Advance Praise

This important volume is a welcome addition to Western-dominated literature pushing top-down approaches to sustainable development. It gives voice to scholars and practitioners in countries like India and Bangladesh who report on promising bottom-up approaches that are suited to a diverse range of local settings.

Riley E. Dunlap
Regents Professor of Sociology, Oklahoma State University, USA

This fine collection of work by insightful scholars adds greatly to our understanding of the challenges of sustainable development. The theme of shifting from top-down strategies to bottom-up runs through the chapters, providing a well-integrated book from a diversity of contributors.

Richard York
Director and Professor of Environmental Studies,
University of Oregon, USA

'Bottom-up' Approaches in Governance and Adaptation for Sustainable Development

Bulk Sales

SAGE India offers special discounts
for purchase of books in bulk.
We also make available special imprints
and excerpts from our books on demand.

For orders and enquiries, write to us at

Marketing Department
SAGE Publications India Pvt Ltd
B1/I-1, Mohan Cooperative Industrial Area
Mathura Road, Post Bag 7
New Delhi 110044, India

E-mail us at **marketing@sagepub.in**

Get to know more about SAGE

Be invited to SAGE events, get on our mailing list.
Write today to **marketing@sagepub.in**

This book is also available as an e-book.

'Bottom-up' Approaches in Governance and Adaptation for Sustainable Development

Case Studies from India and Bangladesh

Edited by

Pradip
SWARNAKAR

Stephen
ZAVESTOSKI

Binay Kumar
PATTNAIK

Los Angeles | London | New Delhi
Singapore | Washington DC | Melbourne

First published in 2017 by

SAGE Publications India Pvt Ltd
B1/I-1 Mohan Cooperative Industrial Area
Mathura Road, New Delhi 110 044, India
www.sagepub.in

SAGE Publications Inc
2455 Teller Road
Thousand Oaks, California 91320, USA

SAGE Publications Ltd
1 Oliver's Yard, 55 City Road
London EC1Y 1SP, United Kingdom

SAGE Publications Asia-Pacific Pte Ltd
3 Church Street
#10-04 Samsung Hub
Singapore 049483

Published by Vivek Mehra for SAGE Publications India Pvt Ltd, typeset in 10/12 pt Adobe Garamond by JMV Design Solutions, Chandigarh 31D, and printed at Chaman Enterprises, New Delhi.

Library of Congress Cataloging-in-Publication Data
Names: Swarnakar, Pradip, editor. | Zavestoski, Stephen, editor. | Pattnaik, Binay Kumar, editor.
Title: "Bottom-up" approaches in governance and adaptation for sustainable development : case studies from India and Bangladesh / [edited by] Pradip Swarnakar, Stephen Zavestoski, and Binay Kumar Pattnaik.
Description: New Delhi, India ; Thousand Oaks, California, USA : SAGE Publications India Pvt Ltd, [2017] | Includes bibliographical references and index.
Identifiers: LCCN 2017011398 | ISBN 9789386446046 (hardbound) | ISBN 9789386446060 (e-pub) | ISBN 9789386446053 (e-book)
Subjects: LCSH: Sustainable development—India. | Sustainable development—Bangladesh.
Classification: LCC HC440.E5 B66 2017 | DDC 338.954/07—dc23
LC record available at https://lccn.loc.gov/2017011398

ISBN: 978-93-864-4604-6 (HB)

SAGE Team: Supriya Das, Alekha Chandra Jena, Shobana Paul, and Ritu Chopra

Contents

List of Tables

List of Figures

Preface and Acknowledgments

"We wrote this book because we got confused."
—Erik Brynjolfsson and Andrew McAfee

We have edited this volume because we intend to address the dilemma of sustainable development. Environmental problems have become persistent and pervasive, affecting almost all life directly or indirectly. The concept of sustainable development evolved out of this ecological crisis. The core idea of sustainable development is directed toward the safety and security of future generations through the judicious use of natural resources in the present. Connecting the past, present, and future regarding the distribution and consumption of natural resources is the fundamental challenge of sustainable development governance. It has been globally accepted that the current crisis is the result of past and present ways of life that, if continued, will jeopardize the future. Now, the pertinent questions are: What to do? How to address this challenge? And most importantly, who is going to do what?

National governments, international economic agencies, and civil society organizations have addressed the issue of environmental crisis like other social and economic problems. Over the years, a range of possible solutions was designed to control the present crisis. Most solutions are directed at those citizens who, defined by their relative poverty, are believed to be the worst affected by ecological disruptions. This process of governance can be summarized as a 'top-down' approach. Due to uneven economic growth and social opportunity, the 'top-down' model failed to reach the bottom of the pyramid. Moreover, many affected people are left out because the economic resources of top-down decision-makers are limited, and they have to prioritize certain issues, communities, and geographic locations.

On the other hand, the people, who are living on the ground, have to encounter the challenges with or without the support of the 'top-down' decision-makers. Their approach is based on their traditional knowledge and limited economic resources. This way of addressing everyday livelihood challenges represents a 'bottom-up' approach. Many climate change adaptation efforts and community resilience in the face of disaster are examples of 'bottom-up' approaches.

This backdrop leads to a dilemma for sustainable development governance. The decision-makers in the 'top-down' model have economic resources and cutting-edge technological solutions. On the other hand, human beings have historically generated extensive local knowledge in response to changing environments. With the help of 'bottom-up' approaches, they have continuously tackled adverse environmental conditions.

This volume is an attempt to showcase instances of both 'top-down' and 'bottom-up' approaches to sustainable development. The book is a collection of various case studies related to diverse environmental challenges by leading scholars from India and Bangladesh. We have focused on micro-level case studies because we believe that case studies are rich in local knowledge, which is difficult to tap otherwise. To elaborate the dilemmas of sustainable development, we have divided the book into four sections. In the beginning, we have tried to summarize the background and context of sustainable development governance. Citing multiple case studies related to forest rights, biodiversity conservation, land rights, and freshwater crisis, section one of this book highlights the challenges of so-called 'top-down' approaches. Section two of this book emphasizes several 'bottom-up' cases on water resource management, flood management, solar home lighting, and ecotourism. We have dedicated section three of this book to the issue of climate change adaptation to highlight the challenges on the ground in actual decision-making procedures. Finally, in the concluding chapter, we have proposed a new approach to address the dilemma of integrating 'top-down' and 'bottom-up' approaches by introducing a new 'middle-out' model. We believe that because of the scale and speed of global climate change, the knowledge being produced locally needs to be aggregated and disseminated as rapidly as possible. And this can be possible with the use of information and communication technologies. With the help of mobile phones and the Internet, a virtual platform can be built on a global scale, where knowledge exchange will be more effective, and resources can be disbursed more judiciously.

This volume would not have been possible without the active support of many individuals and organizations. First of all, we would like to thank our respective institutions for their administrative support to brainstorm and develop this manuscript. Many academic and nonacademic staffs from ABV-Indian Institute of Information Technology and Management, University of San Francisco, and Institute for Social and Economic Change have extended their support to shape the idea, especially Sanjeev Deshmukh, Christopher Brooks, and K.V. Raju. We have organized and reorganized the chapters of this book many times. In this process, we have received valuable feedback from the Helsinki Research Group for Political Sociology (HEPO), the University of Helsinki, and the Finnish Environment Institute (SYKE). In particular, we would like to thank Tuomas Ylä-Anttila, Antti Gronow, Eeva Luhtakallio, Veikko Eranti, Tuukka Ylä-Anttila, and Risto Alapuro from HEPO for their critical comments. We would also like to thank Mikael Hildén and Eeva Primmer from SYKE for extending the opportunity to present our ideas at the sustainability transition workshop at their institute. The idea of this book was benefited by the works on sustainability transition literature. In this regard, we are especially thankful to Anna Wieczorek for her valuable feedback. This volume is the outcome of the cumulative knowledge and wisdom of the contributors. We appreciate their overwhelming support during the publication process. Finally, we would like to thank SAGE Publications for their continuous support at various stages of the publication. We are thankful to Vivek, Sunanda, Alekha, and Supriya.

Reference

Brynjolfsson, E. and A. McAfee. 2014. *The Second Machine Age: Work, Progress, and Prosperity in a Time of Brilliant Technologies.* New York: WW Norton & Company.

1

Introduction: Governance for Sustainable Development in the Anthropocene

Stephen Zavestoski and Pradip Swarnakar

> Because local needs and interests will necessarily vary, sustainable development must be redefined repeatedly, from the bottom up, wherever it is to be put into practice. Sustainable development can have worldwide relevance and appeal, but only if its original purpose of helping the poor live better, healthier, and fairer lives on their own terms is restored.
> —David Victor, 'Recovering Sustainable Development'

Earth systems are facing unprecedented pressure due to anthropogenic activities. So profound are the changes triggered by humans—from the extreme weather and sea-level rise exacerbated by climate change, to ozone depletion, mass extinctions, deforestation, and many irreversible phenomena (Monastersky 2015)—that they are now inscribed in Earth's geological record. In this new Anthropocene epoch, human activity has pushed past several key planetary boundaries. The ensuing global and local environmental changes are happening not only faster than previously predicted but also faster than any planetary-scale changes humans have had to adapt to in the past. National and world leaders must continue to grapple with previous commitments to sustainable development, but they must now do so with approaches to governance capable of confronting a new set of constraints and a heightened sense of urgency imposed by the Anthropocene. These circumstances have led Jeffrey Sachs, Director of the United Nations (UN) Sustainable Development Solutions Network and Special Advisor to former UN Secretary-General Ban Ki-moon, to declare:

> Sustainable development is the central drama of our time. In many ways, humanity has squandered the time it once had to adjust to environmental realities. Now our backs are up against the wall. We are living in history, and our generation's history is the threat of unprecedented, global-scale environmental catastrophe. The starting point is our crowded planet. (2014: 106)

Sustainable development may be "the central drama of our time," but to date it has been a drama without a denouement. Furthermore, it is a drama whose central rhetorical device, as Victor argues, has become meaningless. Finally, it is a drama the conclusion of which will play out in the Anthropocene, a period defined by human impacts on Earth systems having achieved a geological scale. As Kloor explains, the challenge of the Anthropocene is to "meet the needs and aspirations of all of humanity while sustaining the planet's ecology" (2014). This simple equation paradoxically complicates the drama. The Anthropocene erases the society–nature dualism that defined previous sustainable development debates and replaces it with compulsory adaptation to anthropogenic socio-ecological disruption, especially by those in the least developed parts of the world, that risks undermining sustainable development goals (SDGs). To speak of climate change evokes ideas of a depoliticized problem to be tackled by experts applying scientific knowledge toward technological solutions. The Anthropocene, on the other hand, holds discursive power that opens us up to the possibility that economic, cultural, political, and other forms of human social organization may need to be transformed. Within the milieu of the Anthropocene, we must become attentive to alternative approaches to sustainable development that can overcome our past inabilities to bridge the power of top-down approaches with the elements of inclusivity, equity, and justice embedded in bottom-up approaches.

With this book, we aim to direct attention to the need to bridge bottom-up and top-down approaches to sustainable development, which might offer at least one possible approach to resolving the central drama of our time—sustainable development *in the Anthropocene*. To set the stage, in this chapter, we first unpack the history of the concept and the institutional strategies for realizing sustainable development. After contrasting the advantages and disadvantages of both top-down and bottom-up approaches to sustainable development governance, we construct an argument for case study research as a tool for gathering data across contexts that can guide our approaches to innovating new forms of sustainable development governance robust enough to bridge the gap between top-down and bottom-up approaches. Finally, we explain the organization of the book and introduce the chapters.

The Challenge of Defining Sustainable Development

Often little more than a catchphrase to understand the new paradigm of development (Lele 1991), sustainable development has nevertheless become increasingly influential in national and international policy development (for historical and conceptual reviews, see Bolis et al. 2014; Hopwood et al.

2005; Mebratu 1998). The concept first grabbed the world's attention after publication of *Our Common Future*, commonly referred to as The Brundtland Report, by the World Commission on Environment and Development (WCED 1987). WCED defined sustainable development as "development that meets the needs of the present without compromising the ability of future generations to meet their own needs." It contains within it two key concepts: first, "the concept of 'needs', in particular the essential needs of the world's poor, to which overriding priority should be given" and second, "the idea of limitations imposed by the state of technology and social organization on the environment's ability to meet present and future needs" (WCED 1987: 43). This definition has received various criticisms in spite of its wide acceptance (Luke 2005; Redclift 1993, 2005; Seghezzo 2009). The concepts of 'need' and 'intergenerational equity', for example, may be defined differently by different cultures and classes of people.

Many attempts have been made at defining sustainable development (for summaries, see Parris and Kates 2003; Kates et al. 2005; Dernbach and Cheever 2015), yet these efforts often result in additional questions. For example, the Board on Sustainable Development of the US National Research Council (NRC) frames the definitional challenge by asking the following three questions: What should be developed? What should be sustained? Over what period? (NRC 1999; see also Redclift 1993). Looking at the national and international policy apparatus, the NRC concluded that the primary "goals of a transition toward sustainability over the next two generations should be to meet the needs of a much larger but stabilizing human population, to sustain the life support systems of the planet, and to substantially reduce hunger and poverty" (NRC 1999: 5). Thus, under the heading "What Is to Be Sustained," the board identified three major categories: nature, life support systems, and community. Similarly, there were three quite distinct ideas about what should be developed: people, economy, and society (Robert et al. 2005: 11).

Sustainable development definitions are important because they point to the indicators that would need to be measured to determine progress. For example, the NRC's emphasis on human development over economic development points to development indicators such as life expectancy, education, equity, and opportunity. The 2002 World Summit on Sustainable Development (WSSD) aimed to add specificity to the Brundtland definition by identifying the now widely recognized economic, social, and environmental pillars of sustainable development (Osofsky 2003; Wapner 2003; JDSD[1]). The wide range of definitional challenges, as well as related

---

[1] See http://www.joburg.org.za/pdfs/johannesburgdeclaration.pdf, accessed on February 14, 2017.

measurement challenges, inevitably led to criticism of sustainable development policies and strategies. Critics point out the fundamental contradiction between sustainable development's parallel requirements of expansion of economic growth in developing countries, enhanced levels of ecological conservation, and social equity (Agyeman 2003; Pezzoli 1997). Other critics focus on the inattention to power relations among the local-to-global actors and institutions supporting unsustainable development (Lele 1991; Sneddon et al. 2006). Moreover, business organizations' guidelines (UN Global Compact, the OECD Guidelines for Multinational Enterprises, the ICC Business Charter for Sustainable Development, the Caux Round Table Principles, the Global Sullivan Principles, and the Ceres Principles) "tend to emphasize environmental rather than social aspects of sustainable development, in particular to the detriment of the original Brundtland prioritization of the needs of the poorest" (Barkemeyer et al. 2014).

Meanwhile, sustainable development as an organizing concept continued to evolve. The 1992 Rio Declaration on Environment and Development (United Nations General Assembly 1992), in its first three principles, enshrined a human–nature dualism into our conceptualization of sustainable development. As Principle 1 proclaims, sustainable development is human-centered: "Human beings are at the centre of concerns for sustainable development. They are entitled to a healthy and productive life in harmony with nature." Principle 2 then acknowledges states' sovereign rights "to exploit their own resources pursuant to their own environmental and developmental policies," while at the same time imposing "the responsibility to ensure that activities within their jurisdiction or control do not cause damage to the environment of other States or of areas beyond the limits of national jurisdiction." Principle 3 asserts intergenerational equity in a manner that furthers the paradox of human development consistent with environmental protection: "The right to development must be fulfilled so as to equitably meet developmental and environmental needs of present and future generations."

Agenda 21, the plan for action that emerged out of the Rio Earth Summit, endeavored to resolve the conflict and contention between environment and development, in part by prescribing action strategies at local, national, and global levels (Lafferty and Eckerberg 2013: 1). An emphasis on local decision-making devolved responsibility for resolving and managing the intractable problem of balancing environmental protection with human development to communities:

Because so many of the problems and solutions being addressed by Agenda 21 have their roots in local activities, the participation and cooperation of local authorities will be a determining factor in fulfilling its objectives. Local authorities construct, operate and maintain economic, social and environmental infrastructure, oversee planning processes, establish local environmental policies and regulations, and assist in implementing national and subnational environmental policies. As the level of governance closest to the people, they play a vital role in educating, mobilizing and responding to the public to promote sustainable development. (United Nations General Assembly 1993: Section 28.1)

Agenda 21 specifies the importance of local communities' involvement in decision-making, planning, and implementation of processes in local- and national-level governance (Freeman 1996). In the ensuing years, researchers scrutinized sustainable development projects and programs to determine whether the local-level decision-making called for in Agenda 21 was actually occurring (Mehta 1996; Selman 1998; Smardon 2008).

Building on the sustainable development commitments of the Rio Earth Summit, the UN launched a process of identifying a set of development goals for the new millennium—the Millennium Development Goals (MDGs). World leaders signed on the MDGs in 2000, promising to "spare no effort to free our fellow men, women and children from the abject and dehumanizing conditions of extreme poverty" (United Nations 2015), with a target date of 2015. Ensuring environmental sustainability was the aim of the seventh MDG (MDG7). However, this also had inherent problems for developing countries for whom human development continued to be traded against environmental protection (Castello et al. 2010; Sharma 2013). Despite these challenges, by 2015, former UN Secretary-General Ban Ki-moon concluded that "the MDGs helped to lift more than one billion people out of extreme poverty, to make inroads against hunger, to enable more girls to attend school than ever before *and to protect our planet*" (United Nations 2015: 3; emphasis added).

Outside the UN, the MDGs received mixed reviews despite their claimed poverty reduction achievements, not to mention claimed advances in aid from industrialized countries and stakeholder participation (United Nations General Assembly 2012). Critics claimed that the UN's donor-driven approach to addressing the MDGs resulted in neglect of certain issues in developed countries and the failure to consider the real needs in recipient countries, particularly those of marginalized populations (Holland 2008; Miyazawa 2012). Moreover, the seventh of the eight goals, to "ensure environmental sustainability," is presented separately from the other parallel goals that exhibit lack of an integrated and consistent approach to sustainability (McMichael et al. 2003).

After establishing the MDGs in 2000, the next major event on the sustainable development calendar was the WSSD, held in Johannesburg in August 2002. This meeting was intended to assess the progress toward the goals set at the Rio Earth Summit in 1992 and plan further action. However, by some accounts, the most important outcome of the Johannesburg meeting was a sustained critique of UN mega conferences for failing to "incorporate the views of citizens' groups and NGOs, and build on bottom-up activism, at the same time as top-down governmental decision-making" (Seyfang 2003). Scholars and activists pushed for alternative and more inclusive processes and concepts such as 'stakeholder democracy' (Bäckstrand 2006a) and 'just sustainability' (Agyeman and Evans 2004).

Consequently, for many critics, the outcome of the subsequent Rio+20 meeting in 2012 seemed meaningless. Nevertheless, the UN needed to begin preparing for a post-2015 development agenda when the MDGs would 'expire'. A key component of 'The Future We Want' (United Nations 2012), the outcome document of Rio+20, was the mandate to establish an Open Working Group (OWG) to develop a set of SDGs for future consideration. The OWG was formed in January 2013 and concluded its work in 2014 in the *Report of the Open Working Group of the General Assembly* (United Nations General Assembly 2014). Shortly thereafter, the UN General Assembly declared that the OWG's proposal for SDGs would become the basis for integrating SDGs into the post-2015 development agenda.

The OWG deliberations revealed new challenges for the sustainable development agenda. First, members of the OWG heard extensive testimony from scientists regarding rapidly evolving understanding of the planetary boundaries within which humans must operate.[2] These boundaries circumscribe development options in a way that the prior vague and open-ended debates about balancing human development with environmental protection did not. Additionally, having already exceeded three of the nine planetary boundaries, OWG members were warned that anthropogenic global environmental change is already undermining previous development successes. Consequently, the post-2015 development agenda must provide pathways to adapting to these changes that can maintain previous development gains while

[2] A decade of research by Rockström et al. (2009) identified nine "planetary life support systems" essential for human survival. Data were collected to determine how far humans have pushed these systems. For climate change, biodiversity loss, and the biogeochemical flow boundaries, evidence suggests humans have surpassed the boundaries, triggering "irreversible and abrupt environmental change" that could make Earth less habitable. For four other boundaries—ocean acidification, land use, freshwater depletion, and ozone depletion—humans remain in a safe operating space. Data have not been collected on the final two boundaries—atmospheric aerosols and chemical pollution.

Figure 1.1:
Evolution of the sustainable development agenda

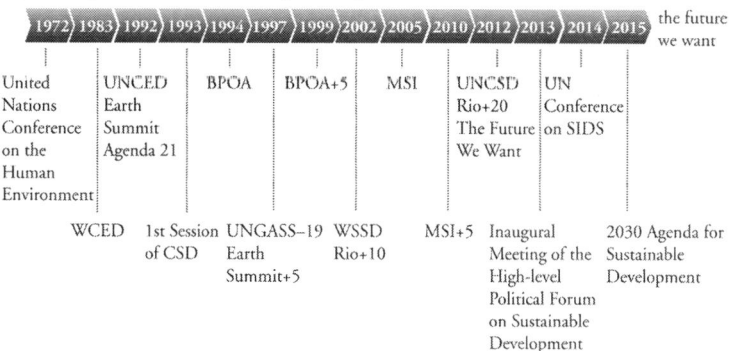

United Nations Conference on the Human Environment	UNCED Earth Summit Agenda 21	BPOA	BPOA+5	MSI	UNCSD Rio+20 The Future We Want	UN Conference on SIDS	the future we want
	WCED	1st Session of CSD	UNGASS–19 Earth Summit+5	WSSD Rio+10	MSI+5	Inaugural Meeting of the High-level Political Forum on Sustainable Development	2030 Agenda for Sustainable Development

Source: https://sustainabledevelopment.un.org/intergovernmental

opening up opportunities for achieving the new SDGs. Second, in the face of the added challenges, social scientists urged the OWG to address the issue of governance, thereby strengthening institutions capable of coordinating global action while also supporting local forms of governance where adaptation is already being managed.

Based on this input, the OWG included among the 17 SDGs several that make explicit reference to ecological limits (e.g., Goal 13 on combating climate change; Goal 14 on conserving ocean resources; and Goal 15 on protecting, restoring, and promoting sustainable use of terrestrial ecosystems and halting biodiversity loss). While less explicit, SDGs alluding to issues of governance were also included (e.g., Goal 16 on promoting peaceful and inclusive societies, providing access to justice for all, and building effective, accountable, and inclusive institutions at all levels; and Goal 17 on strengthening the means of implementation and revitalizing the global partnership for sustainable development). The SDGs, perhaps for the first time in the evolution of the sustainable development agenda (Figure 1.1), prompt us to explore critically and creatively alternative approaches, not merely to governance of sustainable development but more significantly to governance of sustainable development in the Anthropocene.

The Challenge: Sustainable Development Governance

There has never been a shortage of action plans for sustainable development. Actual action, however, has often been paralyzed or ineffectual due to the inability to reconcile human development and environmental protection. As such, there is nothing particularly useful in the UN's intention for the SDGs to "be useful for pursuing focused and coherent action on sustainable development." More importantly, the SDGs represent the first shift from the human-centered development agenda enshrined in the Rio Declaration to a development agenda that attempts to balance more explicitly environmental sustainability with human development, partly by acknowledging ecological limits.

This new focus on what has become known as earth system governance is beginning to generate research and discussions, characterized, for example, by the Earth System Governance Project's series of policy briefs, including one titled 'Integrating Governance into the Sustainable Development Goals' (Biermann et al. 2014). Before returning to the challenge of governing sustainable development in the Anthropocene, we discuss the definitions of global environmental governance. Then, we explain how governing sustainable development in the Anthropocene is further compromised by the ongoing adaptation to global environmental changes such as those to the climate. The challenges posed by integrating adaptation into the goals and practices of sustainable development expose the inadequate synthesis of top-down and bottom-up approaches to sustainable development governance.

Over the last two decades, there has been a gradual evolution from exclusively bureaucratic and managerial-style top-down approaches that treat underdevelopment as a scientific problem to be solved through objective empirical observation and analysis, to a collaborative approach that engages communities in conceiving and implementing sustainable development strategies within acceptable bounds established by the top-down sustainable development apparatus.[3] Both approaches, however, are still very much rooted in the Western notions of 'governance' in which sustainable development is seen as a challenge that must be embedded within a system of global governance. Yet, there is a lack of shared understanding of what is meant by global governance, in general, and global environmental governance (Gupta 2005) or sustainable development governance (Meadowcroft 2007), in particular. Governance, as conceived by Baker et al., is "the institutions, mechanisms or processes backed by political power and/or authority that allow an activity or

---

[3] Victor defines the sustainable development apparatus in terms of the conferences, commissions, and task forces that all find their focus within the UN system (2006: 92). Before Victor, Escobar (1995) referred to a "global institutional apparatus" that aims to manage progress, growth, and efficiency in "underdeveloped" areas according to the norms of development.

a set of activities to be controlled, influenced or directed in the collective interest" (2005: 4).

Global collective interests, such as human rights or a stable climate, require a level of coordination and integration among transnational actors. In its simplest form, this is what is meant by global governance. *Our Global Neighborhood,* the 1995 report of the Commission on Global Governance, advocates going beyond conceptualizing global governances merely as intergovernmental relationships, arguing that "it must now be understood as also involving non-governmental organizations (NGOs), citizens' movements, multinational corporations, and the global capital market." Environmental governance, on the other hand, is "the set of regulatory processes, mechanisms and organizations through which political actors influence environmental actions and outcomes" (Lemos and Agrawal 2006). As global environmental challenges emerged, the concepts of global governance and environmental governance were merged into 'global environmental governance'.

Biermann and Pattberg (2008: 279–280) differentiate between three broad usages of the term global environmental governance: analytical (to make sense of current sociopolitical transformations), programmatic (as a counterweight to the negative consequences of economic and ecological globalization), and critical (as the attempt of the UN and other international organizations to limit the freedom of action of powerful States, in particular the United States). In this context, the roles and responsibilities of nonstate actors are central to global environmental governance. Biermann and Pattberg, who argue that "many vital institutions of global environmental governance are today inclusive of, or even driven by, non-state actors" (2008: 280), identify three elements of this shift. First, the number of organizations and the degree of participation have increased. Second, there has been a diversification of organizations, which goes beyond the government, business, intergovernmental organizations, and NGOs. For example, forest management and biodiversity conservation public–private partnerships emerged as a new mechanism. Finally, traditional organizations have adopted new roles and responsibilities such as NGOs' involvement in agenda setting and policy formulation. The expansion of notions of global environmental governance to include nonstate actors has not, however, necessarily been linked to a shift away from top-down approaches to sustainable development. Strong criticism comes from the global South where the groups gaining access to governance processes are seen as being more favorable to the North's agendas, perspectives, and interests (South Centre 1997). Moreover, the global environmental assessments designed and examined by actors in the North have little influence in developing countries like India (Biermann 2001).

While the SDGs acknowledge the importance of recognizing differing national circumstances, the sustainable development apparatus stops short of promoting bottom-up approaches to sustainable development. They do not, in the words of David Victor, appreciate that "sustainable development must be redefined repeatedly, from the bottom up, wherever it is to be put into practice" (2006: 103). On the other hand, the language of 'good governance' has begun to work its way into the rhetoric used by the sustainable development apparatus. Jeffrey Sachs, in a book that features a foreword by Ban Ki-moon, former Secretary-General of the UN, emphasizes the centrality of good governance to achieving the SDGs: "To achieve the economic, social and environmental objectives of the SDGs, a fourth objective must also be achieved: good governance" (2015: 3). This awareness of the role of good governance, and what it might mean for top-down and bottom-up sustainable development strategies, is discussed next.

Top-down Approaches to Sustainable Development Governance

From the outset, the sustainable development agenda was constructed around an understanding of the challenge as requiring production of knowledge through scientific research and application of that knowledge through the tools of management, all of which happen through an internationally coordinated effort. This understanding of the problem is evidenced in *Our Common Future's* assertion that "ecology and economy are becoming ever more interwoven—locally, regionally, nationally and globally" (WCED 1987: 5). As Bolis et al. point out, "Scientific issues related to sustainability were the basis for the adoption of the term 'sustainable development' in international discussions" (2014: 10). Science, in other words, provided the early evidence justifying discussion of limits to growth and the need to achieve human development goals against a backdrop of limited resources. More recently, international scientific collaboration has been vital in advancing our understanding of global environmental problems and the human drivers of global environmental change, as well as in identifying planetary boundaries and measuring human pressure against them. Reliance on scientific knowledge to form and then drive the sustainable development agenda produced a sustainable development apparatus designed to work from the top down.

Among the problems with the top-down approach of the sustainable development apparatus is its depoliticization of sustainable development. The post-development school of thought (e.g., Escobar 1995), which views the concept of development as bankrupt, has even less kind words to say about *sustainable* development. Prefixing sustainable to development is seen as little more than "the wholesale justification for a new wave of state interventions in people's lives all over the world" (Sachs 2010: 32). The top-down orientation of the sustainable development apparatus emerged out of the managerial focus of the development discourse that preceded it, which allowed for the formation of a consensus around the necessity of global management strategies that produce international declarations, statements, assessments, policies, and interventions. As Pfeifer explains, "the sustainable development discourse de-politicizes the issue of socio-natural relations through the mechanisms of problematization, institutionalization, professionalization and hierarchization" (2011).[4] Put differently, "[t]he production of environmental interventions is intimately connected to the production of environmental knowledge" (Guthman 1997: 45). Yet, in failing to acknowledge that both are intrinsically bound up with power relations, the sustainable development apparatus has historically excluded ways of knowing and forms of knowledge that do not fit its discourse. In examining the major discourses associated with deforestation, desertification, biodiversity, and climate change, Adger et al. find that "in each of the four areas there is a global environmental management discourse representing a technocentric worldview by which blueprints based on external policy interventions can solve global environmental dilemmas" (Adger et al. 2001: 681). Importantly, however, they also observe that "each issue also has a contrasting populist discourse that portrays local actors as victims of external interventions bringing about degradation and exploitation." These populist discourses make up one part of the movement for bottom-up approaches to sustainable development that we discuss next.

Bottom-up Approaches to Sustainable Development Governance

Against the backdrop of a global environmental governance framework dominated by a top-down sustainable development apparatus, there are

---

[4] Pfeifer (2011) nicely summarizes characteristics of post-development thinking as first put forth by Escobar: the interest in alternatives to development, not the interest of alternative development; a fundamental rejection of the classical development paradigm; an interest in local culture and local knowledge; a critical perspective on established scientific discourses; and solidarity for pluralistic grassroots movements.

nevertheless many NGOs and other civil society actors working diligently at much smaller scales, occasionally at the grassroots level, to drive specific sustainable development innovations and initiatives at community or perhaps regional scales. A range of perspectives, some of which are summarized by Brosius, Tsing, and Zerner (1998), try to understand these efforts and their potential effectiveness.

Adaptive management, multistakeholder partnerships (Bäckstrand 2006b), grassroots ecosystem management (Weber 2000), grassroots environmental action (Ghai and Vivian 1995), and other related approaches generally share the underlying belief that through stakeholder participation local and scientific knowledges can be integrated to provide a more comprehensive understanding of complex and dynamic socio-ecological systems and processes (Reed 2008). Research shows that approaches triangulating local and scientific knowledge sources are more robust and have better chances of success (Johnson et al. 2004; Reed 2008).

Some of these approaches have their roots in alternative development models of the 1980s (e.g., Chamber's [1983] work on participatory development) while others draw from theoretical developments such as Sen's (1985, 1999) work on the 'capabilities' approach, which in turn shaped the community capabilities perspective on environmental justice (Schlosberg and Caruthers 2010). Their origins help determine where these various bottom-up approaches might be placed on Hopwood et al.'s (2005) continuum of sustainable development perspectives on the changes required in society's political and economic structures to achieve sustainable development. For example, those viewing bottom-up strategies as crucial to legitimating the sustainable development apparatus fit into Hopwood et al.'s 'status quo' category of actors who believe sustainable development can be achieved within present structures. Others see bottom-up approaches as essential in reforming the sustainable development apparatus while being adequate (Hopwood et al.'s 'reform' category), and the final group fits into Hopwood et al.'s 'transformation' category in its belief that bottom-up approaches are a vehicle for driving transformation of the economic and power structures of society that are at the root of the problem.

An example of the latter category can be found in Ecuador's '*buen vivir*' philosophy, which has been codified in the nation's constitution. *Buen vivir* is an ancient indigenous philosophy striving for harmony and equilibrium among humans and between humans and nature. According to Kauffman and Martin,

> [B]uen vivir breaks with the internationally dominant notion of development as accumulation through economic growth…. Rooted in indigenous

worldviews, buen vivir rejects conventional notions of development based on Western ideals of individualism, a dualism between humankind and nature, and a linear notion of progress rooted in material growth. Proponents see it as an alternative to conventional development that overcomes a false dilemma posed by Western ideals. (2014: 41)

The sustainable development apparatus cannot embrace *buen vivir* as a sustainable development strategy because it cannot be adapted to the dominant development discourse:

Buen vivir is difficult to define because it is not meant to be a preformulated route to sustainable development. Fundamental to the concept is the recognition of rights of a diversity of peoples, deliberation, and respect for many ideas. It involves a dialogue within and among communities to determine the best pathways of sustainable development. Therefore, buen vivir will manifest itself differently in various social and environmental contexts. (Kauffman and Martin 2014: 43)

Even getting the sustainable development apparatus to engage meaningfully at the community level in more conventional ways is challenging. For example, with respect to farmers interacting with environmental regulators, Bartel concluded that "there is an epistemic distance between the bureaucratic knowledge held by government and the vernacular knowledge (place-based knowledge) of heterogeneous environments held by farmers" (2014: 891). Nygren (1999) points to similar challenges in the integration of local knowledge in the sustainable development discourse.

The diversity, range, and uncoordinated nature of bottom-up sustainable development efforts produces an impasse. The top-down orientation of the sustainable development apparatus cannot control the spontaneous and organic efforts happening at the grassroots level, it cannot always understand the forms of knowledge produced and communicated from the grassroots, and it subsequently cannot organize that knowledge into its global management strategy, which hinges on the scaling up of local efforts. For example, agencies or individuals responsible for implementing community-based development projects may discover that carrying out participatory planning processes creates more work for them while making less clear how their work is resulting in the expected outcomes. As a result, rather than transforming the system, community-based approaches may entrench bureaucracies in old ways of doing things.

Where institutions of the sustainable development apparatus have attempted to support bottom-up strategies, they often abandon the work or conclude that it is effective within the metrics established by the apparatus.

Mansuri and Rao (2004) report that the World Bank has spent approximately US$7 billion on community-based and community-driven development projects, despite limited evidence that they are at all effective. Not a single study, report Mansuri and Rao, establishes a causal relationship between any outcome and participatory elements of a community-based development project. Although they uncover "some evidence that participatory projects create effective community infrastructure and improve welfare outcomes," they quickly add that "the evidence does not establish that it is the participatory elements that are responsible for improving project outcomes" (2004: 30). Furthermore, to explain the ineffectiveness of community-based development projects, they engage in blaming the victim. Local governments, they claim, are less accountable and more prone to capture by locally powerful elites, a problem that can be solved by creating close links between higher levels of government and communities.

Bottom-up approaches have their limitations, as do top-down approaches. But whereas top-down approaches generally de-politicize sustainable development by turning it into a managerial task carried out by institutional decision-makers being fed data by producers of scientific knowledge, bottom-up approaches, unless they become co-opted, seem to have greater potential for changing the discourse. Certain types of bottom-up approaches, particularly those working beneath the level of the sustainable development apparatus, have the potential to develop new forms of governance inconceivable to the apparatus. They may also develop promising social and technological innovations precisely because they are more spontaneous, less planned, and at their small scale far less risk-averse than the large institutions of the apparatus. On the other hand, the results of bottom-up approaches are difficult to quantify. They can seem idiosyncratic and anecdotal. Unable to make systematic sense of the wide range of bottom-up approaches, the sustainable development apparatus struggles to understand how to scale up the innovations that work at the community level.

Climate Change Adaptation: A Bottom-up Challenge to Top-down Sustainable Development

The imperative of bridging top-down and bottom-up approaches to sustainable development is clear. The path, however, is less clear and increasingly obscured by the uncertainties introduced by the Anthropocene. When sustainable development discourse could engage with climate change as a future concern, development projects could easily enough integrate notions of mitigation (i.e., reducing greenhouse gas emissions) into their goals. It was possible to embrace optimistic visions of development projects simultaneously creating economic opportunity and reducing emissions. As climate impacts moved from future possibility to present reality, the realization that adaptation to these impacts could hinder or even reverse development efforts began to sink in. Ireland and McKinnon point to an important implication of this shift: "[I]n the face of the uncertainties of climate change, and as yet unknown challenges of a future in the Anthropocene, the development sector must learn ... to act in and for acute uncertainty" (2013: 165). Benson and Craig put it even more starkly: "From a policy perspective, we must face the impossibility of even defining—let alone pursuing—a goal of 'sustainability' in a world characterized by extreme complexity, radical uncertainty, and unprecedented change" (2014: 778).

Others advocate, instead of abandoning sustainable development altogether, developing a new definition for the Anthropocene: Griggs et al. define sustainable development in the Anthropocene as "development that meets the needs of the present while safeguarding Earth's life-support system, on which the welfare of current and future generations depends" (2013: 306). The unpredictability of the Anthropocene combined with the added urgency of taking action to minimize further uncertainty and disruption in the future lays bare the inadequacy of the sustainable development apparatus, particularly its governance approaches. This is best illustrated in terms of how climate change adaptation is being integrated into the sustainable development agenda.

Adaptation poses a host of questions that the sustainable development apparatus has had to address. So far, responses to its answers run to extremes, as explained by Ireland and McKinnon:

> At its best adaptation is focused upon enhancing the adaptive capacity of communities who are facing increasing uncertainties as the climate changes and their environment is transformed. At its worst, adaptation is merely the new catch phrase that is being applied to all kinds of development aid programs, whether they genuinely address climate change or not. Either way, adaptation is being rolled out across the globe. (2013: 158)

Most approaches to understanding adaptation employ a variation of the IPCC's definition of adaptation as "adjustment in natural or human systems in response to actual or expected climatic stimuli or their effects, which moderates harm and exploits beneficial opportunities" (2001). Under this definition, adaptation can be anticipatory or reactive, it can happen spontaneously or in planned ways, and it can be undertaken by private or public actors. In another definition, Nelson, Adger, and Brown define adaptation as "the decision-making process and the set of actions undertaken to maintain the capacity to deal with current or future predicted change" (2007: 396–397).

Interestingly, approaches to understanding and conceptualizing adaptation parallel evolution of the concept of sustainable development. Schulz and Siriwardane, for example, identify three analytically useful categories of adaptation: "(1) 'adjustment' adaptation (also referred to as 'incremental', 'coping', 'resilience' or 'restoration'); (2) 'reformist' adaptation (also called 'systemic', 'transition', 'transitional' and 'more substantial' adaptation); and (3) 'transformative' or 'transformational' adaptation" (2015: 7). The three dimensions of governance—normative change, knowledge production, and social intervention—can then be analyzed for each category of adaptation (see Table 1.1).

As Schulz and Siriwardane (2015: 10) explain, adaptation as adjustment focuses on returning socio-ecological systems to some sort of state of equilibrium, a goal that is achieved through expert-led knowledge production and is then applied in top-down implementation of managerial, technological, and governance solutions. Adaptation as adjustment is equivalent to Hopwood et al.'s 'status quo' category of sustainable development. Reformist adaptation, with obvious connections to Hopwood et al.'s 'reform' category, acknowledges the need for socio-technical system change but aims for incremental rather than transformative change. Coproduction of knowledge and participatory management strategies produce minor modifications of technologies, rules, and decision-making processes. Iterations of small changes, undertaken within the still-dominant managerial, economic, and scientific paradigms, are followed by assessments of outcomes that are then applied in subsequent iterations to result in incremental adaptation. Transformative adaptation, again with obvious linkage to Hopwood et al.'s 'transformation' category, normatively embraces systemic change. Experimentation,

Table 1.1:
Types of adaptation and dimensions of governance

	Adaptation Category		
Governance Dimension	*Adjustment*	*Reformist*	*Transformative*
Normative Change	Equilibrium	Incremental	Systemic
Knowledge Production	Expert-led	Coproduction	Experimentation
Social Intervention	Top-down	Participatory	Intrinsic

Source: Schulz and Siriwardane (2015: 10).

independent of expert status, becomes the knowledge-production paradigm, and social interventions are mostly driven by 'intrinsic' motivations.

Our purposes in discussing the similarities between sustainable development and adaptation frameworks is to point out the risk of adaptation becoming a meaningless aspect of sustainable development discourse, much the way Victor (2006) claims sustainable development itself has lost meaning. Cannon and Müller-Mahn enumerate these risks quite thoroughly, eventually concluding that adaptation and 'adaptive governance' are "embedded in an institutional setting that needs to be critically assessed, especially as they tend to be 'depoliticised' and reliant on systems approaches that play down the significance of self-interested actors who have disproportionate access to and control over ecosystems" (2010: 626). On the one hand, it is impossible to separate adaptation from development (Agrawala 2005; Agrawala and van Aalst 2008). Attempting to do so raises countless questions: "Is adaptation a type of development or something much more? Does development facilitate adaptation? What is meant by development in the context of climate change? Is adaptation the form that development must take under conditions of climate change?" (Cannon and Müller-Mahn 2010). To the extent that answers to these questions are rooted in the same frameworks that dominate sustainable development discourse—that is, an adjustment approach to adaptation that responds to change by seeking to maintain equilibrium, produces knowledge primarily through expert-led science, and is a style of social intervention or governance that is top-down and managerial—we are no closer, and perhaps are further, from dealing with the real challenge of sustainable development in the Anthropocene.

Meanwhile, a diverse range of innovative approaches to adaptation—some more effective than others—unfolds around the world. What would it look like for the institutions engaged in top-down sustainable development (and adaptation) to focus on absorbing lessons from, rather than steering, these processes? What would it look like if, rather than actively incubating or intervening in bottom-up efforts, these institutions allowed space for bottom-up innovations to emerge, identified patterns in the forms of governance linked to successful bottom-up adaptations, and then through collaborative processes evolved new forms of governance to bridge top-down and bottom-up approaches to sustainable development? With few exceptions (Rayner 2010; Urwin and Jordan 2008), attempts to explore these questions have not been made. This volume is a gathering of case study research on adaptation and governance to the potential, as well as the challenges, of building such bridges. Case studies are often criticized as crude tools for evaluating, much less developing, sustainable development strategies. But they hold particularly useful potential for producing the kind of knowledge needed to begin building the top-down/bottom-up bridge so desperately needed.

Case Study Research: Crude Tool or Underutilized Methodology?

Case study methodology is ideal for studying bottom-up approaches. The case study approach examines intensely a single example of the phenomenon of interest within its socio-spatial context. As Ford et al. note, case studies are "commonly used in the social sciences to answer scientific queries related to questions of 'how' and 'why' and have thus been used to address research requiring context-dependent analysis" (2010: 377). The chapters in this volume employ case study methods to explore the 'how' and 'why' of bottom-up approaches to governance and adaptation for sustainable development. In doing so, they reflect some of the benefits of case study methods: deep and rich narrative accounts absent from much quantitative research; careful unpacking of nuances and complexities of a problem; and the organic emergence from field observations of useful analytical concepts as opposed to the typically more deductive approach of quantitative research in which pre-existing theoretical frameworks pre-determine analytical concepts. There is, however, a downside. The extent to which this knowledge is context-specific— a particular challenge for case study research in India where regional and local differences can be pronounced—means that generalizing to other contexts is problematic. For example, as Ford et al. explain, "It has been argued that observations from individual examples cannot be used to infer valid general processes, and that case studies thus have limited value in rigorous scientific analysis" (2010: 377).

Ideally, case study research would be carried out as part of a larger project designed to utilize both of the functions of case study research summarized by Cerceau et al.: "a deductive function that tests theories using case studies to assess a priori models, and an inductive function that generates theories using recurring patterns of case studies to generalize postulates" (2014: 3; Eisenhardt 1989). Over time, the iterative nature of such an intentionally designed case study strategy allows for the emergence of explanatory frameworks and their subsequent testing against additional cases. A case-study-based research agenda could, for example, be aimed at producing a more systematic understanding of the possibilities and limitations of bottom-up governance and adaptation approaches to achieve SDGs. The challenge, however, is that most case study research is conducted not as part of a larger, collective research agenda but rather independently by lone researchers examining a particular aspect of phenomena such as governance and adaptation. Furthermore, these dissimilar case studies might be carried out using discipline-specific methodological traditions, terminology, and interpretive strategies.

One alternative is to carry out meta-analyses of otherwise disconnected collections of case studies. Hoffman, Hinkel, and Wrobel (2011) explain that meta-analysis emerged in the social sciences as a more systematic 'literature review' where criteria for selecting studies and other methodological steps are made transparent. According to Hoffman et al., a meta-analysis consists of three steps: first, selection of relevant studies (with definition of criteria for inclusion and exclusion); second, classification of the information provided by the selected studies in order to translate it into a common language; and third, analysis of classified text. Hoffman et al. (2011) cite a wide range of case study meta-analyses in the context of understanding drivers of environmental change using procedures summarized by Rudel (2008). But they also point to a need for a type of meta-analysis at an intermediate level of abstraction, "between the vague and general concepts such as vulnerability, adaptive capacity etc. on one hand and the specific data, measurements and methodologies applied on the other hand" (Hoffman et al. 2011: 1108).

Our aim in this volume is to hint at the possible benefits of organizing case study research in this way. Although the cases explored vary greatly— from governing freshwater wetlands in Bangladesh to an ecotourism planning process in Goa—by organizing them conceptually around adaptation and governance, we intend to draw out patterns and processes that future case study research can test in the iterative process that gives case study research its power. Ultimately, as we argue in the conclusion, there is much to be gained from more effectively harnessing case study research as a mechanism for aggregating knowledge that is emerging through grassroots and other community-level approaches to adaptation and governance. Organizing this knowledge is the first challenge; linking it to top-down decision-making structures may be an even greater challenge, one explored in the book's third section. Additionally, in the concluding chapter, we offer readers a framework for scaling up the analytically useful yet rather modest power of collecting case studies in an edited volume like this one, into a global database of bottom-up sustainable development efforts. Such a database would produce a type of 'qualitative big data' capable of generating conceptual categories, analytical frameworks, and questions for future research that we were previously incapable of imagining.

To offer a hint at the potential power of a global database of qualitative case study research on sustainable development efforts, we have chosen to focus in this volume on case studies from India and Bangladesh. While many places around the world already feel the pressure to adapt to weather extremes, unpredictable weather, and other global environmental changes, India and Bangladesh are particularly useful locations for examining bottom-up adaptation strategies and the challenges of bridging bottom-up and top-down forms of governance. Both have received extensive attention from the sustainable development community over the last several decades before adaptation became part of the discourse. Today, given their large populations with high levels of vulnerability, India and Bangladesh are ideal places to examine how adaptation is happening beneath the radar of the sustainable development apparatus, as well as the challenges of integrating these efforts into new forms of governance that blur the top-down and bottom-up distinction. In India, Prime Minister Narendra Modi has expressed a vision for "sustainable economic growth without compromising on environmental safety" (Consulate General of India 2015). But broad political brushstrokes from the central government cannot deliver on such promises. Case study research from India and Bangladesh, for example, reminds us that community-level factors like social capital play important roles in the success of sustainable development efforts (Bhuiyan 2011). Wright et al. (2014) draw similar conclusions using case studies from Bangladesh, India, Mozambique, and Uganda. Williams et al. (2016) use case study research from Cambodia, Lao PDR, Bangladesh, and India to demonstrate how livelihood activities and the nonclimate stressors are generally overlooked in the development of adaptation options. The organization of this, which we discuss next, has the modest aim of illustrating how aggregating case studies across a diverse range of researchers can reveal useful insights into sustainable development challenges.

Outline of the Book

The book is organized around four sections. The first two sections point to the limits of top-down approaches to sustainable development (Section 1) and highlight challenges and opportunities revealed in bottom-up experiments in governance for sustainable development (Section 2). Section 3 focuses on climate change adaptation to illustrate the complications to sustainable development introduced by climate change and the potential of bottom-up and hybrid bottom-up/top-down approaches to navigate this complex terrain. Section 4 explored the possibility to synergize top-down and bottom-up approaches.

Section 1, 'Governance I: Questioning the Top-down Approach of Sustainable Development', begins with Jyotiprasad Chatterjee's chapter 'Forest, Adivasis, and the Forest Rights Act (2006): Interrogating Top-down Environmental Governance'. Chatterjee's analysis illustrates the intractability of top-down approaches rooted in the market-oriented managerial framework that has largely dominated sustainable development and conservation discourses. As this chapter shows us, even when nation states advance policies intended to ensure access to resources for people who depend, for example, on forest resources for their livelihood, the uneven implementation and enforcement of these policies constrains people's abilities to meet basic needs and engage in localized economic activity. Chapter 3, 'Science and Politics of Wildlife Enumeration: Questioning the 'Tiger Count' in India', by Jayanthi A. Pushkaran, makes a similar critique but with the target more focused on science and its inability to produce objective knowledge for informed decision-making. The chapter illustrates how what we assume to be science-based policies that are logical and rational can actually be quite political. It is in precisely such instances, according to many of the other chapters, that local-level governance is needed. Chapter 4, 'Urban Land Governance Reforms and Sustainable Development: A Study of Urban Property Ownership Records (UPOR) in Karnataka', suggests that new forms of governance may not necessarily require new institutions. Smitha Kanekanti Chandrashekar and Manasi Seshaiah demonstrate how old institutions can develop new administrative procedures for better coordination of local land use. Whether old or new institutions, Nidhi Yadav and Naresh Chandra Sahu's research, 'Challenges of Sustainable Biodiversity Management: National Chambal Sanctuary in Uttar Pradesh', in Chapter 5, reveals the limits of science-based decision-making when community involvement in decision-making is absent. Their analysis of attempts to sustainably manage the National Chambal Sanctuary in Uttar Pradesh, in fact, reveals negative outcomes for local communities and missed opportunities for economic uplift. In the final chapter of Section 1, titled 'Freshwater Wetlands in Bangladesh: The Need for Alternative Governance', Mohammad Chowdhury introduces the concept of 'comanagement' as a strategy to integrate the state apparatus with local institutions so that community and local governments can more effectively manage natural resources. Chowdhury's discussion of the challenges of sustainable management of freshwater wetlands in Bangladesh highlights the need for innovating new multi-level governance structures.

Section 2, 'Governance II: Experiments with 'Bottom-up' Approaches', covers a range of bottom-up efforts to drive sustainable development. Satabdi Datta's research in Chapter 7, 'Dynamics and Pay-offs in Community-based Water Resource Management: A Case Study from Indian Sundarbans', examines the relationship of pre-existing institutional bodies designed to handle water management with community-led water resource management innovations. In 'Local Solutions to Local Disasters: Governance in Flood Management in Assam' (Chapter 8), Arpita Das and Partha Jyoti Das illustrate how previously existing autonomous councils have taken over decision-making around flood management in Assam. In Chapter 9, 'The Role of Rural Local Bodies in Sustainable Development', James Rajanayagam shows how state agencies are often incapable of providing adequate infrastructure to local communities. Yet, at the same time, rural local bodies lack the resources to take on significant infrastructure projects. The chapter demonstrates a way through the impasse by focusing on strategies for rural local bodies to become entrepreneurs in infrastructure development. Occasionally, top-down efforts, such as electrification of rural villages, begin at the community level. Kartikeya Singh's analysis of outcomes of two electrification efforts in Chapter 10 suggests that without community participation, the benefits of top-down sustainable development projects, even when targeted at the household level, can be mixed. Section 2 concludes with Rohini Fadte's focus on community participation in developing a community-based ecotourism model in Goa (Chapter 11) and confirms what prior research has found: When done well, participatory processes can produce significant outcomes with respect to community empowerment.

Section 3, 'Climate Change Adaptation: A Bottom-up Challenge to 'Top-down' Sustainable Development', offers a selection of case studies illustrating the real-world obstacles of bridging top-down and bottom-up approaches. The section begins with a chapter by Farhat Naz, Marie-Charlotte Buisson, and Archisman Mitra (Chapter 12) that draws on interviews with farmers in West Bengal, India, to identify significant differences in the strategies of smallholder farmers compared to their counterparts farming larger tracts of land. The chapter suggests that not all bottom-up adaptation is equal and that new sustainable development strategies, in fact new forms of governance, may be required to facilitate healthy adaptation strategies that contribute to, rather than detract from, the objective of sustainable development. Toward this end, Chapter 13, 'Adaptive Capacity of Marginalized Urban Women to Climate Change: National Capital Territory of Delhi', by Sakshi Saini and Savita Aggarwal, introduces both a challenge and an opportunity. The challenge is that many people are not consciously aware of the changing climate and so fail to perceive a need to adapt. But communication strategies are being tested and implemented that not only raise awareness but also prompt behavior change. In Chapter 14, 'Community-based Climate Change Adaptation in Coastal India: A 'Bottom-up' Approach Using 3P Model', Rachna Arora, Ashish Chaturvedi, Manjeet Saluja, Nikita Mundra, and Arushi Sen detail a community-based approach to climate change adaptation that aims to understand the perceptions of local communities and how these perceptions can be translated into the planning process through the mediation of pilot initiatives (3P: Perceptions, Planning, Pilots). In the final chapter of Section 3, 'Local Knowledge, Social Capital, and Governance of Climate Change Adaptation in Bangladesh', Md. Masud-All-Kamal explores how local knowledge is converted into social capital to steer climate change adaptation in Bangladesh. Collectively, these chapters point to a sample of the types of adaptation strategies being employed and highlight both the opportunities these strategies present for evolving new forms of governance and the challenges of linking local-level adaptation and decision-making with regional or even higher-level institutions.

In the concluding chapter, 'Neither 'Top-Down' nor 'Bottom-up': A 'Middle-out' Alternative to Sustainable Development', we argue that academics and policy-makers have done an insufficient job of utilizing the extensive knowledge being produced in case studies of sustainable development and climate change adaptation. This is partially due to the nature of case studies that are seen as having limited usefulness because idiosyncratic differences across cases make generalizations problematic. The conclusion draws on the preceding 15 chapters to demonstrate how the power of case studies comes from the accumulated knowledge that is only available when a large number of case studies are aggregated and then analyzed. We call for scaling up the aggregation of case study research to a regional and then global level so that knowledge accumulation and dissemination can happen more rapidly and efficiently than it currently does. The scaling of sustainable development case study findings can generate valuable insights that can subsequently be adapted and implemented through the new forms of local governance described by the authors of the volume.

References

Adger, W.N., T.A. Benjaminsen, K. Brown, and H. Svarstad. 2001. "Advancing a Political Ecology of Global Environmental Discourses." *Development and Change* 32 (4): 681–715.

Agrawala, S., ed. 2005. *Bridge over Troubled Waters: Linking Climate Change and Development.* Paris: OECD.

Agrawala, S. and M. van Aalst. 2008. "Adapting Development Cooperation to Adapt to Climate Change." *Climate Policy* 8 (2): 183–193.

Agyeman, J. 2003. *Just Sustainabilities: Development in an Unequal World.* Cambridge, Massachusetts: MIT Press.

Agyeman, J. and B. Evans. 2004. "'Just Sustainability': The Emerging Discourse of Environmental Justice in Britain?" *The Geographical Journal* 170 (2): 155–164.

Bäckstrand, K. 2006a. "Democratizing Global Environmental Governance? Stakeholder Democracy After the World Summit on Sustainable Development." *European Journal of International Relations* 12 (4): 467–498.

———. 2006b. "Multi-stakeholder Partnerships for Sustainable Development: Rethinking Legitimacy, Accountability and Effectiveness." *European Environment* 16 (5): 290–306.

Baker, A., D. Hudson, and R. Woodward. 2005. *Governing Financial Globalization: International Political Economy and Multi-level Governance.* New York and London: Routledge.

Barkemeyer, R., D. Holt, L. Preuss, and S. Tsang. 2014. "What Happened to the 'Development' in Sustainable Development? Business Guidelines Two Decades After Brundtland." *Sustainable Development* 22 (1): 15–32.

Bartel, R. 2014. "Vernacular Knowledge and Environmental Law: Cause and Cure for Regulatory Failure." *Local Environment* 19 (8): 891–914.

Benson, M.H. and R.K. Craig. (2014). "The End of Sustainability." *Society & Natural Resources* 27 (7): 777–782.

Bhuiyan, S.H. 2011. "Social Capital and Community Development: An Analysis of Two Cases from India and Bangladesh." *Journal of Asian and African Studies* 46 (6): 533–545.

Biermann, F. 2001. "Big Science, Small Impacts in the South? The Influence of Global Environmental Assessments on Expert Communities in India." *Global Environmental Change* 11 (4): 297–309.

Biermann, F. and P. Pattberg. 2008. "Global Environmental Governance: Taking Stock, Moving Forward." *Annual Review of Environment and Resources* 33: 277–294.

Biermann, F., C. Stevens, S. Bernstein, A. Gupta, N. Kabiri, N. Kanie, M. Levy, M. Nilsson, L. Pintér, M. Scobie, and O.R. Young. 2014. "Integrating Governance into the Sustainable Development Goals." POST2015/UNU-IAS Policy Brief No. 3. Tokyo: United Nations University Institute for the Advanced Study of Sustainability.

Bolis, I., S.N. Morioka, and L.I. Sznelwar. 2014. "When Sustainable Development Risks Losing Its Meaning. Delimiting the Concept with a Comprehensive Literature Review and a Conceptual Model." *Journal of Cleaner Production* 83: 7–20.

Brosius, J.P., A.L. Tsing, and C. Zerner. 1998. "Representing Communities: Histories and Politics of Community-based Natural Resource Management." *Society & Natural Resources* 11 (2): 157–168.

Cannon, T. and D. Müller-Mahn. 2010. "Vulnerability, Resilience and Development Discourses in Context of Climate Change." *Natural Hazards* 55 (3): 621–635.

Castello, L.D., D. Gil-Gonzalez, C.A.D. Diaz, and I. Hernández-Aguado. (2010). "The Environmental Millennium Development Goal: Progress and Barriers to its Achievement." *Environmental Science & Policy* 13 (2): 154–163.

Chambers, Robert. 1983. *Rural Development: Putting the First Last.* London: Longman.

Cerceau, J., N. Mat, G. Junqua, L. Lin, V. Laforest, and C. Gonzalez. 2014. "Implementing Industrial Ecology in Port Cities: International Overview of Case Studies and Cross-case Analysis." *Journal of Cleaner Production* 74, 1–16.

Consulate General of India. 2015. *Consul General Participates in the 12th Annual Sustainability Summit and Exposition*. March 4. Available at: http://indianconsulate.com/events/detail/36 (Accessed on November 9, 2015).

Dernbach, J.C. and F. Cheever. 2015. "Sustainable Development and Its Discontents." *Transnational Environmental Law* 4 (2): 247–287.

Eisenhardt, K. 1989. "Building Theories from Case Study Research." *Acad. Manag. Rev.* 14 (4): 532–550.

Escobar, A. 1995. *Encountering Development: The Making and Unmaking of the Third World*. Princeton: Princeton University Press.

Ford, J.D., E.C.H. Keskitalo, T. Smith, T. Pearce, L. Berrang-Ford, F. Duerden, and B. Smit. 2010. "Case Study and Analogue Methodologies in Climate Change Vulnerability Research." *Wiley Interdisciplinary Reviews: Climate Change* 1 (3): 374–392.

Freeman, C. 1996. "Local Government and Emerging Models of Participation in the Local Agenda 21 Process." *Journal of Environmental Planning and Management* 39 (1): 65–78.

Ghai, D. and J.M. Vivian. 1995. *Grassroots Environmental Action: People's Participation in Sustainable Development*. London: Routledge.

Griggs, D., M. Stafford-Smith, O. Gaffney, J. Rockström, M.C. Öhman, P. Shyamsundar ... and I. Noble. 2013. "Policy: Sustainable Development Goals for People and Planet." *Nature* 495 (7441): 305–307.

Gupta, J. 2005. "Global Environmental Governance: Challenges for the South from a Theoretical Perspective." In *A World Environment Organization: Solution or Threat for Effective International Environmental Governance*, edited by F. Biermann and S. Bauer, 57–84. Aldershot, UK: Ashgate.

Guthman, J. 1997. "Representing Crisis. The Theory of Himalayan Environmental Degradation and the Project of Development in Post-Rana Nepal." *Development and Change* 28 (1): 45–70.

Holland, M. 2008. "The EU and the Global Development Agenda." *European Integration* 30 (3): 343–362.

Hopwood, B., M. Mellor, and G. O'Brien. 2005. "Sustainable Development: Mapping Different Approaches." *Sustainable Development* 13 (1): 38–52.

IPCC, 2001. *Summary for Policymakers, Climate Change 2001: Impacts, Adaptation and Vulnerability*. Cambridge: Cambridge University Press.

Ireland, P. and K. McKinnon. 2013. "Strategic Localism for an Uncertain World: A Postdevelopment Approach to Climate Change Adaptation." *Geoforum* 47: 158–166.

Johnson, N., N. Lilja, J.A. Ashby, and J.A. Garcia. 2004, August. "The Practice of Participatory Research and Gender Analysis in Natural Resource Management." *Natural Resources Forum* 28 (3): 189–200.

Kates, R.W., T.M. Parris, and A.A. Leiserowitz. 2005. "What Is Sustainable Development? Goals, Indicators, Values, and Practice." *Environment: Science and Policy for Sustainable Development* 47 (3): 8–21.

Kauffman, C.M. and P.L. Martin. 2014. "Scaling up Buen Vivir: Globalizing Local Environmental Governance from Ecuador." *Global Environmental Politics* 14 (1): 40–58.

Kloor, K. 2014. "Facing Up to the Anthropocene." Collide-a-scape, *Discover* Magazine blog. Available at: http://blogs.discovermagazine.com/collideascape/2014/06/20/facing-anthropocene/ (Accessed on November 13, 2015).

Lafferty, W.M. and K. Eckerberg. 2013. *From the Earth Summit to Local Agenda 21: Working Towards Sustainable Development.* Vol. 12. London: Earthscan (Routledge).

Lele, S.M. 1991. "Sustainable Development: A Critical Review." *World Development* 19 (6): 607–621.

Lemos, M.C. and A. Agrawal. 2006. "Environmental Governance." *Annu. Rev. Environ. Resour.* 31: 297–325.

Luke, T.W. 2005. "Neither Sustainable nor Development: Reconsidering Sustainability in Development." *Sustainable Development* 13 (4): 228–238.

Mansuri, G. and V. Rao. 2004. "Community-Based and -Driven Development: A Critical Review." *The World Bank Research Observer* 19 (1): 1–39.

McMichael, A.J., C.D. Butler, and C. Folke. 2003. "New Visions for Addressing Sustainability." *Science* 302 (5652): 1919–1920.

Meadowcroft, J. 2007. "Who Is in Charge Here? Governance for Sustainable Development in a Complex World." *Journal of Environmental Policy and Planning* 9 (3–4): 299–314.

Mebratu, D. 1998. "Sustainability and Sustainable Development: Historical and Conceptual Review." *Environmental Impact Assessment Review* 18 (6): 493–520.

Mehta, P. (1996). "Local Agenda 21: Practical Experiences and Emerging Issues from the South." *Environmental Impact Assessment Review* 16 (4): 309–320.

Miyazawa, I. 2012. "What Are Sustainable Development Goals." *IGES Rio+ 20 Issue Brief* Volume 1, Institute for Global Environmental Strategies. https://pub.iges.or.jp/pub/what-are-sustainable-development-goals. (Accessed on January 11, 2017).

Monastersky, R. 2015. "Anthropocene: The Human Age." *Nature* 519 (7542): 144–147.

National Research Council (US) Policy Division. Board on Sustainable Development. 1999. *Our Common Journey: A Transition Toward Sustainability.* National Academies Press. Available at: http://rwkates.org/pdfs/b1999.01.pdf (Accessed on November 2, 2015).

Nelson, D.R., W.N. Adger, and K. Brown. 2007. "Adaptation to Environmental Change: Contributions of a Resilience Framework." *Annual Review of Environment and Resources* 32 (1): 395.

Nygren, A. 1999. "Local Knowledge in the Environment-Development Discourse. From Dichotomies to Situated Knowledges." *Critique of Anthropology* 19 (3): 267–288.

Osofsky, H.M. 2003. "Defining Sustainable Development After Earth Summit 2002." *Loy. LA Int'l & Comp. L. Rev.* 26: 111–125.

Parris, T.M. and R.W. Kates. 2003. "Characterizing and Measuring Sustainable Development." *Annual Review of Environment and Resources* 28: 559–86.

Pezzoli, K. 1997. "Sustainable Development: A Transdisciplinary Overview of the Literature." *Journal of Environmental Planning and Management* 40 (5): 549–574.

Pfeifer, E. "De-Politicizing the Environment: An Inquiry into the Nature of the Sustainable Development Discourse." *Global Politics*, July 11. Available at: http://www.globalpolitics.cz/clanky/de-politicizing-the-environment-an-inquiry-into-the-nature-of-the-sustainable-development-discourse?en=1 (Accessed on November 6, 2015).

Rayner, S. 2010. "How to Eat an Elephant: A Bottom-up Approach to Climate Policy." *Climate Policy* 10 (6): 615–621.

Redclift, M.R. 1993. "Sustainable Development: Concepts, Contradictions, and Conflicts." In *Food for the Future: Conditions and Contradictions of Sustainability*, edited by P. Allen. New York, NY: John Wiley.

———. 2005. "Sustainable Development (1987–2005): An Oxymoron Comes of Age." *Sustainable Development* 13 (4): 212–227.

Reed, M.S. 2008. "Stakeholder Participation for Environmental Management: A Literature Review." *Biological Conservation* 141 (10): 2417–2431.

Robert, K.W., T.M. Parris, and A.A. Leiserowitz. 2005. "What Is Sustainable Development? Goals, Indicators, Values, and Practice." *Environment: Science and Policy for Sustainable Development* 47 (3): 8–21.

Rockström, J., W. Steffen, K. Noone, Å. Persson, F.S. Chapin, E.F. Lambin, ... and B. Nykvist. 2009. "A Safe Operating Space for Humanity." *Nature* 461: 472–475.

Rudel, T.K. 2008. "Meta-analyses of Case Studies: A Method for Studying Regional and Global Environmental Change." *Global Environmental Change* 18 (1): 18–25.

Sachs, J. 2014. "Sustainable Development Goals for a New Era." Sustainable Humanity, Sustainable Nature: Our Responsibility, Joint Workshop of the Pontifical Academy of Sciences and the Pontifical Academy of Social Sciences, May 2–6, 2014. Vatican City. http://www.academyofsciences.va/content/dam/accademia/pdf/es41/es41-sachs.pdf dated 11 January 2017. (Accessed on January 11, 2017).

———. 2015. *The Age of Sustainable Development.* New York, NY: Columbia University Press.

Sachs, W., ed. 2010. "Environment." In *The Development Dictionary,* 24–37. London and New York, NY: Zed Books.

Schlosberg, David and David Caruthers. 2010. "Indigenous Struggles, Environmental Justice, and Community Capabilities." *Global Environmental Politics* 10 (4): 12–35.

Schulz, K. and R. Siriwardane. 2015. "Depoliticised and Technocratic? Normativity and the Politics of Transformative Adaptation." Earth System Governance Working Paper No. 33. Lund and Amsterdam: Earth System Governance Project.

Selman, P. 1998. "Local Agenda 21: Substance or Spin?" *Journal of Environmental Planning and Management* 41 (5): 533–553.

Sen, A.K. 1985. *Commodities and Capabilities.* Amsterdam: Elsevier.

———. 1999. *Development as Freedom.* New York: Knopf.

Seghezzo, L. 2009. "The Five Dimensions of Sustainability." *Environmental Politics* 18 (4): 539–556.

Seyfang, G. 2003. "Environmental Mega-conferences—From Stockholm to Johannesburg and Beyond." *Global Environmental Change* 13 (3): 223–228.

Sharma, S. 2013. "Reaching the 7th Millennium Development Goals (MDG) on Environmental Sustainability: The South Asian Response." In *Millennium Development Goals and Community Initiatives in the Asia Pacific,* edited by A. Singh, E.T. Gonzalez, and S.B. Thomson, 69–79. New Delhi: Springer India.

Smardon, R.C. 2008. "A Comparison of Local Agenda 21 Implementation in North American, European and Indian Cities." *Management of Environmental Quality: An International Journal* 19 (1): 118–137.

Sneddon, C., R.B. Howarth, and R.B. Norgaard. 2006. "Sustainable Development in a Post-Brundtland World." *Ecological Economics* 57 (2): 253–268.

South Centre (South Commission). 1997. *For a Strong and Democratic United Nations: A South Perspective on UN Reform.* London: Zed Books.

United Nations. (2012). "The Future We Want: Outcome Document Adopted at Rio+20." Rio de Janeiro, Brazil, 20–22 June 2012. Accessed from https://sustainabledevelopment. un.org/content/documents/733FutureWeWant.pdf (Accessed on January 12, 2017).

———. 2015. *The Millennium Development Goals Report, 2015.* New York, NY: United Nations. Available at: http://www.un.org/millenniumgoals/2015_MDG_Report/pdf/ MDG%202015%20rev%20(July%201).pdf (Accessed on November 2, 2015).

United Nations General Assembly. 1992. *Rio Declaration on Environment and Development.* Report of the United Nations Conference on Environment and Development.

A/CONF.151/26. Vol. I. August 12. Available at: http://www.un.org/documents/ga/conf151/aconf15126–1annex1.htm (Accessed on November 5, 2015).

United Nations General Assembly. 1993. *Agenda 21: United Nations Programme of Action from Rio*. New York, NY: United Nations. Available at: http://www.unep.org/Documents. Multilingual/Default.asp?documentid=52 (Accessed on November 5, 2015).

———. 2012. "Accelerating Progress Towards the Millennium Development Goals: Options for Sustained and Inclusive Growth and Issues for Advancing the United Nations Development Agenda beyond 2015." Annual Report of the Secretary General. A/66/126. New York: United Nations.

———. 2014. *Report of the Open Working Group of the General Assembly on Sustainable Development Goals A/68/970*. August 12. Available at: http://www.un.org/ga/search/view_doc.asp?symbol=A/68/970&Lang=E (Accessed on November 2, 2015).

Urwin, K. and A. Jordan. 2008. "Does Public Policy Support or Undermine Climate Change Adaptation? Exploring Policy Interplay Across Different Scales of Governance." *Global Environmental Change* 18 (1): 180–191.

Victor, D. 2006. "Recovering Sustainable Development." *Foreign Affairs* 85 (1): 91–103.

Wapner, P. 2003. "World Summit on Sustainable Development: Toward a Post-Jo'burg Environmentalism." *Global Environmental Politics* 3 (1): 1–10.

Weber, E.P. 2000. "A New Vanguard for the Environment: Grassroots Ecosystem Management as a New Environmental Movement." *Society & Natural Resources* 13 (3): 237–259.

Williams, L.J., S. Afroz, P.R. Brown, L. Chialue, C.M. Grünbühel, T. Jakimow, I. Khang, M. Mineah, V. Ratna Reddy, S. Sacklokham, E. Santoyo Rio, M. Soeun, C. Tallapragada, S. Tom, and C.H. Roth. 2016. "Household Types as a Tool to Understand Adaptive Capacity: Case Studies from Cambodia, Lao PDR, Bangladesh and India." *Climate and Development* 8 (5): 1–12.

WCED (World Commission on Environment and Development). 1987. *Our Common Future*. Oxford: Oxford University Press.

Wright, H., S. Vermeulen, G. Laganda, M. Olupot, E. Ampaire, and M.L. Jat. 2014. "Farmers, Food and Climate Change: Ensuring Community-based Adaptation Is Mainstreamed into Agricultural Programmes." *Climate and Development* 6 (4): 318–328.

SECTION 1

Governance I: Questioning the Top-down Approach of Sustainable Development

2

Forest, Adivasis, and the Forest Rights Act (2006): Interrogating Top-down Environmental Governance

Jyotiprasad Chatterjee

Introduction

The development initiatives undertaken by the neoliberal state of India since the last decade of the last century have been punctuated by several contestations, claims, and counterclaims. Against the free market proponents' claims of unprecedented economic growth and maturity there are counter-claims and concerns about the rising level of poverty, inequality, under-development, jobless growth, social tensions, and environmental degradation. Perhaps, the greatest challenge that the state of India is facing since its adoption of the new economic policy in the last decade of the twentieth century is to deal with the issue of land acquisition for development projects, especially for the establishment of large industrial enterprises primarily under private ownership. Such industries of scale, under the aegis of capital, national, and/or multinational, necessitate huge amount of land, the acquisition of which brings to the fore the serious issues of dispossession and displacement of the people from their land and property. The indigenous people or the tribes (hereafter adivasis) of India, perhaps, are the greatest victims of these two processes. A report of the Ministry of Rural Development, Government of India (GOI), mentions,

> The tribals have been the biggest victims of displacement due to development projects. Though constituting only 9% of the country's population the tribal communities have contributed more than 40% to the total land acquired till so far. (2009: ii)

It is estimated that adivasis constitute 40 percent of the total people displaced for various development initiatives since 1950 (Fernandes 2005; GOI 2002,

2008). Reddy and Kumar (2010) peg this figure at 55 percent. For the forest-dependent adivasis, displacement also raises serious concern about their livelihood. Besides being antithetical to one of the United Nations Millennium Development Goals pertaining to environmental sustainability, the top-down manner in which development has been pursued appears to be of full potential to increase the vulnerabilities of the adivasis to crises. Viewed from the perspective of sustainable development goals as pronounced by the United Nations, economic development of this nature is not sustainable also (Pojman and Pojman 2012). To 'undo' such 'Historical injustice' meted out to the adivasis, the Government of India has come up with the Scheduled Tribes and Other Traditional Forest Dwellers (Recognition of Forest Rights) Act, 2006 (hereafter Forest Rights Act), which was implemented from January 1, 2009. Since then, the Act so far has failed to achieve its desired goal of protecting the rights of the forest-dwelling communities. The issue here is not to discuss the legal aspects of the Act, but to analyze the paradoxical situation that the enactment of such an Act can develop. On one side of the paradox, there is the state of India, functioning under the neoliberal imperative of 'rational' and 'efficient' economic growth, which is left with very little option but to go for large-scale industrialization necessitating huge amount of land; on the other side, it is the same state that is trying to protect the 'rights' of the indigenous people over their land and forest through legal enactments that will make the process of land acquisition difficult. To unveil such a paradox, a critical look at the rationale behind the enactment of the Act along with its possible impact on the stakeholders, particularly the adivasis (tribes) of India, seems pertinent here. Hence, it is not sufficient to explore the failures of the Act only; rather, the focus should be on the entire discourse of development that attempts to conceal its limitations through such enactments, completely ignoring the indigenous knowledge, cultures, and worldviews. What is of particular importance here is to understand the limitations of top-down development initiatives through a thorough analysis of the Act and its implementation.

Adivasis of India: The Victims of Top-down Economic Development

It is important to have a perception about the precarious situation in which the adivasis of India find themselves in, largely due to the unfolding of the 'development' process. To start with, we must be aware about the analytical distinctions between 'displacement' and 'dispossession'—two consequences contingent on the process of development, especially in the context of the underdeveloped societies. Displacement refers to the unintended consequences of a particular type of development initiative. As a balancing attempt, the Government or other agencies, at least notionally, accepts the right of the displaced persons to be compensated. Hence, the whole discourse of rehabilitation seems to be built into the notion of displacement. Displaced persons can bargain about the rehabilitation package and have the right to mobilize against it if they feel deprived. Dispossession, on the other hand, is a deliberate attempt by which the people are robbed of their resources in a meticulously planned manner, for example, through manipulation of land records and *benami* (illegal) transfers. Unlike the displaced persons, the victims of dispossession do not have any right to seek compensation. The base of collective mobilization by the victims also gets weakened as dispossession often seems to be 'natural' and 'legal' even if it makes use of force. Fred Magdoff while talking about capital accumulation by dispossession maintains, "There have been many variations of means used, including both force and swindling by using a variety of laws and agreements or outright chicanery" (2013: 2). From a sociological point of view, however, such theoretical distinctions notwithstanding, the ground reality of the displaced and dispossessed persons has hardly any difference. This is so because, first, often there is either no proper rehabilitation of the displaced persons or there are plenty of allegations about the discriminatory distribution of rehabilitation packages. In this context, it is pertinent to mention that up to 1994, 75 percent of the adivasis, who have been displaced due to mega investment projects since 1950, were not resettled at all (Fernandes 1994) or were 'still awaiting rehabilitation' (Baviskar 2012: 34).

Then, there is the conviction of the sociologists that it is impossible to rehabilitate the displaced persons because mere resettlement or monetary packages cannot ensure the rehabilitation of the culture and community life of the persons displaced. Often the state and project officials, in charge of the resettlement and rehabilitation package, being outsiders, fail to perceive the cultural distinctiveness of the people displaced. Hence, they tend to "homogenise the displaced and impose their worldview and understanding on them" (Asif 2000: 2007). Instead of being considered as participants of the development process, the displaced, thus, becomes the object of development. Such a 'development' package being managed and implemented completely by the managers and state officials is bound to give birth to a feeling of mistrust and misgivings on the part of the people displaced. To the adivasis, on the face of their centuries-old experience of marginalization, such a feeling gets accentuated many times. Asif quite aptly argues:

> Such misgivings and distrust can be more acute for the members of the tribal communities, especially in cases where the project necessitating involuntary resettlement has resulted in losing not only their material assets but integral elements of their social and cultural life. And if it is followed by forced resettlement in locations where they are vulnerable to exercise of power by others the distrust gives rise to wariness. (2000: 2008)

From this point of view, the social and cultural consequences of both displacement and dispossession are equally grave. It is indeed an irony that the adivasis of India, who are largely out of the development process and whose share in the fruits of development process is dismally low, are badly affected by it. The Foreword of the study report of ActionAid quite succinctly argues:

> After all, much of the development projects that India has witnessed in post-independent era have taken place in tribal regions, especially in regions adjoining the States of Andhra Pradesh, Chhattisgarh, Jharkhand and Orissa. And yet the incidence of poverty, malnutrition, non-literacy is most pronounced in these regions...integration of tribes with the wider world through network of infrastructure, industrial, mineral and other development projects, has not led to corresponding development of the tribal people. Rather, their social and economic situation has worsened. (2008: iii)

Development process has not only marginalized them economically but it has also jeopardized their cultural distinctiveness. Dispossession and displacement of adivasis from their natural entitlements constitute the greatest threat to their distinct sociocultural identity and material existence.

Adivasis of India, especially those of the central Indian plateau region, being highly exposed to such 'development' initiatives, are, perhaps, the worst victims of these two processes. Systematic dispossession and displacement of the adivasis from their land and forest started during the colonial period and has been continuing unabated since then. Independence of India, instead of making any marked shift to their misfortune, has allowed the same odyssey to continue, if not accentuating it. On November 12, 2010, the Minister of State in the Ministry of Tribal Affairs informed the Lok Sabha (Lower House of Parliament) that, as of July 2010, a total of 477,000 cases of adivasi land alienation had been registered, covering 810,000 acres of lands, of which 378,000 cases (79 percent) covering 786,000 acres had been decided by the court. Out of these, 209,000 cases (55 percent) covering a total area of 406,000 acres had been decided in favor of the adivasis while 169,000 cases

(45 percent) had been decided against the adivasis.[1] Although adivasis are being victimized by the state as well as other private corporate interests, the state, perhaps, has been more proactive in this regard, as Guha believes, "the state has more actively dispossessed them—by taking away their forests, by displacing them through dams, and most recently, by subjecting them to the negative externalities of open-cast mining" (2012: 9). Such facts quite glaringly indicate that adivasis are the foremost victims of the processes of displacement and/or dispossession. It is little wonder, therefore, that to the adivasis the whole discourse of development appears to be discriminating. The rhetoric of development germinates, and to a great extent perpetuates, their sense of relative deprivation vis-à-vis the Indian mainstream. Such a feeling of deprivation can be considered as a classic expression of internal colonialism, which in any specific historical juncture may give birth to social tensions and conflict by providing the necessary underpinnings to ethnic exclusiveness. The ongoing stiff challenge raised by the adivasis of central India to the development initiatives of the neoliberal Indian state might have its base on such an emergent feeling of ethnic discrimination.

Interestingly, such an onslaught to one of the most vulnerable marginalized cross sections of Indian society is taking place in spite of a plethora of legal initiatives to check it. The state of India, colonial as well as independent, has enacted a number of laws prohibiting such alienation of the adivasis from their land and forest, the important ones being: Chotanagpur Tenancy Act (1908), Indian Forest Act (1927), Santhal Parganas Tenancy Act (1949), the Constitution of India (5th and 6th Schedule), the National Forest Policy (1952), Forest (Conservation) Act (1980), and the Panchayats (Extension to the Scheduled Areas) Act (1996). These Acts, particularly the forest-related ones, however, attained little success in bringing in significant check in the magnitude of alienation of forest-dependent people, specifically the adivasis, from the forests. The possible reasons for the failure of these Acts are procedural as well as epistemological. At the procedural level, the law makers or the so-called 'experts', being trained and governed by the top-down development ethos, have only considered the adivasis as subjects to be governed. Their active participation in the whole process of governance has never been sought. The problem is there at the epistemological level too. The planners and policy-makers having the 'expert' knowledge about forestry, forest, and environmental management do not regard the indigenous knowledge, practices, and cultures about sustainable forest management to be valid. Hence, policies about forest management and protection are drafted

---

[1] Lok Sabha Unstarred Question No. 831, quoted from IWGIA (2011).

neglecting the knowledge of those who have a lived experience about forest. Louis Bruyere, the President of the Native Council of Canada, describes it quite succinctly: •

> Indigenous peoples are the base of what I guess could be called the environmental security system. We are the gate-keepers of success or failure to husband our resources. For many of us, however, the last few centuries have meant a major loss of control over our lands and waters. We are still the first to know about changes in the environment, but we are now the last to be asked or consulted.... The most we have learned to expect is to be compensated, always too late and too little. We are seldom asked to help avoid the need for compensation by lending our expertise and our consent to development. (1986: 14–15)

Moreover, by considering the adivasis' interests on forests as antithetical to the growth, development, and management of forests, these Acts have farther alienated them. The prioritization of forest as an important resource for commercial and industrial development by the colonial as well as independent Indian state (Arora 1994) may be a possible reason for such a consideration. Fernandes argues in this context, "Forest policy is only a symbol of the present development pattern made for the few powerful, and leading to ever greater inequalities" (1984: 1848). Hence, the forest policies not only marginalized the forest-depending people economically but also did serious harm to the sustainable use and management of forests, a crying need for better stability of the environment. Considering the unsatisfactory performance of all these in redressing the 'historical injustice' meted out to the adivasis, the state of India has come up with its latest initiative by enacting the Forest Rights Act, 2006. Under such a gloomy existential context of the adivasis, the enactment of yet another act raises many questions and concerns. Advocates of the Act may regard it as an important safety net for the most vulnerable segment of the Indian populace provided by the pro-reform state. But the question arises, how safe is the safety net after all? This actually boils down to the point of getting hold of the chief motive of the state, behind the enactment of such a 'landmark' Act. A critical evaluation and analysis of the possible implication of the Act for the adivasis seems necessary here.

Forest Rights Act: Implementation Mechanism

The Forest Rights Act, by distributing title deeds, recognizes and vests forest rights and occupation in forest land to "forest dwelling Scheduled Tribes and the other traditional forest dwellers" to assure their (a) rights over forest land, (b) use rights over forest products, and (c) right to protect and conserve forests. As the present chapter tries to analyze the impact of the Act on the adivasis ('Scheduled Tribe' for the purpose of the Act) exclusively, so the issue of the other traditional forest dwellers will not be taken up as a focal theme here. In the framework of the law, the process of distribution of title deeds starts with the Gram Sabha (a body of registered electorates of a village comprised within the area of the Panchayat at the village level). The claims for such deeds are submitted to Gram Sabha by the claimants. The Gram Sabha constitutes a Forest Rights Committee (FRC) to verify the claims. Verified claims are then passed on to the Sub-Divisional Level Committee for examination, which subsequently forwards it to the District Level Committee for consideration and final approval of the record of forest rights. The entire process of recognition and vesting of forest right is monitored by a State Level Monitoring Committee formed by the State Government.

Challenges from the Environmentalists

The enactment of the Forest Rights Act has faced various challenges from the environmentalists. A section of them who advocate management of forest, wildlife, and other biodiversity with complete exclusion of adivasis, local communities, or forest dwellers have been skeptical about it because to them the involvement of the 'ignorant' adivasis in the system of forest management will be detrimental to the cause of forest protection. This position, however, stands contrary to the Rio Declaration, decisions of the Conference of Parties (COP-V) of the Convention on Biological Diversity (COP 5 Decision V/4 1994), and recommendations of the United Nations Forum on Forest (UNFF), which strongly advocate the active participation of the local communities in the management and protection of forest cover. The UNFF in its Fifth Session, asserting that sustainable forest management cannot be achieved without the participation of indigenous peoples (in the present context, the adivasis), refers to the Corobici Declaration that argues:

> Indigenous peoples provide concrete solutions to many of the issues facing humanity today and by strengthening indigenous peoples' roles through effective participation in areas such as forest management and sustainable development, indigenous peoples can contribute significantly to a sustainable future for all of humanity. (UNFF 2005:11)

The viewpoint of the section of the environmentalists, mentioned previously, is clearly in contradiction with the stand of these international agencies. Such a disagreement may, perhaps, be interpreted as the exhibition of the modernist techno-managerial inclination of the former disproportionately in favor of their anthropologically constructed bias against the 'other'. Strikingly, such a bias is apparent in the outlook of The Ministry of Environment and Forest, Government of India, too, which has criticized such an Act because of the apprehension that it will result in the depletion of the country's forest cover by 16 percent (Chakma 2005). The fact, however, is that over "60% of the country's forest cover is found in 187 adivasi inhabited districts where less than 8% of national population lives" (AITPN 2006). Hence, such an apprehension, perhaps, appears to be ungrounded as the fact clearly shows the expertise of the indigenous people to conserve forest. As a matter of fact, forest occupies such an important place in adivasi culture and worldviews that effective forest management without the necessary application of their everyday knowledge seems impossible. Guha (2012: 6) is also of the same opinion when he argues, "peasants and tribals have a stock of empirical ecological knowledge that can be valuably deployed in the sustainable management of forests and wilderness areas."

The Forest Rights Act and the Political Economy of Development: Issues and Concerns

The Government of India, working under the neoliberal framework since the early 1990s, has been pursuing a development strategy that is based on large-scale industrialization, SEZs, and various mega investment projects. Such projects, quite naturally, necessitate a huge amount of land acquisition, including forest land. It has already been mentioned earlier that to acquire land for these projects of scale a large number of people have been displaced or dispossessed, among which a sizeable section is composed of the adivasis.

If we take the example of the newly formed state of Jharkhand, which is home of a good number of adivasis of India, a clear picture about this may be revealed. Up to August, 2009, Jharkhand has signed 71 Memorandum of Understandings (MoUs) for mega investment projects under private ownership, the total project cost of which amounts ₹2,958.5775 billion. For 55 of these MoUs signed between 2004 and 2006, 98,547 acres of land needs to be acquired (*Business Standard* 2009). Obviously, displacement and land alienation of the adivasis will magnify manifold due to these. The case is not different in other states such as Chhattisgarh, Orissa, and some parts of West Bengal. A study by ActionAid (2008) estimates that, about 1,204,522.64 acres (approximately 488,000 hectares) of land (private and public, including forest) has been acquired in four states alone (Orissa, Andhra Pradesh, Chhattisgarh, and Jharkhand) for projects of various kinds since 1995. In this context, enactment of the Forest Rights Act raises pertinent questions regarding the commitment of the state to protect its most vulnerable section of people, which may stand in the way of achieving its much publicized goal of large-scale economic 'development'. A close scrutiny of the implementation scenario of the Act may provide some valuable insight into the dynamics of this dialectics.

Implementation Scenario of the Forest Rights Act

The Forest Rights Act was enacted in 2006 but its implementation started from 2009. Since then, up to September 30, 2013, more than 3.5 million claims have been filed, out of which more than 3 million (more than 86.96 percent) have been disposed. More than 1.4 million titles (39.75 percent) have been distributed (GOI 2014). From the available data it is quite clear that the overall rate of claim rejection (percentage of claims rejected over number of claims received) is very high, even higher than the rate of claim approval. In the adivasi heartland of central India, the rejection rate is, indeed, alarming. Up to September 30, 2013, the rate of rejection in Chhattisgarh is 55.97 percent, Jharkhand 40.37 percent, Orissa 25.75 percent, and West Bengal 58.19 percent.[2]

The Ministry of Tribal Affairs, Government of India, however, has put forward a number of reasons for such a high level of rejection. The important ones being: 'Nonavailability of written records', 'Nonpossession of forest land', 'Multiple claimants', and 'Doubtful tribal status'.[3] The reasons of claim rejection mentioned by the government may be true, but more often than not are also obvious. As far as nonavailability of written records is concerned, it should be kept into consideration that written records as proof of anybody's property are a condition set or required by the modern discourse of property. The adivasis' concept of property is far off from this. In adivasi

[2] Ministry of Tribal Affairs, for details visit www.forestrights.gov.in
[3] For details visit http://fracommittee.icfre.org/Important%20Order/MOTA%20communication%20July%202010.pdf, accessed on February 16, 2017.

culture and tradition, particularly where the urban influence is insignificant, community ownership of property is in vogue, which, perhaps, depends more on mutual consent, community trust, and the likes than any type of written records. Moreover, viewed from the angle of modernity too, one cannot possibly expect such written documents from the adivasis who simply do not realize the importance of these due to their low level of literacy and poor access to other fruits of development. Coupled with this, in the absence of land survey in many of the adivasi regions in India, availability of written records is a genuine problem. Hence, the conventional mode of top-down governance seems to be incompatible with the adivasi cultural heritage and worldviews. A new form of governance empowered by a thorough understanding of indigenous knowledge and belief system needs to be evolved. Needless to mention that it requires a different management approach marked by the active and meaningful participation of the adivasis. Similar is the case with nonpossession of forest land, if possession or nonpossession depends on written records or any other formal documents of proof. Nonpossession of forest land primarily is due to the forest laws and policies formed during the colonial period and after, formation of Reserved Forests, Protected Forests, National Parks, and Sanctuaries, which have already dispossessed a huge number of forest-dwelling and forest-dependent adivasis. The problem is also with the process of demarcation of forest land. Sundar (2012: 19) argues in this context, "In many places, lands were never surveyed and later arbitrarily declared forest land." The case of Orissa is a pointer here, where during the last revisional land survey and settlement operation the Board of Revenue has issued an instruction that all lands beyond 10-degree slope (on which large numbers of adivasi families live and practice shifting cultivation) should be recorded as state land (Roy Burman 2008). As a result, hardly 1 percent of land under occupation of the various adivasi communities for centuries was recorded in their favor. Sarin points out the hopeless situation of the adivasis in the following words:

> The forest and revenue departments in Orissa "own" 50 to 80 per cent of the land in the State's Schedule V areas, while the vast majority of the tribals are left legally landless. A similar situation prevails in Andhra Pradesh where over 60 per cent of Schedule V areas have been declared Reserved Forests. (2010: 111)

In this fashion, a large chunk of forest land ceased to be as such and passed on to the State Land Revenue Department. If this is the case, even in the scheduled areas designated by the Schedule V of the Constitution of India, where the Scheduled Tribes' interests in land is supposed to be specially protected by the Governor of the respective states, the case of other areas can be understood very easily. Under this overall orientation of state activism toward dispossession of the adivasis of their forest land, the question of furnishing proof of possession of forest land, perhaps, is too much.

In the absence of land records, the issue of multiple claimants, another reason for rejection of claims cited by the government, for a same piece of land is quite natural. This is not only the case with the adivasis exclusively; rather, it is one of the root causes of land disputes throughout rural India. Finally 'doubtful tribal status,' as a reason for claim rejection, deserves some special attention. In India, the designation of any community as tribe is a disputed one since there is disagreement among the scholars and academicians regarding this (Xaxa 1999). Moreover, there are certain subtleties behind the concept of Scheduled Tribe and Tribe. Only those tribes that are listed in the schedule maintained by the President of India statewise are designated to be Scheduled Tribes. Hence, a tribe in any given state may not be treated as such in other states. A good example is that of the Santhal, a Scheduled Tribe in West Bengal, Jharkhand, and Orissa but not so in Assam. But Xaxa argues that the adivasis do not view them according to such administrative characterization, rather, "they view them in the sense of belonging to the same community irrespective of whether a group or segment of it is listed or not listed in the Constitution" (1999: 3595). Finally, the benefits of positive discrimination adopted by the Government of India for the socioeconomic upliftment of some of the disadvantaged social categories of India, including the Scheduled Tribes, often fail to reach most of the extremely marginalized adivasis. As they do not receive these privileges granted by administration, so, they also do not have any perception or experiential knowledge about their administrative categorization. Hence, the point of doubtful tribal status is also obvious.

Interestingly, the government does not make any mention about the nature and functioning of Gram Sabhas, factional feuds at the level of Gram Sabha, which are the crucially important sociological determinants of claim rejection. Chatterjee, while discussing the implementation of the Forest Rights Act in West Bengal, mentions,

> There was complaint about non-holding of Gram Sabha meetings in the proper form.… From these meetings people were supposed to know about the provisions of the Act and the rules. If such meetings were not held properly then the claims submitted without proper knowledge about the Act bore a high risk of rejection. (2012: 17)

Such nonholding of Gram Sabha meetings in the prescribed manner might also be due to factional feuds or criss-crossing political interests that are increasingly making the intended functionalities of almost every democratic institution in India today difficult. Allegedly, in every sphere of public activities, the influence and interference of the ruling party are becoming apparent. Such an imprint of the deteriorating party—government relationship—is, perhaps, making its presence felt at the Gram Sabhas too, the lowest level of decentralized democratic institution of India. The inclusion of the other traditional forest dwellers along with the Scheduled Tribes (adivasis) in the Forest Rights Act, however, has added more complexities into it. It is well known that the adivasis, as an ethnic group in Indian society, designates the nonadivasis as their 'other'. Perhaps the ruthless exploitation and cultural oppression inflicted on the adivasis by the nonadivasis, the so-called 'mainstream' Indian population, during the entire course of 'nation-building' in India may be a reason for this. The history of ethnic movements in India bears ample evidences of distrust and hatred of the adivasis toward the nonadivasis that have often resulted in bitter struggle between the adivasis and the nonadivasis. Hence, there is every reason to apprehend the extent the 'other traditional forest dwellers', the representatives of the nonadivasis, will, or rather, can allow the undoing of the 'historical injustice' meted out to the adivasis, the avowed spirit of the Act. This apprehension might turn out to be the reality in those villages or Gram Sabhas where the nonadivasis outnumber the adivasis numerically. Apart from such a dynamics in the Gram Sabhas, the issues of adivasi culture and worldview, their views about law, and attitude toward forest and its ownership are also crucially important considerations, which also the governmental view fails to address. At this stage, it is not out of place to have a deep introspection in the whole issue of rejection or approval of forest rights claims from a different perspective.

Claim Rejection or Approval: Whose Benefit After All?

It has been witnessed that scholars, intellectuals, NGO workers, planners, and policy-makers have raised their eyebrows over such a high rate of rejection of claims, particularly in the adivasi heartland of central India. But any critic of the neoliberal 'developmental' state policies should examine the whole issue from a different angle and find out the potential beneficiaries of the rejected or the approved claims. In case of rejection, the rejected forest land will be retained by the government. Along with forest-related purposes, according to the Forest Conservation Act 1980 and Forest Rights Act 2006, it can be diverted for nonforest purposes as well. Hence, this may facilitate mining, establishment of SEZs, and other infrastructural development projects under the aegis of capital, public as well as private. The much-debated and contentious process of land acquisition, in this way, may become rather smooth, legal, and, hence, legitimate. Viewed from this angle, the Act at the end might benefit the MNCs, other business houses, and specifically the timber merchants. Perhaps this is what Magdoff (2013) implies when he talks about dispossession by "using a variety of laws and agreements." The other side of the story, so to say the case of claim approval, is also not free from the apprehensions of dispossession. This is because the Forest Rights Act limits occupancy right to maximum 4 hectares (only!), but in actual practice, under community ownership, adivasis used to have access to a much larger area covering the entire village or forest. The Forest Rights Act, thus, poses a great threat to the adivasis' customary access to the Common Property Resources (CPRs). Moreover, experience suggests that the adivasis usually record only their homestead and land of immediate use to their favor in land survey and settlement operations. Hence, a huge amount of forest land, over which the adivasis used to enjoy customary rights, remains unclaimed, and this surplus land passes on to the government. In this fashion, the adivasis' traditional right over the whole forest faces the threat of severe violation. Possibly, the primary reason for this is the adivasis being completely alien to the legal discourse of property ownership. B.K. Roy Burman observed it in Totopara, a tiny settlement of the Toto, one of the so-called 'primitive tribes' of West Bengal (Roy Burman 2009). He narrated very lucidly how the Totos became completely marginalized due to the land survey and settlement operations aimed at their 'development'. They not only lost their land but also lost their home. Thus, naturally, any settlement of forest rights of the adivasis, as envisioned by the Forest Rights Act, may result in the 'legal' acquisition of the 'surplus' land by the government. Hence, approval of the forest rights claim can also result in the passing of a good amount of 'excess' or 'surplus' land to the government, which can be 'legally' diverted for nonforest use and purposes.

It is evident from the foregoing that whether the claims of the adivasis are either accepted or rejected, both ways the national and/or multinational business houses are going to be benefitted. From this account, the accountability of the neoliberal Indian state to the market forces is revealed. The whole rhetoric of granting legal rights to the adivasis over the forest land they inhabit, thus, can be analyzed as a means to pave the way for the integration of the Indian economy with the global economy. But it is indeed very unfortunate that such integration is achieved by dispossessing millions of autochthones of India from their customs, cultures, and livelihood. The ongoing struggles of the adivasis in different corners of India, especially those of the central Indian plateau region, hence, should be interpreted as a desperate attempt on their part to safeguard their identity and existence from the ever-increasing aggression of the state and the profit hungry market forces. In short, it is a clash between two different civilizations envisaged through the most radical differences in their respective social, cultural, economic, and political ethos or worldviews.

Conclusion

The Forest Rights Act, enacted by the neoliberal state of India, raises some serious apprehensions and concerns. It has already been noted that whether the claims of the adivasis are approved or rejected, the threat of dispossession is very much there. Following the Marxist logic of primitive accumulation, it can be said that this will make the wretched existence of the adivasis even more precarious. Apart from its economic dimensions, the dispossession of the adivasis from forest has its cultural dimensions too. To the adivasis, forest is not a mere arithmetic of economic profit and loss. They are, instead, traditionally and in some instances spiritually connected to the forests. It is the means by which they feel an attachment to their ancestors. The overall cultural life of the adivasis materially as well as symbolically veers around the forest. Forest provides subsistence and livelihood to the adivasis. Even more importantly, it imparts a sense of identity to the adivasis. Dispossession is not only a threat to their economic survival but it also leads to their cultural submergence, a well planned attack to their distinct cultural identity. By dispossessing the adivasis, the neoliberal state is not only successful in appropriating the natural resources over which the adivasis believe to have a customary right but it also weakens the possibility of collective resistance of the adivasis by destroying the very base of their cultural identity. Considered from a more general and substantive standpoint, the whole issue is indicative of the failure of the top-down development ideology. Over reliance of this approach on 'bureaucratic and managerial-style,' to quote Zavestoski and Swarnakar ('Introduction' of this volume), prompts it to regard the existential crises of the adivasis and other indigenous communities as an administrative problem to be solved by enacting laws and Acts from above. The implementation story of the Act, as dealt with in the present chapter, has revealed the futility of this effort. Through a number of anecdotal references, the present exercise has made it a point that effective and sustainable forest management is impossible without the meaningful participation of the indigenous communities at all levels, starting from the framing of acts and rules to the implementation phase. Indigenous knowledge, belief patterns, and cultural practices need to be the integral parts of any forest management policy that regards sustainability its primary goal. This may be a pointer to the necessity of the bottom-up approach of governance and development.

References

ActionAid, Indian Social Institute, and LAYA. 2008. "Resource Rich, Tribal Poor: Displacing People, Destroying Identity in India's Indigenous Heartland." New Delhi: ActionAid. Available at: http://www.indiaenvironmentportal.org.in/files/actionaid%20resource%20 rich%20report.pdf (Accessed on May 29, 2014).

AITPN. 2006. "India's Forest Rights Act of 2006: Illusion or Solution?" *Indigenous Rights Quarterly* 1 (2–3): 1–5. Available at: http://www.aitpn.org/IRQ/vol-I/issues-2-3/story03. htm (Accessed on February 26, 2014).

Arora, D. 1994. "From State Regulation to People's Participation: Case of Forest Management in India." *Economic and Political Weekly* 29 (12): 691–98.

Asif, M. 2000. "Why Displaced Persons Reject Project Resettlement Colonies." *Economic and Political Weekly* 35 (24): 2005–08.

Baviskar, A. 2012. "India's Changing Political Economy and its Implications for Forest Users." In *Deeper Roots of Historical Injustice: Trends and Challenges in the Forests of India*, 33–46. Washington, DC: Rights and Resources Initiative. http://rightsandresources.org/wp-content/uploads/2014/01/doc_5589.pdf (Accessed on September 13, 2015).

Bruyere, L. 1986. Quoted in "Our Common Future, Chapter 2: Towards Sustainable Development," 1–20. Available at: http://www.un-documents.net/ocf-02.htm (Accessed on December 20, 2015).

Business Standard. 2009. "Acquisition blues land Jharkhand in MoU heap" *Business Standard,* December 3. Available at: http://www.business-standard.com/article/companies/aquisition-blues-land-jharkhand-in-mou-heap-109120300010_1.html (Accessed on December 15, 2015).

Chakma, S. 2005. "Forest Rights Bill vs Environmental Extremism." *ACHR REVIEW.* Available at: http://www.achrweb.org/Review/2005/71–05.htm (Accessed on May 27, 2014).

Chatterjee, J. 2012. "How Red Is Lalgarh?" *West Bengal Sociological Review* III: 5–26.

COP 5 Decision V/4. 1994. *Progress Report on the Implementation of the Programme of Work for Forest Biological Diversity.* Available at: https://www.cbd.int/decision/cop/default. shtml?id=7146 (Accessed on March 16, 2014).

Fernanades, W. 1984. "The Forest 'Question'." *Economic and Political Weekly* 19 (42–43): 1848.

———. 2005. "The Impact of Displacement on Tribal Women." In *Contemporary Society: Tribal Studies,* edited by G. Pfeffer and D.K. Behera. New Delhi: Concept Publishing Company.

Fernanades, W. 1994. "An Activist Process Around the Draft National Rehabilitation Policy." *Social Action* 45 (July–September): 277–98.

GOI (Government of India). 2002. *Tenth Five Year Plan (2002–2007)*. Vol. II, Sectoral Policies and Programmes. New Delhi: Planning Commission. Available at planningcommission. nic.in/plans/planrel/fiveyr/10th/10defaultchap.htm (Accessed on May 12, 2014).

———. 2008. *Development Challenges in Extremist Affected Areas*. Report of an Expert Group to Planning Commission, New Delhi: Planning Commission. Available at: planningcommission.gov.in/reports/publications/rep_dce.pdf (Accessed on May 12, 2014).

———. 2009. *Draft Report of the Committee on State Agrarian Relations and Unfinished Task of Land Reforms*. New Delhi: Ministry of Rural Development. Available at: http://www. rd.ap.gov.in/IKPLand/MRD_Committee_Report_V_01_Mar_09.pdf (Accessed on May 12, 2014).

———. 2014. "Status Report on Implementation of the Scheduled Tribes and Other Traditional Forest Dwellers (Recognition of Forest Rights) Act, 2006 [for the period ending 30th September, 2013]." New Delhi: Ministry of Tribal Affairs. Available at: www.forestrights. gov.in (Accessed on February 27, 2014).

Guha, R. 2012. "The Past and Future of Indian Forestry." In *Deeper Roots of Historical Injustice: Trends and Challenges in the Forests of India*, 1–12. Washington, DC: Rights and Resources Initiative. http://rightsandresources.org/wp-content/uploads/2014/01/doc_5589.pdf (Accessed on September 13, 2015).

IWGIA. 2011. *The Indigenous World 2011*. Copenhagen: IWGIA, 346.

Magdoff, F. 2013. "Twenty-First-Century Land Grabs: Accumulation by Agricultural Dispossession." *Analytical Monthly Review* 11 (8): 1–17.

Pojman, L.P. and P. Pojman. 2012. *Environmental Ethics: Readings in Theory and Application*. 6th ed. Boston, MA: Wadsworth.

Reddy, M.G. and A.K. Kumar. 2010. "Political Economy of Tribal Development: A Case Study of Andhra Pradesh." Working Paper No. 85. Hyderabad: Centre for Economic and Social Studies. Available at: http://workspace.unpan.org/sites/internet/Documents/political%20 ecy%20cess.pdf (Accessed on May 11, 2014).

Roy Burman, B.K. 2008. "Ambiguities, Incongruities, Inadequacies in Scheduled Tribes and Other Technical Forest Dwellers (Recognition of Forest Rights) Act 2006." *Mainstream* XLVI (15). https://www.mainstreamweekly.net/article607.html (Accessed on February 14, 2015).

———. 2009. "What Has Driven the Tribals of Central India to Political Extremism." *Mainstream* XLVII (44). https://www.mainstreamweekly.net/article607.html (Accessed on February 14, 2015).

Sarin, M. 2010. "Democratizing India's Forests Through Tenure and Governance Reforms." *Social Action* 60 (2)(April–June): 104–120.

Sundar, N. 2012. "Violent Social Conflicts in India's Forests: Society, State and the Market." In *Deeper Roots of Historical Injustice: Trends and Challenges in the Forests of India*, 13–32. Washington, DC: Rights and Resources Initiative. http://rightsandresources.org/ wp-content/uploads/2014/01/doc_5589.pdf (Accessed on September 13, 2015).

UNFF. 2005. "High-level Ministerial Segment and Policy Dialogue with Heads of International Organizations." New York: United Nations Forum on Forest, Fifth Session. May 16–27. Available at: http://daccess-dds-ny.un.org/doc/UNDOC/GEN/N05/338/88/IMG/ N0533888.pdf?OpenElement (Accessed on December 25, 2015).

Xaxa, V. 1999. "Tribes as Indigenous People of India." *Economic and Political Weekly* 34 (51): 3589–3594.

3

Science and Politics of Wildlife Enumeration: Questioning the 'Tiger Count' in India

Jayanthi A. Pushkaran*

Introduction

On March 28, 2011, the Government of India (GOI) released its latest tiger count reporting a 10 percent increase from its 2006 figures. At the International Tiger Conference held at New Delhi, the then Minister for Environment and Forest Jairam Ramesh declared that the population of tigers, the key flagship species for conservation in India, rose from 1,411 to 1,706 since 2006. It was an important moment for the tiger conservationists as the numbers of tigers in India had risen for the first time in a decade. The gain in tiger population once again instilled hopes that India could still save the big cat that suffered a huge decline in the past century. The estimated tiger numbers soon made headlines worldwide. For Project Tiger agencies, these figures were significant at least in two respects. First, the figure was a testimony that their renewed conservation strategies have started bearing fruits. Second, the estimated figures were considered as more accurate than previous ones. Most importantly, the enumeration was carried out by using sampling-based estimation protocol developed by Wildlife Institute of India (WII) rather than the controversial pugmark census method (PCM).

The early discussions on tiger enumeration appeared in the writings of some of the foresters in the late 1960s. However, the strongest critiques of the official method of population estimation emerged in the scientific papers by tiger biologists. In the 1980s, when official reports of the Forest Department

* I would like to thank Dr Rohan D'Souza, Dr Mahesh Rangarajan, Dr Madhav Govind, Dr Bruno Latour, Mr Qamar Qureshi, and Dr Faiyaz Khudsar for their academic guidance, and Upma Manral and Ridhima Solanki for their support.

were extolling the nationwide population boom among tigers, several conservationists in India and across the world called into question the scientific basis for carrying out these tiger counts. As a result, alternative techniques such as camera trapping and radio telemetry were proposed by biologists. However, voices within the Forest Department and the Project Tiger Directorate consistently argued for sticking to the PCM. With the drastic decline in tiger numbers in 2004, the debate had assumed its most stringent tone. The PCM debate serves as a case for understanding the processes, agencies, and actors that were instrumental in generating different narratives of tiger demography. Empirical facts generated through scientific research being subjected to multiple interpretations are not new in ecological studies. An important task then is to understand how such disputes and differences over knowledge, methods, and techniques regarding wildlife research are resolved since decisions-making regarding which problems are important and what needs to be studied is found to be shaped by different actors. This is where factors such as power, interests, and affiliations of these actors enter the debate.

The tiger is recognized as a keystone species. The welfare of many ecologies is dependent on keystone species. Protection of tiger as the apex predator in its ecosystem entails protection of tiger landscapes, which also function as critically important watersheds serving 832 million people. About 71 percent of the tiger landscapes lie in one of the designated 25 biodiversity hotspots of the world (Karanth 2005). Thus actions to protect tigers and other species in their natural habitats are directly related to ecosystem services that are essential for human well-being and sustainability, including the provision of food and water and increasing people's resilience to the impacts of climate change. This is also one of the foundations of the Strategic Plan for Biodiversity of the Convention on Biological Diversity (CBD) (CBD 2010). The Millennium Development Goals (MDGs) prioritize basic needs in global efforts to reduce poverty. The importance of biodiversity for development is recognized under MDG 7 (ensure environmental sustainability) that includes the CBD 2010 biodiversity target to—reduce biodiversity loss (UNDG 2010). Biodiversity has a specific goal in the SGDs agenda (Goal 15) (Nilsson et al. 2013). The worrisome state of the tiger population and the urgency to take action for both present and for future generations, therefore, made it necessary to tackle the problem from a governance perspective in India. In several ways, the tiger became a barometer of how India is doing on the larger question of sustaining environmental quality in the face of ever-increasing demands on finite resources.

By focusing on the early debates of the tiger enumeration exercise and drawing upon the insights elaborated in the field of science technology studies, this chapter contests the received scholarship, which has viewed the disagreements and consensuses as internal contests between conservation sciences. It is argued here that such a simplistic treatment of positions has obscured the various vocabularies of science, politics, and diplomacy used by the state and scientific communities to define the population status of Indian tigers. A number of academic and popular writings on wildlife preservation in India and elsewhere, in recent years, have suggested that conservation efforts must be led by 'good science'. Scientific techniques and ecological data are not only treated as crucial inputs to policy prescriptions but, significantly as well, conservation interventions are often justified principally on scientific grounds. In several ways, therefore, good science is seen as the natural opposite, so to speak, to politics and social complexity. The tiger census debate in India, as I will argue in this chapter, however, shows otherwise. Despite the often times strident claims that the tiger count was a scientific endeavor, subject to rigorous protocols for analysis and objective documentation, the exercise was invariably questioned for its purported subjective biases and institutional prejudices and marred by allegations of being driven by vested interests. Put differently, the 'good science' for conservation can be opened up to reveal complex arrangements of power and knowledges that are inflected and shaped by political, cultural, and social logics.

This chapter is divided into five sections. The first section discusses the logic of quantification in conservation science. The second and the third sections explore the science and politics underpinning the top-down tiger conservation approaches and tiger enumeration methods in India, respectively. The fourth section attempts to understand the sociopolitical construction of tiger numbers and counting methods after the Sariska Tiger crisis of 2005. The last section discusses how the rule of number shaped the knowledge and institutional practices to conserve tigers in India.

Conservation Through Quantification

The emergence of the word biodiversity, one of the most important issues of our times, has redefined the way nature is perceived, treated, and managed. It is not merely a concept that naturalists use to describe connections among organisms and the various components of natural world, but it has also assigned a particular notion of privacy, protection, and control to species and their habitat. The broad definition of the term generally refers to the diversity of life on earth. However, the formal definition adopted in conservation efforts as implied in Article 2 of CBD describes biological diversity as "the variability among living organisms from all sources including, inter alia, terrestrial, marine and other aquatic ecosystems and the ecological complexes of which they are part; this includes diversity within species, between species and of ecosystems" (CBD 2005). According to the CBD, the current gaps in reliable estimate of species can hamper our ability to conserve, use, and share the benefits of our biological diversity. Therefore, it is usually argued that biodiversity has to be identified and counted in order to manage it efficiently.

A major part of the research around this concept is focused on quantification of biodiversity, in terms of its presence, location, abundance, and function. Numbers, graphs, databases, and records become important in biodiversity conservation. They are prepared and used to convey results in an organized and standardized form to summarize the natural world, its complex events and trends, or sometimes to explain how a process was carried out. What is special about the language of numbers? Numbers are considered as a final product of a process or knowledge that is objectively produced. In addition, they impart universality and mobility to the things being counted that can easily be transported across the world, applied to coordinate activities or settle issues. Quantification, Porter (1995) argues, is a "technology of distance" that not only diminishes the need for personal trust but also is a tool for communication that goes "beyond the boundaries of locality and community." In other words, quantification is one of the important means by which scientific knowledge is constructed as an objective, global commodity. Implication for this translation is that the local becomes a subject to a global system of power relations.

Despite its global nature, conservation science can be understood as a situated local practice embedded in its own social context. Some of the recent studies in the field of sociology of scientific knowledge (SSK) and science technology studies (STS) have shown that scientific knowledge, data, and technology, like other social processes, are penetrated by various forms of interests, values, and politics (Haraway 1988; Jasanoff 2004). Studies of the production and use of ecological indicators and its application in policy processes have increasingly been found to be subjected to context-dependent interpretations, negotiations, and institutional interests (Lawrence and Turnhout 2010). Environmental data involve a process of rationalization where practitioners approach nature through their conscious or unconscious selection of species and location, observation, and recording and sharing of results that may seem devoid of motives and values. However, when data are drawn into the political sphere, the values, motives, and interpretations coded in the so-called innocuous knowledge can be discerned. It is relevant to tease out these elements, because often the nonscientific influences get occluded when the end product is portrayed as being natural, objective, and scientific.

Science of Tracking Tigers: Practices and Debates

Tiger is the national animal of India and often figures in the cultural and religious discourses as symbol of power and strength (Jackson 1999; Thapar 1998). As a charismatic species in need for urgent protection, tiger emerged as a symbol for larger national ambitions in India where it gained attention of both federal and state governments (Lewis 2005; Sivaramakrishnan 2011). These cultural and political values are also reflected in conservation discourses, including the notion of flagship species in India and also internationally through projects such as the Global Tiger Recovery program (Seidensticker 2010). In comparison to other felids, tigers also figure notably amongst international funding agencies backing research in India, which is justified on the grounds that it is home to more than 50 percent of the world's tiger population. The dominance of tigers in the wildlife research is further supported by wildlife policies and resources provided by the GOI. In effect, tigers are an important part of the political project of conservation in India operated through scientific discourses, the priorities of funding agencies, and the Indian administration framework, which mediates permits and grants.

Tiger hunting was rampant in the colonial period. Tigers were tracked by foresters and hunters for hunting purpose (Johnsingh and Goyal 2005; Rangarajan 2005). In his several accounts of tiger hunting, Jim Corbett describes pugmarks of the tigers as one of the 'jungle signs' that is examined to interpret gender, age, and sometimes physical condition of the individual tigers (Corbett 1944, 1957). Around the 1930s, two articles appeared in the *Journal of Bombay Natural History Society* that talked about the importance of tiger tracks and its relation to animal locomotion (Pocock 1930, 1939). Champion claimed that the age, sex, and other physical characteristics of a tiger could be determined by studying its tracks (Champion 1929). Soon after these findings, foresters started enumerating large carnivores such as tigers, lions, and leopards by tracing their pugmarks. The first organized census to count the lion population in Gir forest was conducted by the then Junagarh state (located presently in Gujarat) in 1936. The census was carried out by about 700 laborers who counted and measured pugmarks at waterholes. The census was based on the assumption that carnivores visited the waterholes once or more to drink; therefore, the probability of sighting the pugmarks was more around such water bodies (Dalvi 1969). In 1955, the Indian Board for Wild Life (IBWL) in one of its meetings acknowledged the necessity for proper compilation of wildlife statistics. In the same year, the board requested Dharamkumarsinhji, a renowned ornithologist from Gujarat, to prepare a guide for the census operation in India.

The field guide created by him entailed a general introduction of the census methods, their practical application, and detailed notes on 20 different animal species. The tiger census proposed by him included direct and indirect census methods. The indirect method basically involved counting and examining ground evidence such as paw prints or tracks. After ascertaining the tiger locality of the tigers in beats,[1] preliminary census through what he calls 'Track Method' was to be conducted. This method involved collection of the measurements of the diameter of pads and the measures of the rectangular measurement of the front feet to detect the gender of the tiger using bamboo splints (Dharamkumarsinhji 1959). The next stage of the enumeration exercise was direct visual counts that involved fixing of observation posts to observe the tiger using baits. The direct method included visual counts through the vantage points and organized drives.

The methods suggested by Dharamkumarsinhji were based on his experience in the field and the information gained from the expertise of big game hunters and naturalists. These methods, though proposed in the field guide, had not been tried on the field at the time of publication. The guide, however, was the first attempt to codify procedures for scientific enumeration of some of the important wildlife species in India. The so-called scientifically devised methods by Dharamkumarsinhji did not go down too well with the foresters for various reasons. A Conservator of Forests from Gir pointed out that the census methods required a large number of live baits and were too expensive and time-consuming exercises (Dalvi 1969). The strongest critiques of these census methods appeared in the writings of Saroj Raj Choudhury, a forester and senior research officer who also taught the Wild Life Education course at the Forest Research Institution (FRI), Dehradun. His main contention to the method was that despite its huge cost, the method did not yield an absolute count. According to him, the estimate derived through the proposed method was nothing but a 'rational guess' (Choudhury 1970a). He started counting tigers in Orissa and developed a method that he termed as Cooperation Census Method (Choudhury 1970b). He evolved a tool that

---

[1] A beat is an administrative unit, 15–20 sq. km in average size, delineated primarily on natural boundaries.

he named Tiger Tracer to copy the pugmark in the field. It was a colorless, rectangular glass piece of 15 cm by 20 cm with four adjustable props at the corners. The tracing technique was grounded on the assumptions that the track prints of individual tigers are distinctive and an expert can identify them by visually inspecting their pugmarks tracings. The method involved locating the tracks and obtaining either a plaster-of-paris cast or a hand tracing of the tiger's left hind foot from the tracks in the field. The track prints were compared to identify each tiger on the basis of supposed differences in shape and measurement (Sharma and Wright 2005).

A country-wide tiger census using Cooperation Census Method was done in 1972 by Forest Department personnel. Following the commencement of Project Tiger, pugmark method became the standard method to enumerate tigers in Indian reserves. The conservation scenario in India changed considerably after the initiation of Project Tiger in 1972. First, it signified the recognition of wildlife conservation as a priority in official agendas for conservation. Second, the idea of wildlife having an inherent importance in the natural world formed the basis of subsequent conservation efforts. However, what constituted a trophic level for ecologists became a parameter of success for the state, and what represented a lab for the former became an administrative unit for the Forest Department. Within this new configuration of knowledge and power, tiger became a symbol and index of ecological health. For the foresters, tigers in India could be credibly counted by studying their pugmarks. For the biologists, however, such a method was flawed.

Widespread acceptance of PCM within the Indian forestry circle and the population estimates based on it received criticism as more ecological facts about tigers came to be known (Schaller 1967). Some of the earliest ecological and behavioral research on tigers was carried out in 1964 in Kanha National Park by George Schaller, a biologist from New York Zoological Society.[2] In 1972, intensive studies on tigers were taken up in the Chitwan National Park in Nepal by American biologist John Seidensticker using techniques such as radio telemetry that generated ecological data on dispersal pattern, area requirement, and relationship of tiger with other carnivores (Weber and Rabinowitz 1996). Radio telemetry was invented in the United States to study the elusive carnivores. The method is deployed to study tiger behavior such as ranging pattern, use of habitat, and space requirement. Its high cost and logistic complications involved in the capture, handling, and tracking of tigers makes it difficult to apply it extensively over the entire area of interest

[2] Schaller was a biologist from the New York Zoological Society who carried out the first long-term study on tiger in the Kanha National Park in India.

and to cover the animal population (Karanth and Nichols 2010). Meanwhile, Indian Forestry Service (IFS) and the Indian government attempted to reassert national control over the practice of wildlife research and conservation away from the influence of nongovernmental Indian ecologists and their foreign collaborators (Lewis 2003).

Choudury's paper (Choudury 1970b) on the utility of track prints resulted in several investigations being undertaken by tiger ecologists. Based on the ecological studies of prey selection in India, Nepal, and Thailand, biologists demonstrated that densities of tiger in a given area depended on how their prey community was structured (Karanth and Nichols 1998). Ullas Karanth, a carnivore biologist, who studies tigers in India for the World Conservation Society (WCS), started his study in the Nagarhole Reserve in Karnataka. His studies suggested that the population rise reported in the tiger censuses in India was not consistent with the normal population trend of tigers observed in other ranges. Drawing on the findings of tiger ecologists such as Schaller, Seidensticker, Sunquist, and Smith, he derived the prey requirements of each predator and then tested prey biomass in the different parks against the estimated figure of tigers in the same. By showing several inconsistencies between the prey availability and the reported tiger densities, he demonstrated that official censuses overestimated the tiger numbers in many parks (Karanth 1988). He pointed out problems with some of the biological understanding regarding tiger behavior such as its pairing habit and social organization on which the census method was premised. Several other studies stressed on the fact that felids such as tigers, lions, and leopards were difficult to census; therefore, conservation efforts should concentrate on long-term trends rather than absolute numbers.

Ecologists in various parts of the world had begun suggesting that population monitoring methods should focus on deriving reliable indices of tiger density, and assessing prey densities and habitat availability (Wikramanayake et al. 1998). However, wildlife conservation in India largely remained impervious to the newly gained insights on tiger ecology. In fact, during the first two decades of Project Tiger, there was an urgent need for forestry staff trained in wildlife sciences, and the training offered in forest management at FRI had almost nothing on wildlife. Until 1973, wildlife research was not a priority and wildlife sciences had not gained importance as a subject. In effect, members of the IFS who were deployed to conduct research lacked the expertise required for wildlife conservation, and many research posts in Project Tiger remained vacant. In order to generate expertise in the field of wildlife conservation, the Directorate of Wildlife Research and Education started a 10-month postgraduate diploma course for forest officers in 1974.

This Directorate of Wildlife Research and Education became the WII in 1982. The WII was authorized to be a research institution in 1986, and started offering certificate courses and short-term courses in wildlife science to foresters. Despite these courses, only a small number of foresters were trained in wildlife management (Rahmani 2008). As far as wildlife research was concerned, frequent postings coupled with administrative tasks made it difficult for forest officers to conduct any long-term research (see Lewis 2003). Annual tiger census constituted the major chunk of formal research activity for many years.

Tiger biologists came up with another method known as the camera trapping or photographic capture-recapture method. This method used stripe pattern as a feature to identify the individual tiger. When a tiger was photographically captured or marked, the data obtained from subsequent recapture were analyzed within the theoretical framework of the formal capture-recapture theory (Karanth and Nichols 2010). The purpose of the photographic capturing is to build capture histories for each tiger and use them to estimate tiger abundance through statistical models. When tiger biologists recommended this method for tiger monitoring in India, the Forest Department avoided it for the high cost of equipment and risk of their theft (Jhala et al. 2008). Owing to the sustained criticism of PCM, for the very first time in the late 1990s, a brief note was sent to the State Forest Departments to exercise caution while applying PCM (GOI 2005).

Project Tiger: A Numbers Game

Toward the late 1960s, the conservation communities across the world started showing concern about the threat to several wildlife species and their habitat loss in India. Particularly, an influential set of US and Indian biologists and international environmental advocacy organizations such as World Wildlife Fund (WWF) were vocal in making policy recommendations for the conservation of endangered species in India. In the 10th general assembly of International Union for Conservation of Nature and Natural Resources (IUCN), held in New Delhi in 1969, European experts emphasized on the vulnerability of tigers and highlighted that the forests of India, Bangladesh, Nepal, and Bhutan were the last remaining spots for the tiger's future (Seidensticker 1997). These global organizations regularly drew international attention to India and its potential to conserve the diminishing population of tigers. In turn, a special request was made to the GOI to strengthen and formulate legal measures for tiger protection (IBWL 1970; IUCN 1969). The following years saw the Indian government responding to those clarion calls through the launch of Project Tiger. Guised in the ideology of conservation, this project was a product of a distinctive kind of science, diplomacy, and politics. The early 1970s was marked by diplomatic rifts between India and the United States after the Indo-Pakistan War. Around the same time, GOI sought to limit the presence and role of US funds and biologists to the matters related to wildlife in Indian landscape (Lewis 2003). It was mirrored by the emergence of an Indian government-dominated conservation research and planning in place of an internationally oriented and ecologist-dominated conservation of the 1960s.

In April 1972, Guy Mountford, an international trustee of WWF, visited India and met then Prime Minister Indira Gandhi to apprise her of the grim situation of tiger population in India. Shortly afterwards, a group of specialists were appointed by Indira Gandhi to study the situation of the tiger population and to plan strategies to conserve them. From its inception, Project Tiger got the enthusiastic support of Prime Minister Indira Gandhi, renowned for her concern for environmental issues. As a powerful patron, her personal interest in wildlife conservation and her connections with key conservationist in the scientific community and forest bureaucracy proved crucial for Project Tiger (Singh 1972). The project was one of the biggest conservation initiatives of its time. Although a substantial amount of funding for the Project Tiger was provided by WWF, IUCN, and other bilateral organizations, its implementation and planning lay entirely with the Indian Forest Department. By taking up tiger conservation, one of the key concerns for the conservation community around the globe, the GOI was also trying to elevate its international standing.

A five-year moratorium on tiger shooting was placed, and IBWL assembled a Task Force to evolve a project to preserve tigers in the wild. It was also realized that prior to the formulation of any plan, the status of the tiger population in India had to be known. The earlier estimate of tiger population as given by E.P. Gee was 40,000 at the end of the nineteenth century and about 4,000 in 1965. A statewise estimate done by Kailash Sankhla, a forester who later on became the director of Project Tiger, indicated around 2,500 in India. The members of the Task Force suggested organizing a census through a scientific method in order to get a clearer picture of the tiger population. A census team was given a short training, after which the first systematic countrywide tiger census was conducted in 1972. After the entire operation, the team arrived at a total figure of 1,827. Although the earlier figures were considered as overestimates, the 1972 tiger numbers revealed a dismal picture. The Task Force, after studying the field data and tiger census report, prepared a project outline for the conservation of tigers and other wildlife species in India.

In a subsequent series of legal and administrative actions, the GOI enacted the 1972 Wildlife Protection Act (WLPA) and issued guidelines for the project. The Act set out the legal basis for establishing and managing national parks and wildlife sanctuaries in India. It specified legal measures to provide protection to the endangered and threatened flora and fauna, and set restrictions on their possession, hunting, taxidermy, and trade. The Task Force, in its report, indicated that increasing demographic pressures in tiger-occupied regions were leading to the diversion of forest to agriculture, which in turn was causing the shrinkage of the tiger habitat. The Task Force's suggestions were endorsed in August 1972 and Project Tiger was launched in 1973.

The overall plan of the project included setting up several patches of habitat for the tiger that would serve as a "breeding nucleus from which surplus animals can migrate to surrounding forest" (IBWL 1972). The 1972 census had indicated tiger presence throughout India. However, the Task Force selected eight potential areas in eight different states in India in order to include as many biogeographic types of areas as possible and for the ease of implementation.[3] Eight reserves were established in 1973 and a ninth reserve was added in 1975. By 1983, six more reserves were created. The expansion of areas under tiger reserves was paralleled with the forced removal of forest-dwelling rural population. A critical component of the conservation regime was a singular management approach grounded on scientific approach entailing enclosed, regulated, and people-free landscapes. Local forest use was considered as the sole barrier to effective conservation. This hands-off conservation approach was devoid of any mechanism to provide the displaced population livelihood alternatives or to share the benefit of conservation (Kothari 1995). By the late 1980s, conservation efforts evoked strong opposition and resistance from the forest-dwelling rural people who bore the brunt of losses from the exclusion (Rangarajan 2005). During the late 1980s and throughout the 1990s, the management of Project Tiger was on the decline. With the emergence of new political leadership, the political commitment to conservation had decreased considerably and the economic reforms in 1991 facilitated diversion of forest land for other development activities (Rangarajan 2001).

---

[3] The Task Force relied on Resolution 25, adopted at the IUCN general assembly in 1969, that recommended that national parks and reserves should be created in a manner that different biogeographic types were preserved.

In the 1970s, wildlife preservation underwent a major shift with respect to the key actors influencing the conservation agendas and wildlife research. In the 1950s and 1960s, research organizations such as Bombay Natural History Society (BNHS), its US collaborators, and scientists frequently made policy recommendations and suggested changes in protected area management. Some of the major ecological researches carried out in India were being conducted and funded by US institutions such as the Smithsonian and the US Army. However, with the passage of WLPA in 1972, the power to manage protected areas and conduct research for the same was formally given to the IFS.

In 1969, when wildlife experts across the world were trying to find out effective methods to enumerate tigers, the Forest Department of India decided to use the pugmark method to conduct tiger census. When the Smithsonian offered to help IFS by sending a team of US scientist to assist the census team, K.S. Sankhla refused to accept any foreign expertise (Botteron 2001). In effect, the field managers in India continued with the method against the advice of IUCN. Numbers were an important feature of Project Tiger. One of the justifications for the project was the steep difference between the population estimate of 40,000 tigers in 1900 given by E.P. Gee and the 1,800 tigers estimated in the census carried out in 1972. Therefore, the performance of the project was assessed in terms of increased tiger numbers.

Soon after the launch of Project Tiger, the managers of Indian wildlife reported huge improvements in the population of tigers. For instance, describing the triumph of Project Tiger, H.S. Panwar, a forester, stated,

> The tiger population rose from 268 in nine reserves in 1972 to 757 in 11 reserves in 1981. The total Indian tiger population rose from 1827 to 3015 between 1972 and 1979, an increase of 1188 or 62 percent in seven years. The swamp deer, elephant, rhino and wild buffalo also thrived in the reserves, especially in the core areas. (1982: 336)

However, a mid-term review of Project Tiger in 1976 by IUCN raised doubts on the reported increase and recommended field managers to exercise more caution while carrying out census:

> Statements on increase in tiger populations, in particular, are fraught with hazards. In the first place the pugmark census method, whilst almost certainly available for this situation, can only be regarded as an approximation of the true figure and, in any case, few Tiger Reserves have undertaken regular pug mark counts since 1972. Any future census will need to be subjected to the same stringency or the result must be clearly stated to have been analysed and

compiled by somewhat different methods, otherwise the figure will have little comparison value and give quite erroneous impression of the trend. (Holloway et al. 1976: 23)

The rise in tiger population, however, continued in the official data in the later half of the 1970s. This period coincided with the imposition of emergency during Indira Gandhi's prime ministership, when most of the official data reported improvements. A few state governments reported grossly exaggerated tiger population figures in the reserves. Some of the leading national newspapers started questioning these claims of enhanced tiger numbers when a reserve reported as high as 100 percent population rise in just two years. A couple of other reports came out that questioned the unusual increase in tiger populations and raised suspicion over the counting methods (Greenough 2004). An alternative method called camera trapping was proposed by tiger biologists who claimed it to be more effective in tiger monitoring. However, the Forest Department argued that the pugmark method was highly accurate (Pawar 1979).

Around 1988, the overall numbers of tiger started declining. The crisis became apparent when the tigers could no longer be spotted in the Ranthambore National Park in the state of Rajasthan. The park authorities denied these reports, but the raids by local police revealed that at least 20 tigers were killed in Ranthambore since 1989. Rampant poaching of tigers continued in other states of India. Madhya Pradesh was one such state that was affected the worst. Several hundred cat skins were seized from the districts of Jabalpur, Mandla, Balghat, and Satna (Currey 1996). Despite the dismal state of affairs, the forest officials in Madhya Pradesh announced an increase in tiger population even before the completion of the estimation process (Thapar 1998; *Tigerlink* 1997). The biologists once again held the flawed techniques of tiger monitoring as the prime reasons why the huge decline in the tiger population went unnoticed. However, the Project Tiger Directorate lent its support for the PCM. A letter circulated among the Chief Wildlife Wardens in 2001 stated:

Estimation of tiger population, based on pug marks and related evidences, still remains the most cost effective and time tested methodology suited to our conditions, which if carried out due care in a systematic manner, can lead to authentic results. (Gopal 2001)

The next reserve where tigers went extinct was the Sariska Tiger Reserve in Rajasthan. The disappearance of the tigers was first observed in September 2004 by the students of the WII (*Down to Earth* 2005). After initial denial, the authorities confirmed that tigers in the reserve had been poached (Mazoomdar 2005). The local extermination of tigers in Sariska claimed headlines in national and international media. The situation highlighted the failure of the exclusionary reserve management in India that has historically been dictated by the Forest Department without significant reference to both science and public opinion.

Tiger Crisis in India: National Issue of Biodiversity Conservation

Sariska was a highly guarded tiger reserve and was frequented by biologists, bureaucrats, and politicians. Previously a royal hunting reserve, Sariska is spread over 866 sq. km along the Aravalli hill range in Alwar district of Rajasthan, India. With high density of prey base located within 120 sq. km of intact forest, it harbors diversity of predators, particularly tiger, hyena, jungle cat, leopard, and wild dog. It has been regarded as one of the key tiger conservation sites in the semi-arid tracts of northwest India. A number of critical debates raged in the aftermath of the Sariska debacle, especially related to the management of tiger reserves. To further explore the tiger crisis, the GOI constituted the Tiger Task Force (TTF). It recommended increasing the involvement of biologists and wildlife researchers in India to help plan future management of tiger reserves.

The TTF held several consultation sessions with experts and scientists working on population estimation in India. The technical note and the field guide, prepared on the new methodology by the director of Project Tiger and the scientist from WII, were circulated among a team of Indian and international tiger experts for their opinions (GOI 2005). Criticism largely revolved around the technical issues and feasibility of the method. The experts pointed out that the technical note and the field guide were not scientifically peer reviewed and published. One major concern expressed by biologists was the reliability of data obtained by Forest Department staff, who are not trained in wildlife research and in the past had shown lack of ethics in reporting tiger numbers. Experts argued that there was no guarantee that the Forest Department staff would follow the protocol. Given the poor track record of the official tiger population monitoring and hostile attitude of the Forest Department toward wildlife researchers in India, they even went on to suggest that the exercise be entrusted entirely to the biologists (Karanth 2005). Wildlife biologists have often found difficulties with obtaining permission from the Forest Department to carry out research in the protected areas in India.[4] Researchers have bemoaned the delay and lack of transparency in processing and issuing the research permit. Such hindrances, biologists complain, limits the conduct of ecological research, which strictly hinges on time- and area-specific data-collection and is subject to seasonal schedules (GOI 2005; Madhusudan et al. 2006; Shahabuddin 2007).

The Project Tiger Directorate approached WII, Forest Survey of India (FSI), and the Forest Department to refine the estimation procedure and to develop a holistic monitoring program for tigers and their habitat. By 2004, the directorate deployed experts and asked them to develop a monitoring protocol that was simple, practical, and statistically sound so that the frontline staff of the Forest Department could implement it. WII researchers developed a four-stage method aimed at determining spatial occupancy of tigers throughout the potential tiger forest and sampling those forests using camera traps in a statistical framework. It also represented a shift from exclusive concentration on tiger number and protected-area-based approach to landscape-level monitoring. The new protocol included methods of enumeration coupled with assessments of habitat and prey population in order to get unbiased estimates of tiger population within the desired statistical level of accuracy. In 2006, the new methodology was approved by the National Tiger Conservation Authority on the recommendation of TTF. The 2008 All India Tiger Population Estimation was carried out using the new technology. The total country-level tiger population was estimated as 1,411 (mid-value), and lower and upper limits being 1,165 and 1,657, respectively.

Sariska crisis brought to the fore the failure of state-centered efforts to protect wildlife in India. A number of intersecting concerns emerged in response to the situation that included long-term causes of the conservation crisis; lack of transparency in forest management; exclusionary conservation approach; inflated official counts; and outdated techniques of tiger monitoring. Central to these debates was PCM, which came in for criticism from all quarters.

In sum, the process of knowledge production in Project Tiger was embedded in a distinctive mechanism of governance wherein terms were mostly articulated by the Forest Department. Moreover, the construction

---

[4] Wildlife research often requires manipulation of the natural conditions such as capturing of animals for marking and radio tagging, manipulating habitat for experiments, setting fire to small plots of vegetation, and repeting some of these experimental conditions after regular intervals. However, Forest Departments often do not allow any kind of research that entails handling any kind of animal (touching, holding, or even wire tripping them for a camera shot).

of progress was implicated in the incentive mechanism of the Project Tiger agencies. Given a lack of independent scientific estimates of tiger numbers in Indian reserves, the legal and administrative mechanism of forest bureaucracy, in fact, gave rise to an exclusionary conservation regime and politically motivated knowledge about tiger population. PCM, in several ways, gradually became an important tool to achieve such a goal. The biologists' challenge, however, was often compromised by the fact that they were only sparingly given access to the forests in order to conduct wildlife research in India. In other words, the presence of the foresters and forest administration was overwhelming in all the tiger reserves and sanctuaries. In several ways, therefore, many of the biologists were unable to sustain their science and research agendas as transparent, open, and inclusive exercises.

Dialectic of Tiger Conservation: Science and Politics

Conservationists maintain that effective conservation critically depends on sound science. In part, as professional biologists argue, given the complexity of the natural world, effective conservation strategies can only be crafted through rigorous and credible science-based methodologies. Science, hence, institutional or otherwise, should provide the only acceptable formats for producing empirically valid practices and objectively generated data. Such arguments have, in fact, carried the day in most efforts at formulating approaches for wildlife preservation practices and habitat conservation policies. The tiger enumeration debate, however, has shown how official efforts to estimate tiger population in India, perceived as a purely scientific exercise, is actually a wider project of governance entangled in various political and social interests.

The enumeration debate has revealed two important concerns. The first is the manner in which the Indian state has devised a project of nature-making embedded within a dialectic of conservation but linked to a particular constellation of knowledge and power associated with Project Tiger. Integral to this activity is the creation of concepts, objects, and processes that are used to justify key state activities such as formulation of law, policy, and administrative bodies. For such an arrangement, much of the scientific exercise is invariably translated through the prism of top-down power and governance imperatives.

Within this ideological configuration, the tiger at the apex of ecological pyramid is seen as an indicator of ecological well-being and performance of conservation efforts. Specific kinds of environmental narratives were built around the tiger and its ecology to enable conservation through quantification. In doing so, enumeration methods were standardized for the ease of administrative use. Furthermore, the tendency to pursue the 'rule of numbers' was intended to calculate visible and striking outcomes and demonstrate success. The logic of tiger numbers, hence, appeared critical to the legitimacy and funding of Project Tiger. The Forest Department was encouraged to define its ecological goals by the application of such types of 'science' by legal and administrative means, the end product of which was considered as an increase in tiger population. Even when the natural world did not seem to be responding to such inputs, the bureaucracy was often motivated to report improvement. In the absence of a feedback mechanism in Project Tiger, the flawed techniques and imaginary figures were hard to detect. Though individual researchers and different scientific communities pointed out this problem, they were not empowered or recognized within the government-sponsored conservation project in which the power to rule and make decisions lay in the hands of the Forest Department. During the situation of conflict or crisis, for instance in the case of the Sariska debacle, the state carefully assembled various actors, experts, organizations, and its agencies to review its function to build consensus over its strategies. This too was achieved by bringing some shift in the composition and role of existing functionaries and agencies without altering the structure and mode of control.

The second concern is the manner in which the state and scientific elites have constructed the question of tiger enumeration in India. Based on a reliance on biologists and their institutionally derived knowledge, the aforesaid, locally constituted meanings and practices related to the tiger have often been excluded. Such a distinction has been achieved by extracting tigers and the landscapes occupied by them from their historical and cultural contexts. In sum, ideas of scientific elites have taken precedence over any other imagination of tigers and their habitat by any other actors. Such a treatment occluded a range of political and social linkages that were crucial. Before the exclusionary top-down conservation approaches in India, people were connected with natural resources both intrinsically and materially. However, biodiversity conservation focused strictly on species protection without consideration for the humans who inhabited the same ecosystems. As a result, tension between local people and the protection authorities arose, and people lost their interest in conservation since it directly impacted their livelihoods. Studies in Nepal and Russian Far East have increasingly shown that a science-based approach coupled with close collaboration with local stakeholders helps to improve wildlife and habitat management, both within and outside of protected areas. Inclusion of local communities in resolving resource use issues and the application of robust tiger monitoring programs led to sustainable conservation interventions (Chapron et al. 2008; Dhakal et al. 2011; Miquelle et al. 2007).

When we unpack the various components of the tiger debate, it becomes clear that the claims for and against the PCMs were not purely scientific positions, but were rather marked by different kinds of institutional, professional, and individual interests. Hence, the many voices that questioned the biological and statistical basis for tiger population monitoring were not merely challenging the science rooted in conservation claims but were also aimed at criticizing the official conservation approach and its arrangements. Similarly, the methodology and technology used by Project Tiger agencies too were encoded with politics within which they were operated. Seen from this perspective, tiger conservation appears as a stage for the playing out of various kinds of interest and politics for the Indian government, forest bureaucracy, and scientific community.

References

Botteron, C.A. 2001. "India's 'Project Tiger' Reserves: The Interplay Between Ecological Knowledge and Human Dimensions of Policy Making for Protected Habitats." In *Applying Ecological Principles to Land Management*, edited by Virginia H. Dale and Richard Haeuber. New York, NY: Springer Verlag.

Champion, F.W. 1929. "Tiger Tracks." *Journal of Bombay Natural History Society* 33 (4): 284–287.

Chapron, C., D.G. Miquelle, A. Lambert, J.M. Goodrich, S. Legendre, and J. Clobert. 2008. "The Impact of Poaching Versus Prey Depletion on Tigers and Other Large Solitary Felids." *J. Appl. Ecol.* 45 (6): 1667–1674.

Choudhury, S.R. 1970a. "Let Us Count Our Tigers." *Cheetal* 12 (2): 41–51.

———. 1970b. "The Tiger Tracer." *Cheetal* 13 (1): 27–31.

CBD (Convention on Biological Diversity). 2005. *Handbook of the Convention on Biological Diversity.* Montreal: The Secretariat of the Convention on Biological Diversity.

———. 2010. *Strategic Plan for Biodiversity 2011–2020.* Montreal: The Secretariat of the Convention on Biological Diversity.

Corbett, J. 1944. *Man-eaters of Kumaon.* Oxford: Oxford University Press.

———. 1957. *The Man-eating Leopard of Rudraprayag.* Oxford: Oxford University Press.

Currey, D. 1996. *The Political Wilderness: India's Tiger Crisis.* London: Environmental Investigation Agency.

Dalvi, M.K. 1969. "Gir Lion Census, 1968." *Indian Forester* 95 (11): 741–752.

Dhakal, N., K.N. Nelson, and J.L.D. Smith. 2011. "Assessment of Residents' Social and Economic Wellbeing in Conservation Resettlement, a Case Study of Padampur, Chitwan National Park." *Society of Natural Resources Journal* 24 (6): 597–615.

Dharamkumarsinhji, K.S. 1959. *A Field Guide to Big Game Census in India.* New Delhi: Indian Board for Wildlife.

Down to Earth. 2005, March 15. "Maneaten," 26–28.

Gopal, R. 2001, November 27. "Letter to All the Chief Wildlife Wardens." No. 7–1/96 PT. New Delhi: Ministry of Environment and Forest.

GOI (Government of India). 2005. *Joining the Dots: The Report of Tiger Task Force.* Delhi: Ministry of Environment and Forests.

Greenough, P. 2004. "Pathogens, Pugmarks and Political Emergencies: The 1970s South Asian Debates on Nature." In *Nature in Global South: Environmental Projects in South and South East Asia,* edited by Paul Greenough and A.L. Tsing, 201–230. Durham: Duke University Press.

Haraway, D. 1988. "Situated Knowledges: The Science Question in Feminism and the Privilege of Partial Perspective." *Feminist Studies* 14: 575–599.

Holloway, C.W., Paul Leyhausen, and M.K. Ranjitsinh. 1976. *Conservation of the Tiger (Panthera tigris L.) in India: A Report to the Chairman of the Project Tiger Steering Committee on a Mid-term Study of Project Tiger, March/April 1976.* Geneva: IUCN.

IBWL (Indian Board for Wild Life). 1970. *Resolution of the Executive Committee of the Indian Board of Wildlife-Preservation of Tiger.* Delhi: Government of India.

———. 1972. *A Planning Proposal for Preservation of Tiger (Panthera tigris tigris Linn.) in India.* Delhi: Government of India.

IUCN (International Union for Conservation of Nature and Natural Resources). 1969. *The Tiger.* Geneva: IUCN.

Jackson. P. 1999. "The Tiger in Human Consciousness and its Significance in Crafting Solutions for Tiger Conservation." In *Riding the Tiger: Tiger Conservation in Human-dominated Landscapes,* edited by J. Seidensticker, S. Christie, and P. Jackson. London: Cambridge University Press.

Jasanoff, S. 2004. *States of Knowledge: The Co-production of Science and Social Order.* London: Routledge.

Jhala, Y.V., R. Gopal, and Qamar Qureshi. 2008. *Status of the Tigers, Co-predators, and Prey in India.* National Tiger Conservation Authority. Dehradun: Wildlife Institute of India.

Johnsingh, A.J.T. and S.P. Goyal. 2005. "Tiger Conservation in India: The Past, Present and the Future." *Indian Forester* 131 (10): 1279–1296.

Karanth, U. 1988. "Analysis of Predator-Prey Balance in Bandipur Tiger Reserve with Reference to Census Reports." *Journal of the Bombay Natural History Society* 85 (1): 1–8.

———. 2005, July 12. *Scientific and Technical Issues Involved in Monitoring Tiger Population in India.* Submission to the Tiger Task Force. New Delhi: Project Tiger Directorate.

Karanth, U. and J.D. Nichols. 1998. "Estimation of Tiger Densities in India Using Photographic Captures and Recaptures." *Ecology* 78 (8): 2852–2862.

———. 2010. "Non-invasive Survey Methods for Assessing Tiger Populations." In *Tigers of the World: The Science, Politics, and Conservation of Panthera Tigris,* edited by Ronald Tilson and Philip Nyhus. New York: Elsevier Inc.

Kothari, A. 1995. "Conservation in India: A New Direction." *Economic and Political Weekly* 30 (43): 2755–66.

Lawrence, A. and E. Turnhout. 2010. "Personal Meaning in the Public Sphere: The Standardisation and Rationalisation of Biodiversity Data in the UK and the Netherlands." *Journal of Rural Studies* 26 (4): 353–360.

Lewis, M. 2003. *Inventing Global Political Ecology: Tracking the Biodiversity Ideal in India, 1945–1997.* New Delhi: Orient Longman Private Limited.

———. 2005. "Indian Science for Indian Tigers? Conservation Biology and the Question of Cultural Values." *J Hist Biol,* 38 (2): 185–207.

Madhusudan, M.D., K. Shanker, A. Kumar, C. Mishra, A. Sinha, R. Arthur, A. Datta, M. Rangarajan, R. Chellam, G. Shahabuddin, R. Sankaran, M. Singh, U. Ramakrishnan, and P.D. Rajan. 2006. "Science in the Wilderness: The Predicament of Scientific Research in India's Wildlife Reserves." *Current Science* 91 (8): 1015–1019.

Mazoomdar, J. 2005. "Have You Seen a Tiger in Sariska Since June? If Yes Then You're the Only One." *The Indian Express,* January 22.

Miquelle, D.G., D.G. Pikunov, Y.M. Dunishenko, V.V. Aramilev, I.G. Nikolaev, V.K. Abramov, E.N. Smirnov, G.P. Salkina, I.V. Seryodkin, V.V. Gapanov, P.V. Fomenko, M.N. Litvinov, A.V. Kostyria, V.G. Yudin, V.G. Korkisko, and A.A. Murzin. 2007. "2005 Amur Tiger Census." *Cat News* 46: 11–14.

Nilsson, M., P.L. Lucas, and T. Yoshida. 2013. "Towards an Integrated Framework for SDGs: Ultimate and Enabling Goals for the Case of Energy." *Sustainability* 5: 4124–4151.

Panwar, H.S. 1982. "What to Do When You Have Succeeded: Project Tiger Ten Years Later." *Ambio* 11 (6): 331–337.

Pawar, H.S. 1979. "A Note on Tiger Census Technique Based on Pugmark Tracings." *Indian Forester* Special Issue: 70–77.

Pocock, R.I. 1930. "The Panthers and Ounces of Asia." *Journal of Bombay Natural History Society* XXXIV (63–82): 307–336.

———. 1939. *The Fauna of British India, Including Ceylon and Burma, Mammalia.* Vol. II. London: Taylor and Francis.

Porter, T.M. 1995. *Trust in Numbers: The Pursuit of Objectivity in Science and Public Life.* New Jersey: Princeton University Press.

Rahmani, Asad S. 2008. "Indian Conservation Service." *J. Bombay Natural History Society* 105 (3): 246–47.

Rangarajan, M. 2001. *India's Wildlife History. An Introduction.* New Delhi: Permanent Black and Ranthambhore Foundation.

———. 2005. "Battles for Nature: Contesting Wildlife Conservation in 20th Century in India." In *Shades of Green: Movement to Save the Planet,* edited by Christof Mauch, Nathan Stoltzfus, and Douglas R. Weiner. New York, NY: Rowman and Littlefield.

Schaller, G. 1967. *The Deer and the Tiger.* Chicago, IL: Chicago University Press.

Seidensticker, J. 1997. "Saving the Tiger." *Wildlife Society Bulletin* 25 (1): 6–17.

———. 2010. "Saving Wild Tigers: A Case Study in Biodiversity Loss and Challenges to Be Met for Recovery Beyond 2010." *Integr Zoolog* 5 (4): 285–299.

Shahabuddin, G. 2007. "The Endangered Wildlife Biologist." *Seminar* 577. Available at: www.india-seminar.com/2008/581/581_index.htm (Accessed on February 9, 2017).

Sharma, S. and Belinda Wright. 2005. *Monitoring Tigers in Ranthambore National Park: A Report Prepared for Empowered Committee on Forest and Wildlife Management, Government of Rajasthan.* New Delhi: Wildlife Protection Society of India.

Singh, K. 1972. *Foreword to a Planning Proposal for Preservation of Tiger (Panthera tigris tigris Linn.) in India.* Delhi: Department of Environment, Government of India.

Sivaramakrishnan, K. 2011. "Thin Nationalism: Nature and Public intellectualism in India." *Contrib Indian Sociol* 45 (1): 85–111.

Thapar, V. 1998. "Fatal Links." *Seminar* 466 (June): 59–69.

Tigerlink. 1997, June 1. More Paper Tigers, Volume 15, Ranthambore Foundation.

United Nations Development Group (UNDG). 2010. *Thematic Paper on MDG 7: Environmental Sustainability*. New York: UNDG and MDG Task Force.

Weber, W. and A. Rabinowitz. 1996. "A Global Perspective on Large Carnivore Conservation." *Conservation Biology* 10 (4): 1046–54.

Wikramanayake, E.D., E. Dinerstein, J.G. Robinson, U. Karanth, A. Rabinowitz, D. Olson, T. Mathew, P. Hedao, M. Conner, G. Hemley, and D. Bolze. 1998. "An Ecology-based Method for Defining Priorities for Large Mammal Conservation: The Tiger as Case Study." *Conservation Biology* 12 (4): 865–878.

4

Urban Land Governance Reforms and Sustainable Development: A Study of Urban Property Ownership Records (UPOR) in Karnataka*

Kanekanti Chandrashekar Smitha and Manasi Seshaiah

Introduction

Effective urban land governance reforms have been a long pending initiative in India. The smooth functioning of a sustainable economic system requires an accurate and efficient maintenance of the land record system in order to carry out any land-related transactions. The challenges posed by the recent global developments, especially rapid urbanization, unregulated migration to urban centres, ineffective land management practices, institutional fragmentation, and lacuna in the legal system to guarantee land titles, have not only increased the complexity of land administration but contributed to unsustainable economic development. Further, rapid urbanization has created complex land-transactions-related issues such as unnatural increase in the demand for urban property, creation of fake documents, '*benami*'[1] and fraudulent land transactions coupled with insufficient checks/control at various official levels for monitoring irregularities, inappropriate land valuation which have evolved from unclear and inconsistent land laws, policies, and unsustainable management, increasing public insecurity.

---

* This was part of a larger study undertaken by the Centre for Ecological Economics and Natural Resources, Institute for Social and Economic Change, Bangalore, funded by the World Bank on Land Policy and Administration, 2011. Other team members included Professor N. Sivanna, Dr R.G. Nadadur, Professor Chengappa.

[1] With respect to land transactions, *benami* essentially means property without a name. Especially during land transaction, the person who pays for the property does not buy land under his/her own name. The person on whose name the property has been purchased is called the *benamdar* and the property so purchased is called the *benami* property.

A study conducted by Mckinsey (2001) in India showed that India loses 1.3 percent of its potential growth due to poorly maintained land records. Much of the registration process in India does not result in valid ownership titles. Hence, more than 70 percent of legal disputes are regarding land. These inefficiencies have failed to address economic development and alleviate poverty in India. It is with the intention of addressing these problems that the Urban Property Ownership Records (UPOR) has been initiated in Karnataka. The purpose of UPOR, therefore, is to promote efficiency in service delivery and enable citizen interface with the computerized digital management of land records throughout the state of Karnataka in India. The purpose coincides with UN declaration of sustainable goals such as Goal-11,[2] to make cities inclusive, safe, resilient, and sustainable, and Goal-10, to reduce inequality within the cities. Similarly, MGDs Goals 1 and 7 are to eradicate extreme poverty and hunger, and promote 'environmental sustainability'. It was envisaged that UPOR would serve as the main instrument for effective urban land governance and provide secure tenure for sustainable development, and thereby promote 'good governance'. In the UPOR context, 'technology' plays a key role in making a paradigm shift in bringing about effective change.

The chapter critically examines the initiation of the process of UPOR, that is, a computerized process of land title registration system introduced in Karnataka. The chapter is divided into three sections. Section I provides a historical perspective on the concept of city survey in Karnataka. Section II gives details on the process of UPOR initiative and various stages of its implementation. Finally, Section III offers a critique of the top-down approach while implementing UPOR initiative.

Section I

Concept of City Survey: A Historical Perspective

Historically, the first formal and recorded urban mapping or city survey in the country was introduced during the years 1900–10. Such a measurement

---

[2] The SDG-Goal 11 aims to promote safe and affordable housing by 2030 and by 2020, promote human settlements adopting and implementing integrated policies and plans toward inclusion, and resource efficiency. Similarly, the SDG-Goal 10 aims to reduce inequality by making policies universal in principle paying attention to the needs of disadvantaged and marginalized populations.

Table 4.1:
Survey carried during different periods

S. No.	Survey in Different Periods
1	Original survey was carried out between 1863 and 1890. All the documents prepared during original survey are in Modi[3] Marathi language.
2	Re-survey was carried out between 1900 and 1920.
3	Hissa[4] survey was carried out between 1926 and 1940.
4	Second Reclassification was carried out between 1955 and 1966.

Source: Commissioner, Department of Survey Settlement and Land Records, Government of Karnataka, Bangalore.

was carried out in the Bombay Presidency and it later spread to other major cities so that by 1918 the system became a full-fledged one. The Karnataka Survey Settlement and Land Records (KSSLR) Department is one of the oldest survey departments established under the Bombay Revenue Act and Karnataka Land Revenue Act (KLR Act, 1964). At present, urban mapping exists in 48 urban centres of Karnataka, even though the records are not fully updated. Across 42 sectors of Belgaum, it has been completed by continuing the system that prevailed during the Bombay Province. The city survey concept was introduced in six more districts of Karnataka (Bellary, Gulbarga, Kolar Gold Fields [in Kolar district], Mysore, Bangalore, and Davangere) between 1969 and 1975. As many as 112 villages of Belgaum had been notified for mapping and measurement. Similarly, 137 sq. km of Bangalore core city has been measured and mapped, barring new extensions and fringe areas. As per the available records, in Karnataka, survey was carried out during different periods in time (see Table 4.1).

In the past few years, there has been a manifold increase in the property rates in urban India, while urban land management practices have remained unplanned. This has been further accentuated by increasing population, migration to cities, and increase in income levels of the educated and employed urban middle class. Hence, the need for a well-maintained land record system for urban areas needs emphasis, particularly in the context of soaring land market prices and a competitive urban economy that contributes a large share to the gross domestic product (GDP).

---

[3] The Modi script was used to write the Marathi language spoken in the Indian state of Maharashtra. Modi was used until the 1950s when Devanagari replaced it as the written medium of the Marathi language.

[4] When a piece of land which has been assigned a Survey Number is divided into multiple subdivisions, each subdivision is assigned what is known as a 'Hissa number'.

Recent Developments in Urban Areas of Karnataka

In recent years, though the state of Karnataka has experienced a slower growth rate, the state's GSDP (gross state domestic product) is estimated to grow at 6.4 percent (according to the Economic Survey of Karnataka) in the year 2011–12 and reach ₹2,979.64 billion as compared to ₹2,799.32 billion in 2010–11. This can be attributed to the growth in manufacturing (4.1 percent) and service sectors (10.6 percent) in 2011–12.[5] Since two decades, manifold increase in public sector investment followed by private sector investment as a result of liberalization and introduction of urban sector reforms[6] has impacted substantially on land rights of urban citizens. As a result of such macropolicies, there is a need for proper implementation of Karnataka's urban land legislation and regulations.

Rapid Urbanization

The process of rapid urbanization[7] in Indian cities has shaped sustainable economic growth and registered massive influx of migrants[8] from rural areas to urban centres between the 1990s and 2000 and continues till date. Therefore, cities have become major centers of industrial development and have contributed to boom in construction and service sector industries. The expansion of these sectors has created enormous job and investment opportunities and urban sprawl. By 2025, 50 percent of Karnataka's[9] population (40 million) is expected to live in urban areas. Urban poverty constitutes 32.6 percent in Karnataka which is high as compared to the other states. The fundamental issue is to accelerate an orderly urban growth for sustainable development in towns and cities, other than Bangalore, to

---

[5] *The Businessline* (2012). Karnataka's economic growth to slow down to 6.4 percent in 2011–12, March 20th. www.thehindubusinessline.com

[6] Urban sector reforms are initiated by the government of India through JNNURM Project (2005)—Jawaharlal Nehru Urban Renewal Mission. The mission has inserted optional land reforms for both rural and urban areas through computerization.

[7] Nearly 28 percent of total population lives in urban areas in India (UDP 2009: 3).

[8] Migration toward the higher order of urban centres is with 66 percent of the urban population concentrated in 23 Class I cities. Mysore and Hubli-Dharwad constitute emerging metropolitan cities in Karnataka (UDP 2009: 3).

[9] Urban settlement structure in Karnataka is highly dense with 67 percent of urban population living in 24 cities (UDP 2009: 5).

accommodate waves of migration and regulate the allocation and use of land in a fair and equitable manner.

Private Investment and Demand for Urban Land

Demand for urban land is the result of certain factors, namely macrolevel liberalization policies promoting private investment in the delivery of basic services, the development of service sector in the form of establishing IT (Information Technology)/BT (Biotechnology) industries, and rural–urban divide resulting in mass exodus of migrants from rural areas to cities. As a result, great demand for urban land is created by both informal communities and investors, giving rise to conversion of land from agricultural land to nonagricultural uses. As this is done without clear land titles, dubious land transactions and land litigations have changed the concept of land use (Wadhwa 2002). Therefore, there is a lacuna in the existing legal framework concerning ownership of land. Hence, the need for a system of state guarantee of land title or property which can ensure and protect local land rights is very important. Further, decentralized reforms through 73rd and 74th Constitutional Amendments have facilitated a comprehensive land policy in India.

Urban Land Governance Practices

Effective urban land governance is critical to the good governance and sustainable development. Absence of well-planned urban land management resulted in spiraling land prices, speculation and inflation, and concomitant growth of informal settlements adversely impacts social welfare and overall sustainable development. Inadequate land management in urban areas also results in the violation of land use regulations and encroachments (Dowall and Clark 1996: 5; UDP 2009: 9). Unfortunately, since past three decades, the issue of sustainability has not received the much-needed attention with respect to urban land governance. Typically, urban land regulations are influenced by zoning, regulated densities, building bylaws, master plans, and comprehensive city development plans.

Violation of urban land-use regulations and practices is due to the lack of valid cadastral, registration, and approved records that results in haphazard and unconstrained land use seriously impacting on urban governance. Further, an increase in the volume of urban land transactions due to rapid industrialization, and changes in land-use pattern have further complicated the formal process of land registration leading to unregulated urban sprawl.

The situation is aggravated by informal settlements such as slums[10] and encroachments occupying urban land without any formal security of land tenure, causing disjunction between planning and affordable housing (Ramanathan 2009). Lack of transparency further accentuated unsustainable urbanization.

Lacuna in Legal System to Guarantee Land Titles

Time and again, uncoordinated, disintegrated, and outdated land-related legislation has led to critical situations in many developing countries (Burns 2007). Lack of effective implementation and predominance of deeds[11] system over title registration in the existing land governance in India means that the title is not backed by the state guarantee (Wadhwa 2002). There are instances of land-related conflicts leading to court litigation involving social cost and disrupting societal harmony.

Discrepancies in legal provisions have contributed enormously for lack of sustainable practices in land governance. Though legal provisions guaranteed the right to property as a fundamental right, later, with the 44th Constitutional Amendment in 1978, the right to property was reduced to the status of legal right and is no longer a fundamental right with constitutional remedies. In Karnataka, land use is regulated by Karnataka Land Reforms Act of 1964 and Indian Registration Act of 1908. While these acts envisage the compulsory registration of sale of land, they do not provide for registration of title, but only result in a deed of transaction. Registration process merely acts as a fiscal instrument of the state to collect a 'fee', but does not provide statutory guarantee[12] to the land title. Further, Section 18 of the Registration Act does not require compulsory registration of all land-related transactions[13] (Ramanathan 2009; Wadhwa 1989). As a result, existing legal titles do not reflect any changes resulting in succession,

---

[10] Nearly 7.8 percent of the urban population lives in slums in 35 towns of Karnataka state. According to the NSSO, 58th round (2002), the number of slums in Karnataka is estimated to be 1983 with 483,828 households (UDP 2009: 10).

[11] A deed does not in itself prove title; it is merely a record of an isolated transaction.

[12] State laws do not entertain any suit against the state government or any officer of the state government in respect of a claim to have any entry made in any record or register maintained by the government or to have any entry omitted or amended (Wadhwa 2002).

[13] Such as state acquisition of land, court decrees, land orders, partitions, mortgages, agreements to sell the land, etc.

transfer or, more broadly, the customary rights recognized in the community (like slums/informal settlements) and these differences add to the existing complex urban land dynamics.

Section II

Initiation of 'Urban Property Ownership Records' (UPOR) in Karnataka

Urban local bodies (ULBs)[14] in Karnataka have been maintaining property tax details through a property tax record called 'Khata'.[15] 'Khata' document is presumed as a property ownership record by majority of the property owners. Besides, as ULBs[16] are not authorized to create and maintain property ownership records, they maintain 'Khatas' for the purpose of tax collection from urban citizens, which is not a Record of Rights (RoR) issued in respect of revenue or agricultural land. Therefore, having Khata per se does not indicate legal status.

To address the lacuna in urban land governance, a unique urban land reform project called Urban Property Ownership Rights (UPOR) was introduced in December 2009 by the Government of Karnataka (GoK) (Chawla 2004; Pathak and Prasad 2005). As per the KLR Act 1964, the Survey, Settlement and Land Records (SSLR) Department introduced UPOR, a comprehensive framework for the creation and management of urban property records in December 2009.

The purpose of UPOR is to promote digitalization of urban land records for greater transparency and openness. Further, UPOR aims to modernize land records management, minimize scope for land disputes by enhancing transparency, and eventually facilitate toward conclusive land titles in the urban context of Karnataka.

---

[14] There are 213 ULBs in Karnataka varying in size based on population from 10,000 to almost a million. Bangalore city, the state capital, has almost reached 9 million.

[15] Khata means an account maintained by the city municipal council (CMC) for collecting property tax from property owners and it is not an ownership document.

[16] There are 214 ULBs in Karnataka, India.

Mandates of Survey Settlement and Land Records (SSLR)[17]

The SSLR Department, GOK, as per the KLR Act of 1964, and KLR Rules[18] of 1966 has been mandated to prepare, maintain, and preserve spatial and nonspatial data relating to ownership of land for urban properties in the state of Karnataka. It is essential for every land owner to have proper and legally acceptable documents related to property. Though 'Khata' is issued by a municipality or city corporation for the purpose of paying revenue tax, it is considered as property right certificate by the citizen and the fact is that it does not legally constitute a document regarding ownership of property. Property cards, therefore, under UPOR are proposed to be issued to all land owners, and will be legal ownership records as per Section 133 of KLR Act 1964, replacing the existing records of presumptive ownership. The records will be maintained in the form of maps and sketches depicting boundaries, and the extent of individual properties and text relating to ownership, land use, and other land related particulars.

Although SSLR maintains urban property records in Bangalore, Mysore, and 41 other towns of Karnataka, due to insufficient manpower and financial crunch, the records were not updated on a regular basis. Arising out of the successful implementation of '*Bhoomi*'[19] in rural Karnataka the need for similar property records system was felt for urban centres also (Mukerji 2011). Form 13 Property Card (PC) comes under the statutory provision and is mandatory for transactions such as bank loans or selling property. The UPOR property ownership records is expected to legally authenticate documents, finalized through a statutory enquiry process as per law ascertaining the ownership of properties and measurements, used as legal and primary references for resolving any kind of legal disputes related to ownership, location, dimensions, and areas of urban properties.

Initially, UPOR is being implemented in five cities, namely Mysuru, Shimoga, Hubli-Dharwad, Bellary, and Managalore of Karnataka. It is a process of confirming the presumptive property title and providing a conclusive property title of all 'urban properties' as mandated under the Land Revenue

---

[17] Based on the extracts from the Concept Note of UPOR.

[18] The specific sections of the KLR Act 1964 governing the implementation of UPOR in Karnataka include (a) Chapter 1—Section 1(2) and Section 2(2, 6, 38), and (b) Chapter XIII—All Sections (148 to 156) popularly known as 'City Survey'.

[19] *Bhoomi* (meaning land) is the project of online delivery and management of land records in Karnataka. It provides transparency in land records management with better citizen services and takes discretion away from civil servants at operating levels.

Act 1964 and Rules 1966. The project aims to create a comprehensive database, both spatial and nonspatial, of all properties of urban areas in Karnataka.

Objectives of UPOR

The main objective of UPOR is to issue ownership title to urban households and the creation of a fresh data base for urban mapping. It is important to note here that the generation of revenue is not the sole purpose of urban mapping under UPOR. The sub-objectives of UPOR include:

- Measurement and mapping of all nonagricultural lands and urban properties
- Creating and maintaining RoR for all nonagricultural lands
- The preservation of existing land records by the revenue department

A robust system of UPOR is created for every property with a view to documenting accurately both spatial characteristics of properties and RoR data in respect of (a) land parcels, (b) structures/buildings, and (c) roads, etc. This property record will serve as a trusted record for all land related transactions promoting transparency, openness, and efficient service delivery. The PC thus created under UPOR would serve as evidence for property ownership for regulatory and legal purposes. The PC issued will continue to remain valid and will not become obsolete or inaccurate (UPOR webpage).[20]

Outline and Scope of the UPOR[21]

Under UPOR, property records will be recreated for urban areas under the provisions of KLR Act 1964. UPOR was initiated to create and maintain property ownership records in five urban bodies within a framework for management of land records in urban centres. Five cities were selected on a pilot basis, which include Mysore, Mangalore, Bellary, Hubli-Dharwad, and Shimoga. The major components of the UPOR are computerization of

---

[20] UPOR objectives are listed out in City Survey Department-UPOR webpage www.upor. karnataka.gov.in

[21] UPOR outline and scope are presented in Request for Proposal (2009).

all land records including mutations, digitization of maps and integration of textual and spatial data, survey/re-survey and updation of all survey and settlement records including creation of original cadastral records wherever necessary, computerization of registration and its integration with the land records maintenance system, development of core Geospatial Information System (GIS), and capacity building. Another interesting feature of the project is that UPOR is being implemented through Public–Private Partnership (PPP) mode, in particular Build, Own, Operate, and Transfer (BOOT).[22] GoK has teamed up with private sector partners in the area of land administration and records. The UPOR authorities, led by the deputy commissioner of the concerned district have set up citizen facilitation centers for receiving relevant documents from property owners for ascertaining their ownership claims.

Under UPOR process, the task of creating and maintaining the property records for urban areas involves four distinct steps.

1. Creation of property records which includes (a) survey and mapping of all properties in the cities and (b) verification of ownership claims.
2. Continuous management and maintenance of property records includes (a) creation of property records for new extensions of cities, (b) documenting changes in property records in the context of new constructions, and (c) updating changes in land use and mutation of property records in case of sale/partition.
3. Operation and maintenance of service delivery channels to 'Bangalore One' centres to delivery various property record related services to citizens. These service centres are established by the Survey Department. The above three tasks are performed by service provider (SP), and finally.
4. Creation of IT infrastructure that will enable the storage of property records and delivery of various property record related services such as (a) title enquiry, (b) certifying transactions, and (c) changes in property records. The IT infrastructure comprises (a) software application; (b) various COTS software products; and (c) servers, storage, and other IT hardware. These activities are carried out by technical service providers (TSPs).

The first three tasks are performed by SPs, while the fourth by TSPs.

---

[22] BOOT model followed for the implementation of UPOR.

List of Services 'Expected to Be Delivered' by UPOR[23]

The following services are expected to be delivered by the UPOR project for promoting efficiency, transparency, and openness in the land transactions.

Information Services which includes (a) Search for various types of properties and related documents like sale deeds, (b) History of transactions of various properties post UPOR project, (c) Various thematic layers of Maps (for example roads, public amenities, government properties, etc.), and (d) Search for the entire city-based list of properties based on the parameters of the property database. In addition, following services are delivered.

1. Information to owners/power of attorney holder or lease, bank, etc. on transaction initiated in respect of property.
2. Search and Print (Certified Copies) which includes (a) Certified copies of ownership record and (b) Certified copies of transaction recorded in the UPOR database.
3. Map Services include (a) Maps of the properties with the adjacent properties mapped, (b) Maps of the locality like sector 7 Hosur Road Sarjapur layouts, and (c) Division of the city into different squares of 1 km to 1 km and mapping of all properties in the maps.
4. Transaction Services include (a) Transfer of ownership provide details on: through registration and without registration, (b) Charge details consist of (i) Creation of Charge; (ii) Release of Charge, and (iii) Reducing charges on the repayment of the loan by the recipient; (c) Details on recording of any transactions include: Power of Attorney, long term lease; and (d) Details on subdivision of property.
5. Survey Management Process for (a) resurvey, (b) subdivision, (c) amalgamation cases—these could include special cases such as (i) the acquisition of land by government for public works and (ii) purchase of two adjacent properties by a private individual and the creation of a single commercial/residential building—and (d) *hudbust-*refixation of properties, and (e) additional services/survey on demand/instructions from courts/agencies and other government departments.
6. Appeal provision includes handling cases related to courts and court commissions/partitions.

---

[23] Notes collected from RFP (2009a: 9–10).

Services related to the municipal department include: The Department of Survey Settlement and Land records while networking with urban development departments and the City Municipal Corporation (CMC) maintains details on property taxes, accepts applications for building plan permits, and grants permission. While delivering such services through UPOR, the following services are offered: (a) statements of tax dues on property, (b) retrieval of building plan on the property, (c) application for building plan permission, (d) payment of property taxes, (e) preparation of demand notices for the municipalities, (f) generation of overdue notice for the municipality, (g) any change in the land use pattern, and (h) layout approval.

Implementation of UPOR: Public–Private Partnership (PPP) Model

The PPP model has been followed for city survey, where private agencies are selected through bidding, for implementing the UPOR activities in a phased manner across five pilot cities (Mysore, Mangalore, Bellary, Hubli-Dharwad, and Shimoga) in Karnataka.

The implementation work has been sub-contracted to two categories of vendors, namely SP and TSP.[24] While SP is in charge of the creation of both spatial and nonspatial database under the supervision of the department staff, TSP is in charge of the creation of software, not only for pilot cities but also for the entire state. TSP is responsible for storage and continuous online update of property details. The time frame fixed for the project completion was six years for both SP and TSP, starting 270 days from the issue of order to the service provider.

Governance Structure: Project Monitoring

The progress of each UPOR project was to be monitored by different committees constituted for the purpose. The monitoring activity comprises of four prime execution stages of the UPOR project they include constituting: (a) Project Monitoring Committee, (b) Project Management Unit (PMU), and (c) Inter-departmental coordination and coordination between vendors.

---

[24] Notes from UPOR-Concept Note, page 5.

The City Survey

The city survey work under UPOR relates to both spatial and nonspatial data. The spatial data involves the establishment of control networks via primary, secondary, and tertiary control points and maps, while the nonspatial data includes (a) collection of data and verification of documents, (b) measurement of properties, (c) preparation of index mapping, (d) creation of a master data base (data entry), (e) collection of original documents, (f) scanning of the documents, and (g) preparation of draft property cards.

The four stages of city survey include (a) establishing control networks, (b) conducting a detailed survey of each property, (c) title enquiry process and preparation of property cards, and (d) citizen service delivery.

Establishment of Control Networks[25]

One of the major tasks of the project is to establish control networks which include (a) primary control point, (b) secondary control point, and (c) tertiary control point. These control points are established for fixing coordinates, linked to each other for accuracy. The Town Planning Authority (TPA) respective to cities has already established Primary Control Points (PCPs) and Secondary Control Points (SCPs). But the area of interest in UPOR was to establish a robust control-point network covering points 100 percent of developed area as well as 50 percent of buffer area considering future developments additional PCPs (one for every 4 km) and SCPs (for every 1 sq. km) were to be established. The coordinates for these points are satellite based obtained using DGPS, and observation is carried out as per the accuracy points. The obtained coordinates are processed in house. Based on the instruments of Town Planning of Mysore, the department of Survey of India (SOI) and TPA has established 67 PCPs, 160 SCPs, and 8,340 TCPs in Mysore city.

Use of Electronic Total Stations (ETS) Instrument

A Base Map on a scale of 1:20,000 was compiled for Mysore city, using SOI Maps and updated Satellite Image. The Base Map is on WGS 84 Datum

---

[25] Regarding control points, a note is taken from UPOR—Concept Note and UPOR webpage www.upor.karnataka.gov.in.

and Universal Traverse Mercator (UTM) Projection. SOI established PCPs as well as additional PCPs established subsequently have been plotted on the Base Map. The established PCP and SCP points have been further extended to Main Line Traverse (MT) and Secondary Line Traverse (ST) by providing Tertiary Control Points (TCPs). MTs commence from PCP and SCP, and converge on closed in PCP and SCP using ETS, for achieving high degree of accuracy. TCPs having 3 cm positional accuracy have been established on permanent structures (Figure 4.1).

Figure 4.1:
UPOR second stage—Implementation process in Mysore city

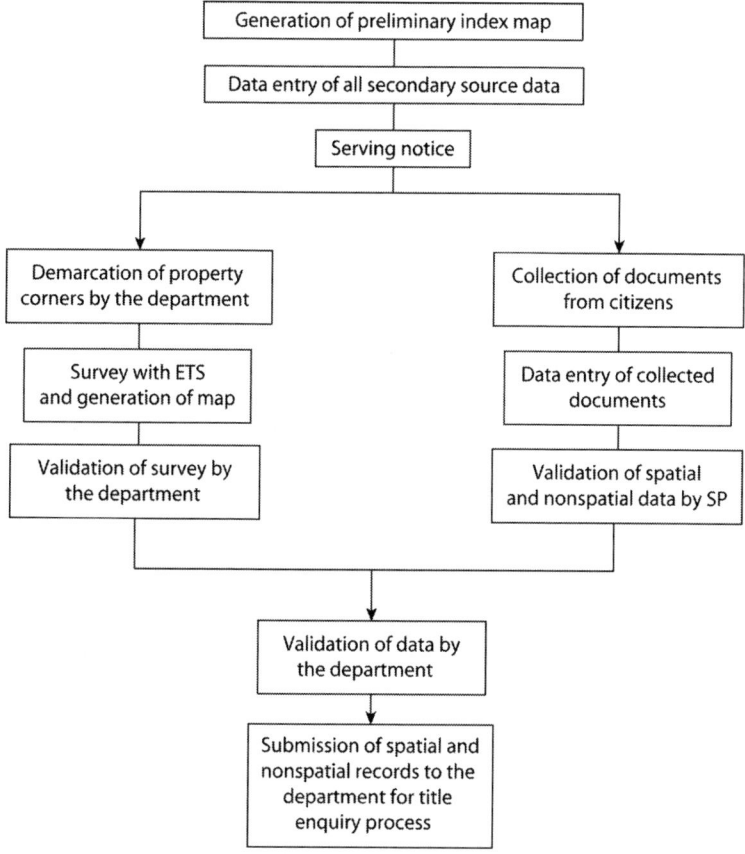

Source: UPOR office, Mysore.

All the TCPs have been established in such a way that there is at least one TCP within a range of 100 meters from the other one, with a fair inter-visibility, using TCPs. The total station readings are taken as per the requirements.

INDEX MAPPING: UPOR

After establishing control networks, the next stage is index mapping. There are three steps involved in index mapping. They are (a) dividing the entire city area into sectors and zones, (b) giving numbers to all the properties, and (c) geo-referencing of village maps.

Survey: Nonspatial Survey

The second and third stages of UPOR implementation relate to a nonspatial survey. There are seven steps involved in a nonspatial survey. They are (a) collection of data and verification of documents, (b) measurement of properties, (c) preparation of index mapping, (d) creation of a master data base, (e) collection of original documents, (f) scanning of the documents, and (g) preparation of draft property cards.

SECOND STAGE OF UPOR IMPLEMENTATION—ACTIVITIES INVOLVED

During the second stage of UPOR implementation, as the first step, different sets of stakeholders are informed of the survey through a notice from the department of city survey. Stakeholders are requested to[26] produce relevant ownership documents for scanning and verification. The list of documents depends on property type and category of stakeholders. Table 4.2 presents details of stakeholder's categories and list of documents collected by the SSLR in Mysore city. The UPOR office in Mysore city has received about 90,000 documents (files).[27] The second step is for detailed spatial planning to be carried out for demarcating and categorizing the properties in the city (Table 4.2).

As each individual property is unique, in terms of coordinate[28] measurements, the entire city is divided into zones, sectors, and village properties. The city of Mysore is divided into 11 zones, 50 sectors, and 42 villages. As

---

[26] Quantifiable details are collected from UPOR office, Mysore.

[27] Quantifiable details presented in this chapter for Mysore are up to December 2011.

[28] Coordinates for each property is fixed based on easting, westing, southing, and northing measurements in the city.

Table 4.2:
Stakeholder categories and a list of minimum documents collected for Mysore

Category Type	List of Minimum Documents
Mysore Urban Development Authority (MUDA)	(a) Sale deed from MUDA to allottee (b) In case the present owner is not an allottee, the sale deed* from original owner (c) In case some of the sale deeds are missing, then EC to cover the period of missing deeds (d) Possession certificate issued by MUDA (e) Approved layout plan (f) Khatha certificate (g) Latest tax paid receipt
Gramathana Ashraya layouts, Janatha Colony properties	(a) First sale deed after the conversion of land (b) Second sale deed from private person/agencies to the site allottee, EC (c) Present owner's sale deed* (d) Possession certificate (Hakkupatra) (e) Tax paid receipt (f) Khatha certificate
Layouts developed by private persons/Agencies	(a) First sale deed after the conversion of land (b) Second sale deed from private person/agencies to the site allottee (c) Present owner's sale deed*, EC (d) Conversion order (if conversion order for the layout already exists then this document is not required) (e) Approved layout plan (if a layout plan for the layout already exists, then this document is not required) (f) Municipality Khatha (g) Tax paid receipt
Housing Board Layouts (KHB)	(a) Allotment letter (b) Sale deed from KHB to allottee (c) In case the present owner is not allottee, the sale deed* from the original owner to the present owner (d) In case some of the sale deeds* are missing, then EC to cover the period of missing deeds (e) Possession certificate issued by KHB (f) Khatha and tax paid receipt
Properties in alienated lands but layout plan not approved	(a) RTC in the name of the owner (b) Mutation and present RTC, EC (c) Conversion order (d) Conversion Sketch (e) Sale deed from land owner to the first buyer site (f) Latest sale deed* if the present owner is not the first buyer (g) Khatha and tax paid receipt

(Table 4.2 continued)

(Table 4.2 continued)

Category Type	List of Minimum Documents	
Properties in agriculture survey numbers-revenue sites	(a)	RTC in the name of the owner
	(b)	Mutation and present RTC
	(c)	Sale deed* from the owner as per RTC to the present owner of the site
Properties in government land with survey numbers	(a)	Grant order in the name of original grantee
	(b)	Saguvali chit in the name of the original grantee
	(c)	Conversion order
	(d)	Conversion Sketch
	(e)	Sale deed* from grantee to first buyer after conversion
	(f)	Latest sale deed* in case multiple transactions have taken place and EC
Agriculture lands	(a)	Latest RTC in the name of the owner
	(b)	Mutation

Source: UPOR office, Mysore.
*Note:** Or all other supporting documents showing property ownership.

the survey knowledge of SPs is inadequate, 32 surveyors have been appointed by the department for the measurement of individual properties in the city.

In the third step, government surveyors identify property corners which are critical to determining the boundaries of properties. The discretion of private agency is not considered for this critical activity. Here, all the three control points (PCPs, SCPs, and TCPs) network and fix a value for each property. Values are fixed for all the properties including roads, lakes, and public establishments in the city for measuring individual plots/sites and households. As the fourth step, boundaries are fixed using Electronic Total Stations (ETS) to arrive at coordinates.

As the work proceeds, in the fifth step, using UPOR vectorization software, property maps are prepared also referred to as 'Index Mapping'. Maps are generated for the entire city survey jurisdiction. A detailed survey map is generated depicting roads, public establishments, and households. Complete details are collected block, compound, and plinth area-wise. A relevant notice is issued regarding the survey date to the individual property owners in the city. However, it is not final because the survey has to be validated by the SSLR department.

DEPARTMENT VALIDATION OF SURVEY

The department selects 5,000 properties randomly (20 percent of the work) from each sector and check for validation. Going by the coordinates drawn using ETS, variation between 1 percent minimum and 2 percent maximum is ignored. However, if the difference is 10 percent for each sector, then the entire sector is resurveyed.

After mapping of the city properties, in the sixth step, a master data base is generated by collecting documents from urban authorities, municipalities, and individual property owners. For example, municipal data, alienation data, government land data and city survey details (Mysore, Bellary, and Hubli-Dharwad). Vendors (SPs) collect relevant documents from property owners (citizens) in the city. These documents are scanned using UPOR software before porting the same. Households without documents are left out. Finally, exception reports are prepared.

Third and Final Stage of UPOR Implementation: Title Enquiry Process (TEP)

In the third and final stage of UPOR implementation, title enquiry (TE) process is initiated. The title enquiry process is in progress in three cities out of five UPOR cities. Under the TE process, four steps are involved for verifying the ownership claims of properties and the preparation of draft property cards (DPC). They include:

1. Based on the data collected during first and second stages, the title enquiry process gets initiated by TE team.
2. Confirmation of tenure and boundaries of the properties.
3. Preparation of DPCs.
4. Preparation of final property card, enquiry register, and property maps.

Privacy Protection of UPOR[29]

There were physical barriers to accessing property records in the pre-UPOR phase. However, with the computerisation of property records through UPOR project, SSLR aims to protect the confidentiality of property ownership details which could be shared only by property owners. Accordingly some essential security measures have been taken up by both SP and TSP for maintaining the property confidentiality of records through biometrics/

---

[29] Source: RFP (2009a: 20).

PKI. SSLR shares the cadastral data with the public agencies on request on free of cost basis or on case-to-case basis.

UPOR Project[30]—Expansion Portfolio

In case of any expansion of portfolio or adding more services to the UPOR project, both SP and TSP are to comply with providing additional services without any additional charges. Further, both SP and TSP are to recover the charges only through user charges during provisioning of services, In case of any services that cannot be related to the existing category of services, the charges paid to both SP and TSP are decided by the PMC on 75:15 share basis.

Expected Services from UPOR

It is expected that Services under UPOR Project will be delivered from Service Centers and have been set up by SSLR in all study cities (Figure 4.2).

Figure 4.2:
Functional architecture of integration process in respect of urban property records project, Karnataka

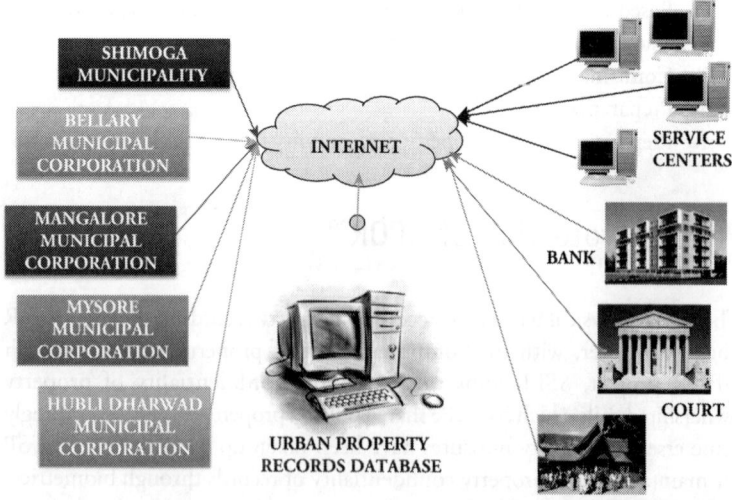

Source: RFP (2009a: 11).

---

[30] Source: RFP (2009a: 23).

Experiences—UPOR Project

UPOR project has commenced in five cities of the state. Progress achieved in Mysore with almost 90 percent of the 2.75 lakh properties identified and the survey completed. Other cities, Hubli-Dharwad (50 percent) and Shimoga (70 percent),[31] have registered good progress. The experiences gained in Mysore, Hubli-Dharwad, and Shimoga are found relevant for the present analysis as it is a considerably bigger operation as compared to other cities (Figures 4.3 and 4.4).

Figure 4.3:
Stages of UPOR implementation process

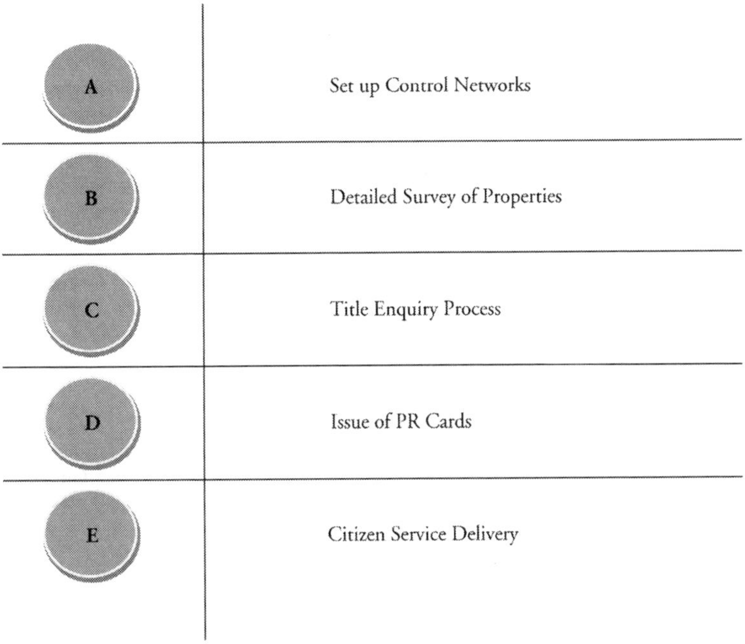

A	Set up Control Networks
B	Detailed Survey of Properties
C	Title Enquiry Process
D	Issue of PR Cards
E	Citizen Service Delivery

Source: UPOR office, Mysore.

---

[31] Based on discussions with officials in 2011.

Figure 4.4:
UPOR implementation process

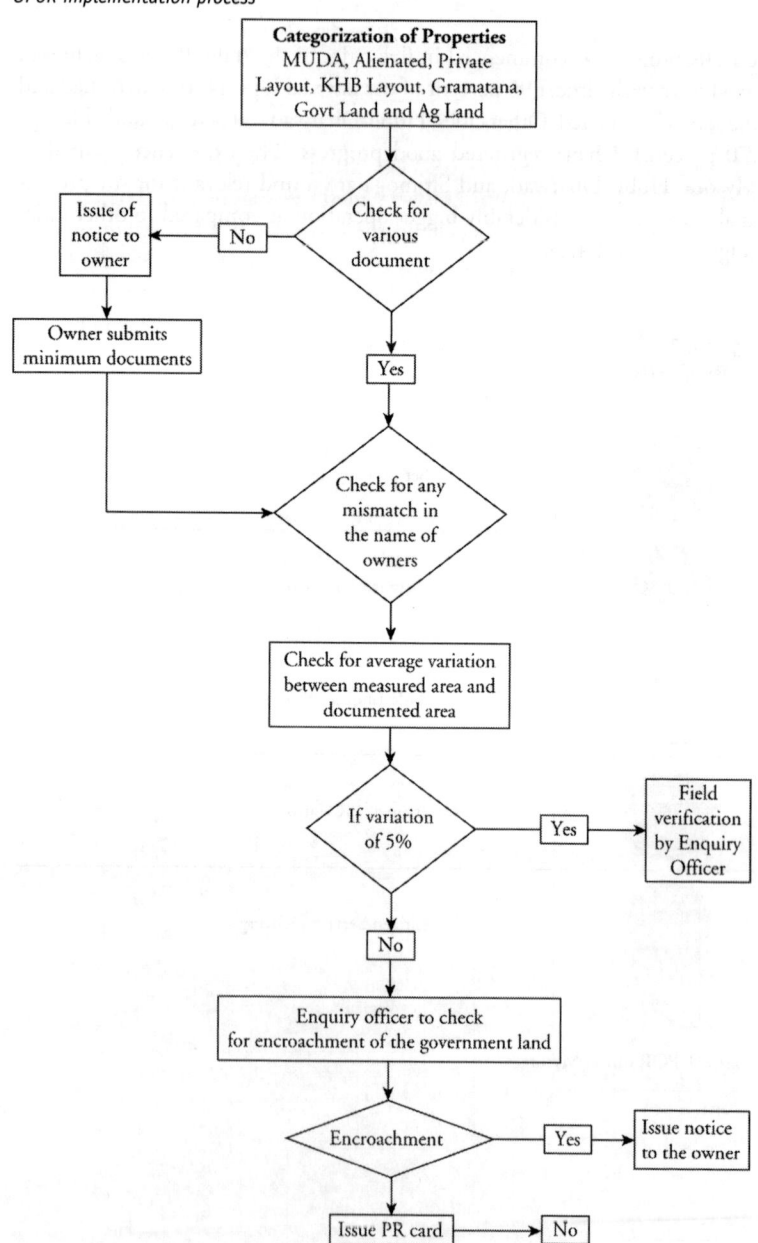

Source: UPOR office, Mysore.

Section III

Key Stakeholders

The present UPOR project has been launched to aid citizens. Key stakeholders identified under UPOR project is presented (see Figure 4.5). Therefore, an attempt is made here to understand the expectations of different types of users and also to identify the potential users, especially of TEP data base. These institutions across society will ultimately play an important role in determining the design of a pricing strategy, extent of coverage, compliance from SPs for the UPOR project and the associated database to remain an economical proposition from the investors' point of view. As a primary stakeholder, the department has prepared a Request for Proposal (RFP 2009b) for SPs and TSPs of the UPOR task, detailing the roles and responsibilities, functional and technical specifications, commercial terms and bid documents,

Figure 4.5:
Classification of stakeholders

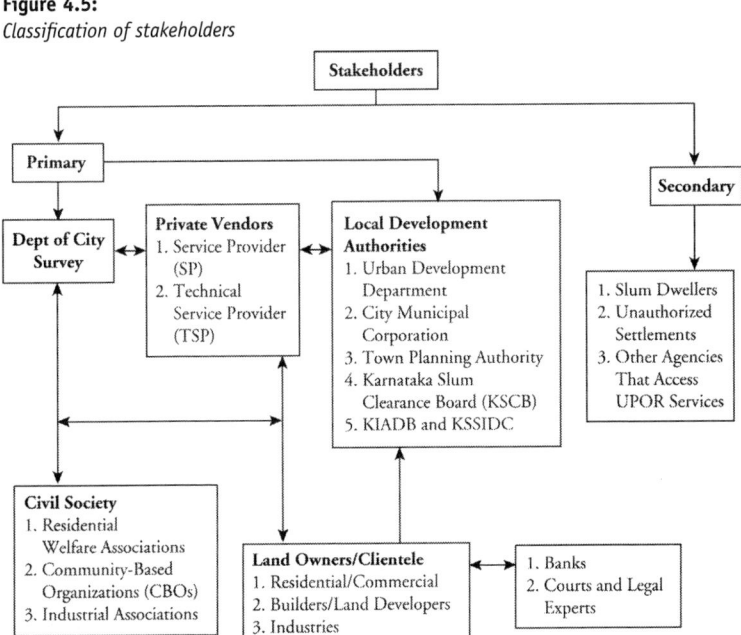

Source: Authors, based on field study.

Master Service Agreement, etc. Properties belonging to the ULBs,[32] UDA, and KHB, in addition, banks, courts and lawyers, builders, private agencies, etc. are major end users of property data.

Implementation Constraints of UPOR

The implementation of UPOR project in five pilot cities of Karnataka has encountered several process related obstacles and challenges.

Management Constraints

In order to ensure the successful and timely implementation of a project like UPOR, the presence of efficient manpower is significant. Constraints encountered are summarized in Table 4.3.

Table 4.3:
Constraints encountered—Project implementation

	Hubli-Dharwad	*Mysore*	*Bellary*	*Shimoga*
Service Provider				
No pre-project plan			√	
Inefficient project manager			√	
Manpower shortage			√	
Financial instability			√	
Poor awareness campaign	√	√	√	√
Department				
Shortage of manpower			√	
Technical constraints	√	√	√	√
General				
Poor coordination between Department and SP			√	
Poor confidence among public	√	√	√	√

Source: Observations of ISEC team, during field visits.

---

[32] Urban Local Bodies (ULBs), Urban Development Department (UDD), Karnataka Housing Board (KHB).

Operational Constraints: There were several operational constraints that affected implementation. Pockets of central old city areas in cities such as Hubli-Dharwad and Bellary do not consist of sketches/maps to validate the properties and their measurements. It became difficult to identify the exact location of sites for fixing ownership title. UPOR implementation process is time consuming and requires enormous time in conducting survey of properties, document collection, fixing ownership titles amidst all inconsistencies. During the collection of property documents, people were skeptical to submit property documents in spite of awareness creation programs. The delay in collection/submission of documents by the citizens burdened the SPs financially. Some of the properties were involved in disputes and litigations, hence, confirming the titles was an issue.

Institutional and Structural Constraints: The institutional and structural constraints faced during the implementation of UPOR affected progress significantly. Owing to lack of monitoring on the part of institutions, matching of mapping process was done with the details of documents made available from other departments (matching spatial and nonspatial). The details on records were not matching with the field details collected by the survey team resulting in frequent resurveys. Multiple agencies fix coordinates which do not match with each other and clash with the UPOR projections.

Technical Problems: There were issues at various levels inclusive of (a) Missing traverse points (MTP), (b) Difficulty in mapping multi-storey buildings, (c) Problems associated with identifying the backside corners of the plots, *Gramathana*[33] and congested properties. Similarly, multiple dimensions with measurements in the form of *guntas*[34] was an issue as most revenue lands have irregular units, (d) Problems with Open Site/Rented Properties, and (e) Errors during survey of properties.

Issues in practical application of hatching tool in the field were constraining. Hatching tool (software TE process tool) automatically clips the physical area of properties if it is more than two feet toward the public domain assets such as roadside. Thus, more than 10 percent of the sample properties were marked for re-notice. For instance, any difference in length/area/breadth of a single property might entail resurvey for the entire block.

---

[33] *Gramathana* sites are areas within village, which are reserved for residential purposes and for expansion. These areas are converted from agricultural land to residential land.
[34] *Gunta* is a measure of an area. This unit is typically used to measure the size of a piece of land. For instance, 1 *guntha* = 33 ft (10 m).

Besides, processing of UPOR data were due to hardware and software issues. Since server was slow, scanning and uploading of data took more time. Errors in data entry were an issue; out of 43,823 properties in Mysore, 7,127 cases (16 percent) had data entry mistakes.

Delays in TE process as notices are not served for two kinds of properties: (a) vacant sites/plots and (b) door locks (DL). The service provider, therefore, contacts land developers to trace residential addresses before issuing notice to them for submission of property-related documents, but as an issue of concern.

Lack of Information/Awareness: Under UPOR, the Department of SSLR is primarily responsible in popularizing the urban property survey work and in this regard wide publicity was undertaken. Awareness campaign includes (a) issue of notice, (b) banners/pamphlets, (c) auto announcements, and (d) creation of citizen facilitation centres. However, across all cities, it was observed that people had not understood the rationale underlying the UPOR initiative. There was lack of awareness which may be due to properties located on revenue land, owners' fear of losing it if they reveal the details, and lack of trust to submit property documents.

Slums and UPOR

One of the critiques of UPOR project has been following top-down approach and lack of equity concerns incorporated into the project. Under UPOR project, slum areas have been marked and individual-level titling process is yet to commence. Only survey is conducted in slums marking them as individual blocks. Therefore, UPOR has not processed any individual documents belonging to authorised slums. Most of the slums are located either on (a) private, (b) vacant, or (c) government land. The slums located on government land will be authorised under CMC or urban development authority. Only authorized slums receive 'Title deeds' or 'Possession certificates' and these documents are issued on lease for 20 years. The number of slums on private land or unauthorized are more than the slums on government land with authorization. Most of the slums located on private land and vacant lands are subject to disputes and recognized as unauthorized or illegal occupants. Authorised slums do not possess documents to validate their stay in the city. Therefore, UPOR has processed only documents and is not able to extend the property rights to the slum dwellers.

Conclusion

The case of UPOR implementation in Karnataka serves to understand how urban land governance reforms contribute to more inclusive governance and sustainability. Globally, land administration has been streamlined, particularly developed countries such as USA, France, and Nordic countries; as a result, violations are minimal. Hence, India's attempt in improving land records management was much called for. The chapter has brought out the highlights of the process and implementation of UPOR besides critical issues and concerns in its implementation and, thus, acts as lessons for upscaling the initiative in other cities of India. Land being key for achieving broader sustainable goals, the UPOR project has facilitated access to accurate and transparent information, hassle-free transfer of land title, clean and updated records, easy access to information, quick retrieval, minimal visits, and security against tampering, overall leading toward transparency in urban governance in Karnataka. Therefore, this new solution to the existing chaotic land governance practice would certainly support urban land reforms and thereby promote sustainable development.

References

Burns, Tony. 2007. "Land Administration Reform: Indicators of Success, Future Challenges." Agriculture and Rural Development Discussion Paper No. 37 Washington, DC: The World Bank.

Chawla, Rajiv. 2004. "Online Delivery of Land Titles to Rural Farmers in Karnataka, India." Paper presented at seminar titled "Scaling Up Poverty Reduction: A Global Learning Process and Conference," May 25–27, Shanghai, China.

Dowall, David E. and Giles Clark. 1996. *A Framework for Reforming Urban Land Policies in Developing Countries*. Washington, DC: The World Bank.

McKinsey. 2001. *India the Growth Imperative*. Washington, DC: Mckinsey Global Institute.

Mukerji, Anirban. 2011. *Urban Property Ownership Records: Background Document*. Available at: http://www.slideshare.net/anirmukerji/urban-property-ownership-records-background-document (Accessed on November 20, 2011).

Pathak, R.D. and R.S. Prasad. 2005. "Role of E-governance in Tackling Corruption and Achieving Societal Harmony: Indian Experience." Paper presented at workshop on "Innovations in Governance and Public Service to Achieve a Harmonious Society," Beijing: NAPSIPAG.

Ramanathan, Swathi. 2009. "Security of Title to Land in Urban Areas." In *Indian Infrastructure Report: Land—A Critical Resource for Infrastructure*. New Delhi: Oxford University Press/3iNetwork and Infrastructure Development Finance Company.

RFP (Request for Proposal). 2009a. *Selection of Service Provider for Urban Property Ownership Records Project*, vol. I A Bangalore: Commissioner Survey Settlement and Land Records.

————. 2009b. "Selection of SP for Urban Property Ownership Records Project, Volume II (Commercial Terms and Bid Documents)." Reference Notification No. STR: 05 TCS: 4: 09–10, Bangalore: Commissioner Survey Settlement and Land Records.

UDP (Urban Development Program). 2009. *Urban Development Policy for Karnataka.* Bangalore: Urban Development Department.

Wadhwa, D.C. 1989. "Guaranteeing Title to Land: A Preliminary Study." *Economic and Political Weekly* 24 (41): 2323–2331, 2333–2334.

————. 2002. "Guaranteeing Title to Land." *Economic and Political Weekly* 37 (47): 4699–4722.

5

Challenges of Sustainable Biodiversity Management: National Chambal Sanctuary in Uttar Pradesh

Nidhi Yadav and Naresh Chandra Sahu

Introduction

Environment degradation is an appalling problem that the world faces today. If it continues, it can lead to food shortage, water scarcity, and other severe problems that have an effect on all living species on the earth (Homer-Dixon 1994; Saleth and Dinar 2004). In a nutshell, the sustainable management of natural resources and biodiversity is very important (Balmford et al. 2002). The continued loss and deprivation of biodiversity has compelled researchers and policy-makers athwart the globe to revamp the existing natural resource management practices and discover different approaches to promote sustainable resource use and to prevent further ecosystem degradation (Shahabuddin and Rao 2010). To control the overexploitation of natural resources and environmental degradation, sustainable development has been proposed to follow and is defined as "development that meets the needs of the present without compromising the ability of future generations to meet their own needs" by the World Commission on Environment and Development (WCED 1987). Sustainable development approach has become the most important debate in the present time. In the current phase of rapid development, environmental degradation and pollution have attributed as a hindrance to economic development in developing countries.

Environmental sustainability has been recognized as one of the eight international Millennium Development Goals (MDGs) that were established following the Millennium Summit of the United Nations (UN) to be targeted for the 15 years from 2000 to 2015 (UN 2015). The MDGs focus on widespread public concern about poverty, hunger, disease, unmet schooling, gender inequality, and environmental degradation (Sachs 2012). However, the main incentive behind the MDGs was to reduce poverty from the world, and the MDGs helped also to lift more than one billion people out of extreme poverty from the world, but the progress has been uneven (Sachs 2012). To reduce this inequality over the different developed and developing countries and to solve the dangerous climate change problem, the urgent need of sustainable development has been felt along with the poverty-reduction objectives. One hundred and ninety-three countries of the UN General Assembly adopted the 2030 development agenda titled 'Transforming Our World: The 2030 Agenda for Sustainable Development'. The agenda outlines the 17 Sustainable Development Goals (SDGs) and its related 169 targets for next 15 years that include ending poverty and starvation, improving health and education, making cities more sustainable, fighting climate change, and conserving natural resources such as oceans and forests (UN General Assembly 2015; Sachs 2012). The implementation of post-2015 SDGs for 15 years from 2015 to 2030 needs to eradicate poverty, protect the planet, and make sure thrive for all as a part of a new sustainable developmental agenda.

Indeed, the world is fighting with poverty and climate change. It is crucial to reduce and control both problems. Furthermore, the creation and effective management of a comprehensive Protected-Areas (PAs) system, covering the world's highest-priority conservation areas, is vital for the protection of biodiversity and the numerous market and nonmarket benefits that integral nature provides (Pimm et al. 1995; Rodrigues 2004). Dudley (2008: 8–9) defines PA as: "A protected area is a clearly defined geographical space, recognized, dedicated and managed through legal or other effective means to achieve the long-term conservation of nature with associated ecosystem services and cultural values." Therefore, significance and application of these PAs remain in the protection of key biological resources, endangered species along with the scope for sustainable development initiatives that will strengthen local livelihoods (Chape et al. 2005; Dash and Behera 2012). India is one of the 17 mega-diverse countries of the world in terms of biodiversity and natural resources (MDGs 2014). It has 2.4 percent of the world's land area, 16.7 percent of the world's human population and 18 percent livestock, but contributes about 8 percent of the known global biodiversity (MoEF 2002). In India, 166,352.63 sq. km which is 5.6 percent of the total land of the country, has been declared as national parks, wildlife sanctuaries, conservation reserves, and community reserves. There are 689 PAs which include 102 national parks, 526 wildlife sanctuaries, 57 conservation reserves, and 4 community reserves (MDGs 2014; WII 2013). People and PAs are dependent on one another for their survival. The mutual interdependency of PAs and people has created conflicts and issues between local people and forest department officials (Kothari et al. 1995). However, these conflicts and issues mostly exist in developing countries (Rout 2008). The priority of livelihood support, development, and climate change concern depends on the socioeconomic conditions of a country. In the case of developing countries like India, local communities are more susceptible to losing their livelihood after the conversion of natural resources areas into PAs. But in the case of developed countries, analysts and policymakers focus more on social impacts of large-scale environmental and climate change as the priority is higher than development. In developing countries, a huge population is already suffering from shortages of food, land, water, forests, and fish. Therefore, it is very difficult to consider the social effects of climate change and ozone depletion in developing countries like India. Providing basic livelihood support to local people is the major concern in developing countries. The achievements of PA management are difficult to assess in case of developing countries due to lack of an established system for monitoring trends in natural resources (Danielsen et al. 2000).

In India, local people accrue benefits from PAs in forms of land use, livestock grazing, and collection of fuel wood and other nontimber forest products (NTFPs) (Dash and Behera 2012; Udaya Sekhar 2003). However, numerous issues and challenges have been observed in PAs. When we talk about wildlife conservation, we find that there is a severe problem related to local people's livelihood. In the case of developing countries like India, there are various issues involved when a PA gets established. After the MDGs, when the world is following SDGs, the current study focuses on the challenges in sustainable biodiversity management in National Chambal Sanctuary (NCS). NCS is a kind of PA, which had been established in 1979 by the Uttar Pradesh (UP) forest department by converting the Chambal river region of UP and its forestry ravine area into protected wildlife area. In this sanctuary area, both poor socioeconomic conditions and environmental degradation are main challenges for the sustainable management. It is well known that a poor person cannot think about climate change and environmental issues without fulfilling his/her basic needs. In this context, the present study analyzes and discusses critical issues and challenges faced by NCS in UP through undertaking a field survey of surrounding villages of NCS. The main objective of this study is to understand how to sustainably manage the biodiversity of NCS in UP by bridging the gap between top-down approach and bottom-up approach. Apart from it, the study reveals how community involvement in the management of NCS would more effectively achieve the goals of conservation and sustainable livelihood support. In addition, secondary information collected by NCS, review of the literature, and our interaction with local people living in and around NCS during the field visit in December 2013–January 2014 has also been used for the present study. Apart from the introduction, the rest of this chapter is given as follows. The section 'Description of the Area' is related to the description of the study area. The section 'Importance and Current Status of NCS' describes the importance and current status of NCS. The section 'Institutional Dynamics and Management of NCS' explains the institutional dynamics and management of NCS. The management issues and challenges with the findings of the study have been discussed in the section 'Current Management Issues and Challenges'. Finally, conclusions have been drawn in the section 'Conclusion' focusing on the importance of local community involvement in the biodiversity management by using both top-down and bottom-up approach.

Description of the Area

The Government of India conveyed administrative approval for the establishment of the NCS in Order No. 17–74/77-FRY (WL) dated September 30, 1978 declared under Section 18(1) of the Wildlife Protection Act (WPA), 1972 (Kumar 2010). A 600 km stretch of the Chambal river has been protected as the NCS for the conservation and management of wildlife species (Hussain 2013). The Government of UP, vide its order no. 7835/14–3-103–78 dated January 1, 1979, issued the proclamation under Section 18 of the WPA, 1972, declaring the area given in the proclamation as "NCS" including river, forest, and private land. The area, as per notification, of the sanctuary in NCS, UP, lies between latitude N 24°55′ and N 26°50′ and longitude E 75°34′ and E 79°18′. The NCS covers an area of 635 sq. km in UP (Kumar 2010). The headquarters of the NCS, UP, is at Agra with subordinate offices at Etawah and Bah.

Four villages, namely Senhson, Bhareh, Nandgaon, and Pinahat, are located in and around NCS of UP region. In all four villages, there are entry points for tourism purpose. During the survey, it has been found that only one place, that is, Nandgaon, is opened for tourism purpose. Originating in the Vindhayan ranges in Madhya Pradesh (MP), the Chambal river snakes its way through the states of MP, Rajasthan, and UP before finally meeting the Yamuna in the Etawah district of UP. Chambal was selected as one of the most important habitats for the reproduction of Gharials bred in captivity at Kukrail in Lucknow and Deori in Morena.

Importance and Current Status of NCS

The Chambal is a persistent river recognized for its pure and uncontaminated water. It is home to an affluent biodiversity of flora and fauna. Uncontrolled poaching and unsystematic fishing have led to the decrement of Gharial population in India. Therefore, a captive breeding and reintroduction program was started in the 1970s to bring this species back from the edge of extermination. The Chambal river supports the largest population of Gharials in the wild. It has been found that NCS is among the most important and significant habitats where numerous worldwide endangered fauna still survive (Nair and Krishna 2013). NCS is listed as an important bird area (Zafar-ul-Islam and Rahmani 2004). NCS is also proposed as a Ramsar Site (NCS 2014). A wetland site, which is chosen of international importance under the Ramsar Convention, is called Ramsar Site. "The convention has provided guidance to Contracted Parties on the management of Ramsar Sites, in addition to its guidance on the wise use of all wetlands" (The Ramsar Convention and Its Mission 2017). Apart from having a strong position for World Heritage and Ramsar Convention listings, the NCS is also an area under discussion in international treaties like the Convention on the Conservation of Migratory Species of Wild Animals (Nair and Krishna 2013). It constitutes Asia's only river-based ecosystem sanctuary and only landscape-level tristate riverine PA in India.

Protection of the river was the main activity in the sanctuary after the declaration of NCS. The main objective of NCS is the preservation of biodiversity, the maintenance of its ecological processes and functions for the benefit of the nation and the humanity (Kumar 2010). It has been found that illegal fishing, sand mining, rampant felling of trees, and poaching are the main activities occurring in NCS. The sanctuary has been under severe pressure from local people. The vegetation in NCS area can be classified as ravine and thorn forest. Prosopis juliflora is widespread in the region. It has been found that in NCS there are 147 fish species, 56 reptile species, 320 bird species, and 60 mammal species from the region including 6 critically endangered, 12 endangered, and 18 vulnerable species, as categorized by the International Union Conservation and Nature (IUCN) (Nair and Krishna 2013). NCS is a natural habitat for Gharials and several different species of reptiles, mammals, and birds. In addition, 8 rare turtle species are found in NCS out of 26 turtle species of the country. Dolphins (national Gangetic animal) are also spotted in NCS.

Table 5.1:
Demographic and infrastructure status of NCS surroundings in 2010

Items in No.	Bah Range	Etawah Range	Total
No. of Villages	119	156	274
Village Population	121,360	108,409	229,769
Cattle Population	80,104	97,709	177,813
People Having Firearms (Living Inside NCS)	838	1,098	1,936
Forest Blocks	31	19	50
Gram/Gram Panchayat	28	19	47
Forest Roads	32	34	66
Temples	31	19	50

Source: Office of the Deputy Conservator Forest, Agra and National Chambal Sanctuary Project, Bah Range, Agra, UP.

This is also a well-known sanctuary where the nesting of Indian Skimmers is recorded in large numbers. The recent demographic and social overhead capital profile is explained in Table 5.1.

More than 60 percent of the people living in and around NCS depend on NCS for their livelihood. They follow ancient cultures and social customs. Around 55 percent people are literate. People are very poor and 90 percent depend on farming and labor. However, they have little agriculture land that is neither irrigated nor fertile. Due to poverty, villagers commit crimes such as illegal fishing, poaching, sand mining, robbery, and kidnapping in the local area. People spend too much on social and religious rituals and customs, for example, birth and thread ceremony, marriage, Indian funeral traditions, and special worship on different occasions. To meet their needs, they are involved in illegal felling of trees for fuel wood. Poor and landless people are depending on the following activities: collecting fuel wood, grazing, sand mining for construction, NTFPs, grass collection, etc. All the activities give rise to the conflict between villagers and PA managers.

The territorial part of NCS is located between the Yamuna and Chambal rivers. During the field survey, it has been found that the most of the local poor people were not satisfied with the protection of the Chambal river in the form of NCS. According to them, they have lost their livelihood and agricultural income because of restriction in the NCS area, which was the only source of their livelihood before the establishment of the sanctuary. The major problems that have been observed in NCS are illegal fishing, dams, poaching, sand mining, stone mining, infrastructural development, water pollution, riparian chemical agricultural-related activities, human–wildlife conflict, and local people–NCS officials' people conflict (Nair and Krishna 2013). However, it is found that these issues and conflicts are related to socioeconomic and developmental status of people. Therefore, socioeconomic status is given in Table 5.2.

Table 5.2:
Socioeconomic and human status of people in and around NCS

Issues	Current Status
Socioeconomic Profile	People are marginal farmers, labor, and off-farm employee as semi-urban workers. Majority of SC people are poor agriculturists with small-scale land holdings, and largely unskilled workers.
Employment	Employment opportunities are very less in NCS area. Most of the people are dependent on fishing and agriculture. The poor opportunity of employment and lack of education cause the problem of poaching and illegal activities. It has been found that 36.03 percent people have unemployment problem (WWF & CCF 2008).[1]
Livelihood Security	A general estimate by NCS management shows that more than 50 percent of the periphery area dwellers depend on NCS for fuel wood collection, fodder, grazing, and NTFPs.
Sanitation	There is a severe problem in NCS area during the summer season of drinking water. People and cattle use river water for drinking purpose, which creates human-wildlife conflict. 79.27 percent people have water problem (WWF & CCF 2008).
Communication, Electricity, and Solar Light	According to a survey by (WWF & CCF 2008), it has been found that 14.41 percent people face problems related to bad roads and 49.94 percent face electricity problem. However, there is an extensive network of road connectivity inside NCS area in both Etawah and Bah range.
Education	55 percent people are literate. Hence, existing illiteracy in and around NCS is the main cause of higher dependency on sanctuary for the livelihood of local people.
Social Overhead Capital (SOC)	There are total 47 gram panchayats and 274 villages in and around NCS. In one gram panchayat, there is one primary school and one junior high school. It has been found that 24.37 percent people do not have access to medical facilities and 13.51 percent people do not have access to schools. It has been found 65 percent villages are having primary schools facility in NCS area.

(Table 5.2 continued)

---

[1] World Wide Fund for nature (WWF-India) and Chambal Conservation Foundation (CCF) have conducted a field survey to know the people perception about the problems of people and Ghariyal status is NCS in 2008.

(Table 5.2 continued)

Issues	Current Status
Self-Help Groups (SHGs)	It has been found that there are three SHGs in one gram panchayat. These SHGs work for conducting community welfare programs and provide alternative employment opportunity. There are active SHGs only in 10 percent of NCS villages.
Aganwadi	There are aganwadi centers in-charge of mid-day meal at schools and organizing child welfare and health programs.
Asha Karmi (Health Workers)	There are only three hospitals in and around NCS. However, health workers are appointed in each gram panchayat for health facilities. It has been found that health facilities are not good in surrounding villages of NCS.

Source: Office of Forest Range Officer in Bah Range, NCS, UP Project; World Wide Fund for nature (WWF)-India; Chambal Conservation Foundation (CCF).

The increasing livelihood dependency of local people on the sanctuary poses a number of challenges to sustainable management and biodiversity conservation in NCS. It has been found that 7 major, 12 medium, and 134 minor irrigation projects are being operated in the Chambal river basin. These projects have really reduced water flow in the river (Hussain and Badola 2001; Nair and Krishna 2013). Table 5.2 clearly indicates that local livelihood in NCS is also under severe pressure. In the rural areas of India, people are still uneducated and unemployed due to the lack of resources, motivation, and agricultural land. They further involve in illegal activities to meet their needs by extracting more resources from such protected areas. Table 5.3 shows the major threats to biodiversity in NCS due to increasing impact of humans.

Table 5.3:
Major threats to biodiversity in NCS due to human interventions

Key Threats to Biodiversity	Current Status
Population Growth	Total population in surrounding villages of NCS shows an upward trend. It has increased from 159,766 in 2000 to 229,090 in 2010. This increased population puts direct pressure on forest resources of NCS. It has given rise to severe exploitation of forest resources.
Illegal Fishing and Poaching	After the declaration of NCS, fishing is banned in the sanctuary. It has been found that Etawah range is more sensitive for illegal fishing while Bah range for turtle poaching.

Key Threats to Biodiversity	Current Status
Extraction of Forest Produce	People living in and around NCS still depend on NCS to meet their needs for firewood and NTFPs.
Livestock Grazing	Livestock graze in and around of NCS. It is one of the important problems faced by the management of NCS that is affecting the Gharial habitat.
Livestock Population	Livestock population of NCS area has increased two times during 2000 to 2010. This increased population puts direct pressure on NCS.
Sand-Mining	Large-scale sand-mining destroys the sandy banks that are very harmful to Gharials and turtles nesting. It is considered one of the major threats to their natural habitat.
Human-Wildlife Conflict	Crop damage by blue bull is a common problem faced by local people in NCS surrounding area. During the last couple of years, it has been found that because of frequent intervention to the natural habitat, people have been killed by crocodile's attack. This gives rise to frequent conflicts between local people and protected area.
Forest Fire	It has been found that forest fire is happening every year causing serious damage to the wildlife. This is considered as threat faced by NCS. There are three cases in 2011; seventeen cases in 2012 and seven cases in 2013 of forest fire.
Encroachment of Forest Land	Illegal felling of trees, illegal removal of NTFPs, and other illegal activities are major problems for the management of NCS. Lack of staff, incomplete rehabilitations, lack of monitoring and an inadequate number of staff are the main reasons for all the problems.

Source: Kumar (2010).

Perusal of Table 5.3 reveals that NCS is under severe threats and is facing several challenges from different quarters to achieve sustainable management. The potential threats highlighted in Table 5.3 have resulted in the depletion and degradation of the rich flora and fauna of NCS. To accomplish sustainable management of NCS, the management of the sanctuary should implement various programs and policies to minimize the various threats. Furthermore, loss of livelihood of local people due to the PA also creates conflicts. Therefore, effective management of the NCS toward biodiversity conservation and livelihood improvement has become necessary by central/ state government and local institutions.

Institutional Dynamics and Management of NCS

In developing countries like India, forest and natural resource management policies have been conventionally characterized by general doubt of local people's capability to manage and protect the environmental resources on which they depend (Heltberg 2001). It is found that after the declaration of NCS as a sanctuary, there have been frequent conflicts of interest between sanctuary management and local community people in the river area. The agreement on what constitutes as efficient sustainable forest management is still a controversial issue (Dash and Behera 2012). It has been stated that effective conservation in any PA can be sustained only with proper management initiatives and with the local people's participation (Hetlburg 2001). There is a requirement to provide gainful employment to the local people to secure wildlife animals and to stop them from dying after being entangled into fishing nets hidden illegally in the river. To resolve the problem of conflicts between local people and NCS officials, it is required to create an environment for the proper involvement of local participation interest toward conservation, development, and management by the sanctuary. For proper natural resource management, developing countries are adopting the participatory approach (Heltberg 2001). The main purpose is to encourage local people's active involvement in the management of PAs and other environmental resources (Kiss and Mundial 1990). But in the case of NCS, there is basic and traditional resource dependancy for the livelihood of the local people settled inside or on the periphery of the NCS. The rehabilitation and resettlement (R&R) were not carried out properly in the NCS area. The majority of the local people feel that they are having their own private land. They also attempt to indulge in illegal activities. There are many bricks kilns all along the boundary of the sanctuary. Prosopis is being used for burning the chimneys that is harmful to wildlife habitat. These are the major challenges for the sustainable management of NCS. WPA, 1972, does not permit any kind of resource use by the villagers. The people living inside the NCS, UP, are still dependent on the River Chambal for their water need in the form of irrigation, drinking, and watering cattle. The components of management include creating new livelihood opportunity for the local people, empowering them, ensuring representation and equity, strengthening resource security, or providing property rights and broad-based participation in decision-making (Singh and Borthakur 2015). There are several institutions, which are working for wildlife and environmental protection as well as in collaboration with the sanctuary for its wildlife conservation and to sustainable ecosystem management over the last couple of years. These have been explained in Table 5.4.

Perusal of Table 5.4 reveals that institutions and NGOs are designed to encourage biodiversity conservation and livelihood for local people. During the World Bank Forestry Project (1982–2002), ecodevelopment as a strategy was applied in many PAs including NCS. It has been found during the study

Table 5.4:
Institutions working for the development and protection of biodiversity in NCS

Name of Institutions	Functioning/Role
Ecodevelopment Committees (EDCs)	The basic theme of EDC is to get people's active support in reducing their dependency on the sanctuary area, create employment for local people, and uplift the living standards by trying to solve the basic problems by also involving the other developmental agencies.
Turtle Survival Alliance (TSA)	TSA is an NGO that is committed to zero turtle extinctions in the 21st century. It is also working on turtle in NCS area over the last couple of years.
Wildlife Institute of India (WII)	It works toward wildlife research in different areas like biodiversity conservation, rare species, wildlife protection and management, wildlife forensics, spatial modeling, ecodevelopment, and climate change etc. with policy implications.
Madras Crocodile Bank (MCB)	The main objective of MCB is to promote the conservation of reptiles and amphibians on the Indian subcontinent.
The Gharial Conservation Alliance (GCA)	The main objective of GCA is to conserve the Gharial. It seeks to find out the current status of the Gharial (Gavialis gangeticus) all over its range, identify the pressure to the species, and to establish conservation programs to ensure the Gharial's survival into the future.
Chambal Conservation Foundation (CCF)	CCF is an NGO. It works to create sustainable projects, to an improvement in the lives of the local people while building support for the natural resources and the beauty of NCS.
Society for Conservation of Nature (SCoN)	It is working for wildlife and nature conservation. The main objective of SCoN is to manage and protect the environment through effective networking, confrontation, education awareness, and conservation programs.
World Wide Fund for Nature (WWF-India)	WWF-India deals with nature conservation, environmental protection, and development related issues in the country. Its main objective is to stop the environmental degradation. It is also working in NCS to conserve the nature and wildlife and to make local people aware toward nature conservation.

Source: Kumar (2010).

that there were six EDCs in NCS, UP. One EDC, namely Kassauaa, in Etawah range and five EDCs, namely Guha, Dhadhupura, Nadgaon, Jhirna ka pura, and Shivlal ka pura, were constituted in 2001–02 on the basis of proximity to the PA. UP government has invested the fund in EDCs in the installation of hand pumps, laying of brick-roads, digging of ponds, construction of the earthen dams, installation of solar lights, development of pasturelands etc. However, it has been found that due to many inherent reasons, the concept of having a self-sustainable revolving fund could not be successful. Currently, WWF is working in NCS villages to motivate the local people for biodiversity conservation and to promote self-generated employment like handicraft with the collaboration of UP forest department.

It has been found that NCS is facing severe problems related to management of its large area due to lack of fund and insufficient personnel. Some efforts are being made to create awareness related to biodiversity conservation, wildlife education, research, and training of the common people by different government programs and NGOs. In spite of efforts to promote local people participation in biodiversity conservation and increasing the livelihood sources to local people, still poverty and unemployment persist in the NCS. There is a need to relook the failure of EDCs and restart it to make sustainable biodiversity management effective and successful in the sanctuary.

Current Management Issues and Challenges

There are many issues and challenges that need to be taken care by UP government, forest department, and local people. Some NGOs also work to conserve the natural habitat, wildlife, biodiversity, and turtle, crocodile, and other endangered species. Development work in NCS should not be based on the cost of a villager's life. All the members of a community must be equally benefited for proper development of effective resource management institutions. Without the support and development of villagers and local people, it is very difficult to manage the NCS. The vision of NCS is the long-term biodiversity conservation for prosperity and the maintenance of its ecological processes and functions for the benefit of the nation and the humanity. Conservation is not an easy task for any PA without the support of local people. Despite being one of the last relic rivers in the greater Gangetic drainage basin to have retained significant conservation values, the Chambal river faces rigorous extractive and edifying pressures for resources (Hussain and Badonla 2001; Nair and Krishna 2013). There are some other issues of land ownership. Some people claim the ownership of land right over the river bed. Such claims must be settled to secure a disturbance-free habitat for the Gharial and other sympatric species. There are some specific issues identified in NCS that need to be solved, such as dependence for NTFPs on NCS, fuel wood collection, fishing, grazing, and dependence of water. It has been found that more than 50 percent people in and around NCS depend on the sanctuary for these activities. Illegal activities such as fishing, poaching, and sand mining are area-specific and their impact extends 1–5 km radially from settlements toward the river. The major emerging issues are increasing construction and area development activities, lack of government incentive, private tourism, continuous population growth, undervaluation of ecological services, poverty, unemployment, inadequate knowledge about the use of corridors by the wildlife animals, lack of awareness, construction of dams and barrages, etc.

It has been found that the Chambal Gharial population has experienced two dramatic declines: the first, during 1999–2003, resulting in a status change to 'Critically Endangered', and the second, the unexplained winter die-off of 2007–08. Over 100 sub-adult and adult Gharial perished in the winter die-off of 2007–08. Available evidence from postmortem examinations indicates that the affected Gharials died of kidney failure and an unidentified toxicant was suspected (Choudhury et al. 2012). Forest offenses during 2009–10 have been given in Table 5.5.

Perusal of Table 5.5 reveals the illegal activities in NCS. It is very difficult to conserve the sanctuary without the support of local people. The pattern of staff by hierarchical level to manage the sanctuary is given in Table 5.6.

Perusal of Table 5.6 reveals that the existing staff is not sufficient for the sustainable management of NCS. There are only seven forest guards in NCS, UP, that covers 635 sq. km. It is very difficult to manage such a large area with few forest guards. The government has invested funds to enhance the employment opportunity through starting small income-generating activities with the help of EDCs. But the employment opportunity remains the same prior and post ecodevelopment project. Some of the entry point activities

Table 5.5:
Forest offenses in NCS

Range	Illicit Felling	Poaching/Hunting	Mining	Other	Total
Bah	11	04	22	18	55
Etawah	32	07	00	06	45

Source: Kumar (2010).

Table 5.6:
Existing staff of NCS

S. No.	Post	Bah Range	Etawah Range
		Number	Number
1	Wildlife Warden	–	1
2	Range Forest Officer	1	1
3	Dy. Ranger	–	1
4	Forester	3	5
5	Research Asstt.	–	4
6	Wildlife Guard	6	8
7	Forest Guard	3	4
8	Guard	3	2
9	Asstt. Wildlife Warden	1	1
10	Field Assistant	2	–
11	Boat Man	4	5
12	Security Guard	–	2
	Total Staff	23	37

Source: National Chambal Sanctuary, Management Plan for 2010–11 to 2019–20.

were successful and highly appreciated by the people. It was successful in terms of generating goodwill for the sanctuary by making direct interventions to meet community needs in a participatory and transparent manner. There was no improvement in generating stall-feeding practices. After some years, EDCs got stopped and could not be successful. There is also no system for joint forest management (JFM) activities for promoting the people's participation in NCS management. However, it is found that animal husbandry is practiced in the entire sanctuary area due to natural geographical conditions and zero investment in cattle rearing. Overall, there is no proper participation of local people in management, protection, and development of NCS.

It is found that NCS has the huge potentiality for attracting tourists. Historically, the name Chambal was a dreaded name as the entire Chambal zone in all the three states was bandit intensified (Kumar 2010). The perception of poor law and order situation discourages the flow of tourism infrastructure. Poor infrastructure is the main cause of poor ecotourism in NCS. However, in spite of having four entry points in NCS, UP, only one is opened for tourism activity, that is, managed by Chambal Conservation Foundation (CCF). The tourists' data of Keoladeo National Park (KNP) reveals that the tourists' inflow was 130,000 during 2012, whereas it was 750 in NCS, UP during 2011–12. In comparison of KNP, the tourist inflow in NCS is very low. The opportunity of ecotourism has not been explored properly. All the threats and issues faced by NCS hamper the sustainable management of the sanctuary. Further, these issues impose a serious challenge toward the long-term conservation of representative biodiversity of NCS and the maintenance of its ecological process.

Conclusion

It has been found that NCS is facing severe problems due to lack of sustainable management framework. In addition, local institutions have failed in NCS management and biodiversity conservation due to lack of people's participation. In spite of having the ecotourism potential in NCS, tourism is not developed due to lack of publicity, security, funds, and infrastructure. However, NGOs are playing a major role to develop ecotourism and to create awareness among local people about importance and significance of conservation in the region. In this respect, both the state and central government should focus more on sustainable development, management, and protection in NCS area by encouraging the local people's participation. Without the active involvement of local people in the sustainable biodiversity management of NCS, the protection of the sanctuary is not successful by the forest department. It indicates that to achieve sustainable biodiversity management of NCS, we need to follow not only top-down approach but also bottom-up approach with proper involvement of local communities. In the absence of local community participation, forest department officials are not able to protect sanctuary because poor socioeconomic condition made the communities more vulnerable. In spite of protection in the NCS boundaries, still people try to exploit the forestry areas for their basic needs and create an insecure environment for the tourists as well as for the forest department officials. Hence, lack of infrastructure, funds, and staff etc. put pressure on the management of NCS. On the other hand, without the local people's participation in the management of NCS, top-down management approach has failed. Therefore, we suggest that government should promote local people's participation in the environmental protection and management by supporting their livelihood system and making them aware of the importance of biodiversity conservation. In this way, for the effective conservation and management of natural resources or PA such as NCS, policymakers should use both top-down approach and bottom-up approach in a collaborative way. The implementation of both approaches will lead a sustainable biodiversity management and will be proved as an effective policy. Apart from that, there is a need to develop appropriate infrastructure, communication facilities, transportation, and some initiatives for the proper development of ecotourism. Increasing tourists' activities will enhance employment and will reduce anthropogenic pressure on NCS. The revenue from ecotourism can be reallocated to develop ecodevelopment committees. Based on the literature survey and field study of villages near NCS, we would like to suggest that 25 percent of ecotourism revenue from NCS can be sent to forest department revenue sector and 75 percent of revenue can be spent on ecodevelopment committees for development of the surrounding area and for local people's participation for nature conservation and protection. The government has to take the incentive to make aware the local people toward the importance of biodiversity. For environmental governance, it is suggested that forest department of UP should come up with employment-generating programs such as establishing small-scale cottage industry, agro-forest handicraft, and plantation and promoting self-help groups (SHGs). Therefore, government has to formulate and implement effective policies to resolve the conflict between local people and management of NCS by creating scope for people participation in the management and development of NCS. A participatory management strategy of PAs, ecodevelopment aims at protecting the biodiversity by including both the impact of local people on the PAs and the impact of the PAs on local people. In order to enhance partnership and to apply effective policies, EDCs should be promoted again around NCS. In addition, JFM also encourages partnerships in forest movement involving both the state forest departments and local communities, should be implemented in NCS.

References

Balmford, A., A. Bruner, P. Cooper, R. Costanza, S. Farber, R.E. Green, ... and R.K. Turner. (2002). "Economic Reasons for Conserving Wild Nature." *Science* 297 (5583): 950–953.

Chape, S., J. Harrison, M. Spalding, and I. Lysenko. 2005. "Measuring the Extent and Effectiveness of Protected Areas as an Indicator for Meeting Global Biodiversity Targets." *Philosophical Transactions of the Royal Society B: Biological Sciences* 360 (1454): 443–455.

Choudhury, B.C., P. Gautam, and T. Nair. 2012. Generic Tri-state Management Plan— National Chambal Sanctuary. National Tri-State Chambal Sanctuary Management and Co-ordination Committee (NTRIS-CASMACC). New Delhi: Ministry of Environment & Forests, Government of India.

Danielsen, F., D.S. Balete, M.K. Poulsen, M. Enghoff, C.M. Nozawa, and A.E. Jensen. 2000. "A Simple System for Monitoring Biodiversity in Protected Areas of a Developing Country." *Biodiversity & Conservation* 9 (12): 1671–1705.

Dash, M. and B. Behera. 2012. "Management of Similipal Biosphere Reserve Forest." *Advances in Forestry Letter (AFL)* 1 (1): 7–15.

Dudley, N., ed. 2008. *Guidelines for Appling Protected Areas Management Categories.* Gland, Switzerland: IUCN.

Heltberg, R. 2001. "Determinants and Impact of Local Institutions for Common Resource Management." *Environmental and Development Economics* 6 (2): 183–208.

Homer-Dixon, T.F. 1994. "Environmental Scarcities and Violent Conflict: Evidence from Cases." *International Security* 19 (1): 5–40.

Hussain, S.A. 2013. "Activity Pattern, Behavioural Activity and Interspecific Interaction of Smooth-Coated Otter (Lutrogale Perspicillata) in National Chambal Sanctuary, India." *IUCN/SCC Otter Specialist GROUP Bulletin* 30 (1): 5–17.

Hussain, S.A. and R. Badola. 2001. "Integrated Conservation planning for Chambal River Basin." Paper presented in the National Workshop on Regional Planning for Wildlife Protected Areas, August 6–8, 2001 1–20. New Delhi/Dehra Dun: India Habitat Centre/ Wildlife Institute of India.

Ingram, P. and V. Neel. 1998. "Embeddedness and Beyond: Institutions, Exchange, and Social Structure." In *The New Institutionalism in Sociology,* edited by M.C. Brinton and V. Nee. Stanford: Stanford University Press.

Kiss, A. and B. Mundial. 1990. *Living with Wildlife: Wildlife Resource Management with Local Participation in Africa,* Vol. 23. Washington, DC: World Bank.

Kothari, A., S. Suri, and N. Singh. 1995. "Conservation in India: A New Direction." *Economic and Political Weekly* 30 (43): 2755–2766.

Kumar, N. 2010. *National Chambal Sanctuary Management Plan.* Lucknow: Forest Department, Uttar Pradesh.

MDG (Millennium Development Goals). 2014. *India Country Report.* Available at: http://mospi. nic.in/Mospi_New/upload/mdg_2014_28jan14.pdf (Accessed on December 23, 2014).

MOEF (Ministry of Environment and Forest). 2002. *Protected Area Network in India.* Available at: http://moef.nic.in/sites/default/files/protected-area-network.pdf (Accessed on February 14, 2017).

Nair, T. and Y.C. Krishna. 2013. "Vertebrate Fauna of the Chambal River Basin, with Emphasis on the National Chambal Sanctuary, India." *Journal of Threatened Taxa* 5 (2): 3620–3641; doi:10.11609/JoTT.o3238.3620–41.

National Chambal Sanctuary (NCS) Agra, Etawah, UP. Available at: http:// nationalchambalsanctuary.in/ (Accessed on January 30, 2014).

Pimm, S.L., G.J. Russell, J.L. Gittleman, and T.M. Brooks. 1995. "The Future of Biodiversity." *Science-New York Then Washington* 269 (5222): 347–350.

The Ramsar Convention and Its Mission. 2017. Available at http://www.ramsar.org/about/ wetlands-of-international-importance-ramsar-sites (Accessed on March 27, 2017).

Rodrigues, A.S., H.R. Akcakaya, S.J. Andelman, M.I. Bakarr, L. Boitani, T.M. Brooks, and X.I.E. Yan. 2004. "Global Gap Analysis: Priority Regions for Expanding the Global Protected-area Network." *BioScience* 54 (12): 1092–1100.

Rout, S.D. 2008. "Anthropogenic Threats and Biodiversity Conservation in Similipal Biosphere Reserve, Orissa, India." *Tiger Paper* 35 (3): 22–26.

Sachs, J.D. 2012. "From Millennium Development Goals to Sustainable Development Goals." *The Lancet* 379 (9832): 2206–2211.

Saleth, R.M. and A. Dinar. 2004. "*The Institutional Economics of Water: A Cross-country Analysis of Institutions and Performance.* Washington, DC: World Bank and Cheltenham, UK: Edward Elgar.

Shahabuddin, G. and M. Rao. 2010. "Do Community-conserved Areas Effectively Conserve Biological Diversity? Global Insights and the Indian Context." *Biological Conservation* 143 (12): 2926–2936.

Singh, B. and S.K. Borthakur. 2015. "Forest Issues and Challenges in Protected Area Management: A Case Study from Himalayan Nokrek National Park and Biosphere Reserve, India." *International Journal of Conservation Science* 6 (2): 233–252.

Udaya Sekhar, N. 2003. "Local People's Attitudes Towards Conservation and Wildlife Tourism Around Sariska Tiger Reserve, India." *Journal of Environmental Management* 69 (4): 339–347.

UN (United Nations). 2015. The Millennium Development Goals Report. Available at: http://www.un.org/millenniumgoals/2015_MDG_Report/pdf/MDG%202015%20rev%20(July%201).pdf (Accessed on December 26, 2015).

UN General Assembly. 2015. *Transforming Our World: The 2030 Agenda for Sustainable Development.* Available at: https://sustainabledevelopment.un.org/content/documents/21252030%20Agenda%20for%20Sustainable%20Development%20web.pdf (Accessed on December 24, 2015).

WCED (World Commission on Environment and Development). 1987. *Our Common Future.* Oxford: Oxford University Press.

WII (Wildlife Institute of India). 2013. *Protected Areas of India.* Available at: http://wiienvis.nic.in/Database/Protected_Area_854.aspx (Accessed on January 29, 2014).

Zafar-ul-Islam, M. and A.R. Rahmani. 2004. *Important Bird Areas in India: Priority Sites for Conservation.* Mumbai: Bombay Natural History Society.

6

Freshwater Wetlands in Bangladesh: The Need for Alternative Governance

Mohammad Abu Taiyeb Chowdhury

Introduction

Wetland is one of the most important components of ecosystems in Bangladesh, having enormous ecological and economic significance (Khan et al. 1994). Developing successful management arrangements that ensure wise use[1] of aquatic resources and meet the needs of resource users (RUs) and other stakeholders is, however, a big challenge. Bangladesh's innovative pilot program in the people-led freshwater wetland management system demonstrates how sustainable enterprise can lift people out of poverty and protect the environment in the process. Even a decade ago, many of Bangladesh's wetlands were devastated—an exploding population, poor government policy, increased deforestation, and environmental pollution, all had taken their toll on these ecosystems. But when the Government of Bangladesh changed the way how it grants access to freshwater fisheries through a comanagement (CM) mechanism in three major watersheds,[2] the fisheries were restored along with the rural communities surrounding these (Thompson et al. 2003). CM is basically a strategy to integrate the state apparatus with local institutions so that communities and local governments

---

[1] As defined by the Ramsar Convention, wise use is the "sustainable utilization of wetlands for the benefit of mankind in a way compatible with the maintenance of the natural properties of the ecosystem." In the case of Bangladesh, wise use of wetlands primarily denotes to provide a safe and sound habitat for water birds.

[2] Names of the three major watersheds are: (a) Hail Haor (located in northeast Bangladesh and is typical of deeply flooded basins in that region known as *haors*); (b) the Turag-Bangshi site (located just north of Dhaka and is typical of most low-lying floodplains of Bangladesh, and (c) the Kangsha-Malijhi site (located in north-central Bangladesh in Sherpur Sadar and Jhenaigathi Upazila of Sherpur District).

can more effectively manage their natural resources (NRs). An attempt has also been made to explore the potential of CM with a focus on the development of institutions. The institutional aspects demonstrate how their approach of CM was applied to wetland governance, in general, and their impact on the flood plain area of Bangladesh, in particular.

The need for alternative governance in natural resources management (NRM) arises from a debate whether the development path should be driven either by a 'top-down' or a 'bottom-up' approach or by a combination of both. This chapter focuses on the need to bridge the 'bottom-up' and 'top-down' approaches to sustainable development, a bridge that might offer at least one possible approach—CM (Borrini-Feyeraband et al. 2007). The history of CM is rooted in decades of field-based and theoretical efforts made by individuals and groups concerned with social justice (Scott 1985). In the past, many traditional societies formed relatively closed systems in which NRs were managed through complex interplays of reciprocities and solidarities. These systems were fully embedded in the local cultures and accommodated for differences of power and roles—including decision-making—within holistic systems of reality and meaning. Dialogues and discussions among interested parties on the basis of field experience (what is referred to as 'CM' today) were widely practiced in some of these societies.

However, the historical emergence of colonial powers and nation states, and their violent assumption of authority over most common lands and NRs, led to the demise of traditional NRM systems virtually everywhere. The monetization of economic exchange weakened local systems of reciprocity and solidarity, as did the incorporation of local economies into increasingly global systems of reference. In addition, the rise in power of modern, expert-based, 'scientific' practices induced severe losses in local knowledge and skills. This generalized breakdown of local NRM systems finally resulted in the disempowerment and 'de-responsibilization' (see Banuri and Amalrik 1992) of local community.

The theme of this chapter is relevant to the Millennium Development Goals (MDGs): Goal 1 (to eradicate extreme poverty and hunger) and Goal 7 (to ensure environmental sustainability by 2015) that were established following the Millennium Summit of the United Nations in 2000. Most recently, these goals have found a place in the United Nations Sustainable Development Goals (SDGs), officially known as 'Transforming Our World: The 2030 Agenda for Sustainable Development'—an intergovernmental set of aspiration goals outlining the 17 SDGs and their associated 169 targets. Among the 17 SDGs, several make explicit reference to ecological limits, for example, Goal 15 on protecting, restoring, and promoting sustainable use of terrestrial ecosystems and halting biodiversity loss. The goals are contained in paragraph 51 of the United Nations Resolution A/RES/70/1 adopted in New York on September 25, 2015. The resolution is a broader intergovernmental agreement that, while acting as the Post-2015 Development Agenda (successor to the MDGs), builds on the principles agreed upon under Resolution A/RES/66/288, popularly known as 'The Future We Want' (Redclift 1987; Sachs 2012; WECD 1987).

Natural Resources Management: Concepts, Approaches, and Context

One good example is CM. This may also be called participatory, collaborative, joint, mixed, multi-party, or roundtable management. CM is a situation in which two or more social actors negotiate, define, and guarantee amongst themselves a fair sharing of the management functions, entitlements, and responsibilities for a given territory, area, or set of NRs. As such, CM is more a pluralist approach to managing NRs, incorporating a variety of partners in a variety of roles, generally to the end goals of environmental conservation, sustainable use of NRs, and the equitable sharing of resource-related benefits and responsibilities. It may also be regarded as a political and cultural process par excellence: seeking social justice and 'democracy' in the management of NRs (Murphree 1994). The main CM values and principles are recognizing different values, interests, and concerns involved in managing a territory, area, or set of NRs, both outside the local communities and within them; being open to various types of NRM entitlements beyond the ones legally recognized (such as private property or government mandate); seeking transparency and equity in NRM; and allowing the civil society to assume ever more important roles and responsibilities. These lead to NRM partnerships, harnessing the complementarity of the capacities and comparative advantages of different institutional actors, linking entitlements and responsibilities in the NRM context. Despite the fact, CM is a complex, often lengthy, and sometimes confused process, involving frequent changes, surprises, sometimes contradictory information, and the need to retrace one's own steps (Fisher 1995).

Participatory decision-making is not a new phenomenon in Asia. It has long been a part of development thinking in the region (Agarwal 2001). Local rural communities in the humid tropical Asia have a long tradition of managing wetlands to secure their livelihoods with local institution playing a central role in NRM practices. After the World War II, however, local communities have been systematically excluded, opposed, or marginalized from taking part in the decision-making process due mainly to top-down, expert-driven, and command-controlled management regimes. As noted above, in public policies, economic development has been given priority, which has influenced NRM systems—maximizing economic benefits from NRs. At times, NRM policies have tilted toward centralized, command-control, logical, left-brain service-based approaches (Khan 2011).

Since the participation of village communities has been widely accepted as an institutional imperative in NRM, community-based management (CBM) of NRs is now receiving attention as a potential mechanism for increasing the efficacy, legitimacy, and sustainability (Basnet 1992; Western and Wright 1994). As such, the term 'community participation' means the involvement of the local people in the NRM system. The core principle of CM is to share management decisions and responsibility between RUs and government, thus improving the quality of decisions and local compliance with management plans. CM emphasizes the development of local institutions (changes in institutions, behaviors, and attitudes to support communities and local government in the management of natural aquatic resources). In an attempt to find new solutions to problems resulting from top-down model addressing resource conservation and sustainability, community-based CM recognizes that local community should have direct control over the utilization, benefits, and management of common resources.

Objectives and Methodology: Bangladesh— A Case Study

This chapter deals with sustainable natural resources management (SNRM) in Bangladesh from a policy perspective with a focus on the freshwater wetlands—locally known as *haors*.[3] The major thrust of the study is twofold:

---

[3] The *haor* basin of the northeastern part of Bangladesh encompasses the floodplains and swamps of the Meghna River systems—tributaries, covering an area of approximately 24,500 sq. km (Kabir and Amin 2007). It is characterized by the presence of numerous large, deeply flooded, bowl-shaped (extensive) natural depressions often tectonic in origin, found between the natural levees of rivers. These are subjected to seasonal (monsoon) flooding every year and

(a) to identify the factors causing the degradation of freshwater wetlands in Bangladesh, including environmental impacts and (b) to explore the best approaches or practices in their governance. The chapter mainly discusses about what strategy may help the policy as well as decision-making to achieve the goals of enhancing wetland resource base. The underlying objective there has been to find out ways how to alleviate poverty through efficient wetland resource management, adopting good governance toward conservation based on wise-use principle, as well as participatory and sustainable management practices.

The study is centered on the critical review, analysis, and synthesis of empirical findings that were drawn from four major case studies including Tanguar *haor* (Kabir and Amin 2007), Hakaluki *haor* (Khan 2011), Hail *haor* (Thompson et al. 2003), and Khaliajuri *haor* (Roy 2012) representing the northeastern *haor* region (Greater Mymensingh—Sylhet Basin) of Bangladesh. The literature survey process involves browsing over websites on issues relating to wetland resources management and collection of relevant information from various published/unpublished sources including major original comprehensive investigations—case studies, theses, reports, etc.

An attempt has been made to examine major issues facing the sector today including promises and challenges that lie ahead, especially in the early twenty-first century. An attempt has also been made to explore the potentials of CM with a focus on the development of institutions. The institutional aspects demonstrate how its approach was applied to wetland governance and its impacts in the entire flood plain area. The chapter then draws some lessons learnt from a mega project entitled 'Management of Aquatic Ecosystems through Community Husbandry (MACH)' implemented over nine years (1999–2007), while best practices from previous experience are widely adopted. In the MACH case, the community-based conservation of wetland resources was aligned with CM as the guiding principle of the project. The rationale there has been to demonstrate democratic mechanisms and decision-making processes as a way of SNRM in the country while taking into consideration the issues of governance, public policy, and environmental challenges–climate change.

remain under water for several months—7 to 8. The basin includes about 47 major *haors* and some 63,000 *beels* of varying sizes (Khan 1994).

Importance of Wetlands in Bangladesh

International Concern

This chapter stems from a genuine concern of the Convention on Wetlands of International Importance (1971), popularly known as Ramsar Convention.[4] Bangladesh demonstrated its concern for wetland conservation through the National Environment Policy and became a signatory of the Ramsar Convention in May 1992, followed by convening a workshop on wetlands. As a signatory of the Ramsar Convention, Bangladesh is now obliged to manage its wetlands and wetland resources on a sustainable basis (GoB 2002; Kabir and Amin 2007), requiring participation of local communities in decision-making. The government has already suggested a CBM approach of *haor* resources in order to foster sustainable uses. The first project of this kind was implemented by the International Union for the Conservation of Nature (IUCN)-Bangladesh in 1998 with financial support from the Ministry of Environment and Forest.

National Significance

Bangladesh is a country of wetlands. More than two-thirds of the country fall under wetland category[5] (Table 6.1) as defined[6] in the Ramsar

---

[4] The Convention on Wetlands of International Importance (especially as Waterfowl Habitat) is the first global intergovernmental treaty that reflects emphasis on wetland conservation. The convention was adopted in February 1971 in the city of Ramsar, Iran, and thus is popularly known as Ramsar Convention. The goal of the convention was to stop the progressive encroachment and loss of wetlands worldwide. The underlying objective was to recognize the fundamental ecological processes of wetlands and their social, economic, aesthetic, and recreational values.

[5] The total area of wetlands in Bangladesh has been variously estimated at 7–8 million ha, encompassing a wide variety of dynamic ecosystems ranging from rivers and streams, seasonally inundated extensive floodplains and flooded paddy fields, estuarine systems with extensive tidal mangrove swamp forest, natural freshwater lakes and marshes, man-made water reservoirs, fish ponds, and tanks (Table 6.1), creating a unique mosaic of habitats for diverse flora and fauna (Khan 2011).

[6] IUCN adopted the following working definition of wetlands at its first convention (1971) in Ramsar, Iran. According to Ramsar Convention (1971),

> [T]he areas of marsh, fen, peat-land or water, whether natural or artificial, permanent or temporary, with water that is static of flowing, fresh, brackish or salt, including areas of marine water, the depth of which at low tide does not exceed sic meters can be defined as wetlands.

Table 6.1:
Distribution of wetlands in Bangladesh

Type	Area (Hectare)
Permanent rivers and streams	48,000
Estuaries and mangrove swamps	61,000
Shallow lakes and marshes	120,000
Large water storage reservoirs	90,000
Small tanks and fish ponds	150,000–180,000
Brackish water/shrimp polders	90,000–115,000
Seasonally flooded flood plain	577,000

Source: Nishat et al. (1993).

Convention (1971). About 50 percent area of the country is comprised of floodplains that are regularly inundated, forming one of the most important wetland areas of the world—fertile and productive ecosystems with rich biodiversity. These are critical landscape units both economically and ecologically: for human settlements, agriculture, fisheries, navigation, communication, recreation, ecotourism, and biodiversity that serve as a source of income and nutrition for millions of people (FAO 1988).

Floodplain fisheries, for example, play a vital role in cushioning rural poverty and supplying animal protein to the poor. Although fish and rice are considered the main wetland resources, these also provide millions of poor families with numerous other resources such as freshwater, crops, water fruits, reeds, vegetables, grasses, grazing land, peat coal, fuel, fertilizer, honey, medicinal plants, birds, and wildlife (CNRS 2009). They are rich in biodiversity—flora and fauna composition—provide habitat for a variety of resident birds and migratory waterfowl of regional and global significance, and serve as nursery and important breeding grounds for fish production (Rahman 2005). In fact, the lives and livelihoods of the majority of Bangladeshi people, particularly the poverty-stricken and the landless groups, are intricately linked to these freshwater wetlands, providing sources of income and nutrition for millions of poor Bangladeshi (Chowdhury 2012).

Factors Causing Degradation of Wetlands

Freshwater wetlands in Bangladesh are continued to decline in terms of size, shape, and area, regardless of all beneficial functions. Various natural processes and human activities have significantly damaged and altered the system. Whatever remains are devastated, degraded, or being threatened. Environmentalists (Nishat et al. 1993) have identified the following factors behind the degradation of wetlands in the country:

- Immense population pressure on limited land;
- Expansion of human settlements;
- Overexploitation of wetland resources;
- Overfishing and associated disturbances;
- Overgrazing by livestock;
- Enormous changes in land-use patterns, for example, expansion of winter agriculture;
- Subsequent conversion of wetlands through drainage into rice fields;
- Increased deforestation, that is, overfelling of wetland trees;
- Excessive siltation due to degradation of watershed areas;
- Flood control embankments and irrigation projects (drainage schemes) for enhancement of agricultural productivity;
- Landfills for various purposes—residential, commercial, and industrial development;
- Pollution of water due to urban, industrial, agro-chemical, and other sources;
- Indiscriminate control/regulation/use of water flows of main river systems in upper riparian area;
- Rural infrastructures such as poorly planned roads, sluice gates, and narrow culvert; and
- Poor government policy (Nishat et al. 1993).

Environmental Impacts

Degradation of wetlands in Bangladesh has created several environmental impacts (CNRS 2009; Khan et al. 1994; Nishat et al. 1993) as follows:

- Loss of biodiversity—Extinction and reduction of wildlife, including birds and reptiles;
- Loss of soil nutrients—Decline in many indigenous aquatic plants, herbs, weeds, and shrubs;
- Extinction of many wild and domesticated varieties of rice with the propagation of high-yielding varieties;

- Overharvesting of fisheries and aquatic resources;
- Serious reduction of fish habitat, fish population, and diversity;
- Shrinking of natural water reservoirs and of their resultant benefits;
- Reduced dry season standing water and river flows;
- Disappearance of natural connections between floodplains and rivers;
- Increased sedimentation and pollution, frequent occurrence of flooding, and deterioration of living conditions; and
- Degeneration of wetland-based ecosystems, occupations, socioeconomic institutions, and cultures.

Major Issues and Challenges

Flood embankment and water control structures have already blocked many fish migration routes in the *haor* region. Irrigation facilities have expanded winter rice cultivation at the expanse of the surface water that sustains aquaculture life in the dry season. Industrial development has caused severe local pollution, killing breeding fish population during the dry season. Modern agriculture (with the use of agro-chemicals such as synthetic fertilizers, pesticides, and herbicides) has also affected wetland habitats tremendously. Deforestation and poor land management has caused high rate of siltation, filling freshwater wetlands in the dry season that serve as fishing nursery and breeding grounds during a crucial time of the year. Furthermore, more and more people fish destructively by dewatering using fine mesh nets.

Furthermore, there is a jurisdictional ownership problem of wetlands as well. The following is a good example of top-down power diminishing the bottom-up capacity. The resources of the wetlands belong to the state, and access to and control over wetland resources are determined by the existing top-down, command-and-control, bureaucratic management regimes. The government leases out fishing rights in public water bodies (*Jalmahals*— fishery estate) to vested interest groups while encouraging the leases for maximum benefits or exploitation (without giving incentives to protect wetland resources for the future generation). Despite recent changes in national policies that call for an end on overexploitation of resources, wetlands continue to be encroached for a number of economic activities including agriculture, aquaculture, industry, and brickfields with no sign of abatement. Grounded solely in the economic aspects of NRs exploitation, the wetland management objective of the government has frequently been focusing on rent-seeking mechanism to maximize revenues and other economic incentives or benefits. It is widely held that governance of wetland management in favor of the vested interest group is one of the main factors responsible for marginalization of local communities related to wetland because of their limited right and access in managing those resources (Ahmed et al. 2008; Khan 2011).

Wetland Conservation and Management

Comanagement in Wetland Conservation

As has already been recorded, wetland ecosystems are threatened mainly due to excessive and inefficient exploitation (pressures or stresses) of wetland resources and impacts of human intervention in the form of so-called development activities. The participation of local communities in the protection of their own environment, particularly in the conservation of their wetland resources, has, therefore, become an urgent issue to slowdown the rate of loss of biodiversity, and thereby to minimize the damage to the wetland ecosystem. In fact, active participation of the local people is crucial for the sustainable use of wetland resources. According to many commentators, management plan carried out on wetlands without the active participation of the local people is more likely to be unsuccessful (Borrini-Feyerabend 1996; Freeman 1996; Dugan 1990).

Community-based natural resources management (CBNRM) in Bangladesh has demonstrated its prospects in some development projects such as Sustainable Environment Management Program (SEMP). However, there have been arguments that CBNRM is more effective in decision-making, distributional equity/implications, coping with uncertainty and risks, learning and adaptation, and sustainability compared with many other approaches such as top-down, command-control, systematic science, technology-based management system, etc. (Agarwal and Gibson 1999; Barton et al. 1998; Halder and Thompson 2003; Western and Wrights 1994). Despite its advantages and effectiveness, the CBNRM approach has been adopted only in experimental development projects and is yet to be mainstreamed as a formal management approach (IUCN 1980, 2004). Although the Government of Bangladesh has undertaken some projects toward this end, the 'community participation', which is one of the main guidelines of Ramsar Declaration, is yet to be implemented at some proposed Ramsar sites, for example, Tanguar *haor* (Kabir and Amin 2007).

Khan (2011) investigated options for institutionalizing participation of stakeholders in freshwater wetland resource management, particularly in the Hakaluki *haor* area. The field study method selected three development initiatives in the Hakaluki *haor* for critical assessment, with the intention to seek alternatives to the state-governed management approach (SMA) and find a means of governance that would encompass multistakeholders in the management of NRs. As an added dimension, the study also attempted to examine the weakness of SMA and appreciated the role of local community participation and their deliberation in decision-making processes.

The research findings of Hakaluki *haor* have revealed that the community-based organizations (CBOs) were quite capable of contributing effectively to the community-based and CM approaches in wetland resource management. What was greatly in need is the establishment of a multilevel stakeholder governance system as an institutional structure and process, necessary to sustain CBOs operations in decision-making. According to the study, the participation of local RUs in Hakaluki *haor* would require an appropriate degree of integration of the 'bottom-up' and 'top-down' approaches to include all relevant stakeholders in the decision-making processes at multiple levels of social organizations. This alternative approach could be an effective instrument to facilitate the deliberations of stakeholders and to strengthen institutional linkages to engender benefits to the local RUs (Khan 2011).

The following section summarizes the institutional aspect of the MACH project undertaken in Hail *haor* as a good example or a success story. The experiment fits well within a broader context of CBM of common property and conservation. It has also the added dimension of targeted livelihood development, in terms of both alternatives and enhanced income from wetland resources (Berkes 1989).

Managing Sustainable Development in the Wetlands

The MACH Approach

Recognizing the need for new and better approaches to wetlands management, the Governments of Bangladesh in cooperation with the United States formulated the MACH project, which was implemented between 1998 and

2009. Funded by the United States Agency for International Development (USAID), and implemented by a group of partner organizations such as Winrock International, Bangladesh Centre for Advanced Studies (BCAS), Centre for Natural Resource Studies (CNRS), and Caritas, in close collaboration with the Department of Fisheries, Government of Bangladesh, since 1998, this project developed an innovative approach to watershed management, centered on a community-based CM mechanism, applied over an entire wetland ecosystem (covered about 32,000 ha in the wet season, directly involving about 184,000 people over 110 villages), including the *haor* basin in Bangladesh.

The foundation of the MACH project is CM—the transition of which is comprised of the following three key elements: (a) working with local communities and government to develop CM organizations; (b) building the capacity of those institutions to manage themselves and empowering them to conserve wetland ecosystems and use resources sustainably; and (c) providing support to improve the livelihoods of poor people dependent on these wetlands. The project has been able to simultaneously improve wetland ecosystems and the livelihoods of local people through a community-centered, integrated ecosystem (watershed) management approach,[7] as opposed to one that focused solely on fisheries management. One distinctive feature of MACH project is its decentralized and democratic approach to CM that focuses on collaboration with local government. Reviving fish production aside, MACH also includes supplemental income-generation activities and focus on the hardcore poor who would be denied access to fishing (restriction imposed) for a specified period to restore resources. The MACH program has indeed improved the lives of nearly 200,000 of Bangladesh's poorest citizens (Thompson et al. 2003).

Institution and Organization

MACH emphasized the development of local institutions to support communities and local government in the management of natural aquatic resources (Figure 6.1). The project helped develop resource management organizations (RMOs), that manage NRs in the target areas, and resource

---

[7] An integrated ecosystem (or watershed) approach considers all biodiversity, functions, and human communities within the watershed as opposed to a fisheries management approach that focuses more narrowly on fish resources.

Figure 6.1:
Institutional arrangement—MACH CM mechanism

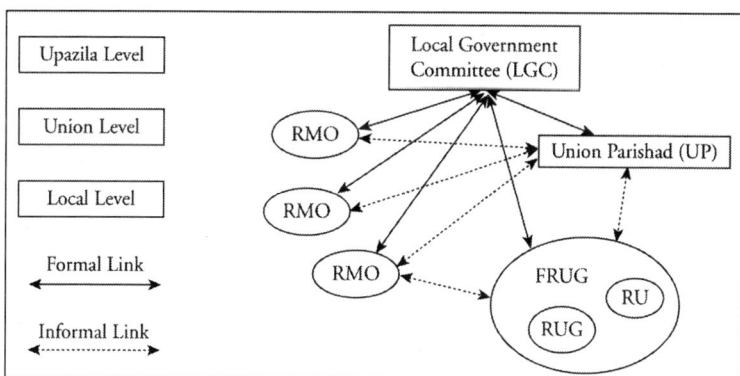

Source: Halder and Thompson (2006).

user groups (RUGs), formed to diversify and enhance livelihoods. The RMOs and RUGs were linked to the Central government through local government committees (LGCs—Upazila Fisheries Committee [UFC]). The committees include officials, elected representatives, and community leaders. Together, the LGC members oversee wetland management and make CM decisions.

Key Lessons Learnt

Building community resource management organizations/institutions (RMOs): The development of community organizations for wetland and wetland resource management empowered local bodies to take responsibility for decisions and actions to restore and sustain wetlands. Evidence shows the merits of establishing sanctuaries for fish stock (e.g., Baikka beel, Hail *haor*) during the dry season. The RMOs played a key role in this regard, addressing problems identified by the communities, protecting water bodies, and restoring wetland habitats. The RMOs adopted regulations covering their wetland resource areas they directly controlled, such as fishing times, means of harvesting, and plans for physical interventions.

Developing CM linkages: The sharing of responsibilities between RUs and government (UFC) commonly devolves greater management responsibilities to the community. While local government is a key in all development work at the grassroots level, MACH placed a greater emphasis on the strengthening of community linkages with local government (LGC) and formalizing the status of the RMOs.

Empowering and enabling the poor: Since wetlands provide a variety of resources, stakeholders use these resources for income and subsistence. MACH's comprehensive approach involved all communities neighboring the wetlands and ensured the participation of the poor in LGCs. Constitutional arrangements (eligibility for different posts, roles of leaders, term limits) governing the operation of the RMOs promoted pro-poor participation.

Ensuring women's participation: An outstanding achievement of MACH project has been the empowerment of women, since the project has been operated in conservative rural areas, where women have traditionally had few rights and little power over their lives or livelihoods, according to a study conducted by the USAID in 2006 by external experts. By insisting that a proportion of positions in RMOs be filled by women, and by setting up RUGs for women, the project has forced the pace of social change. Roughly two-thirds of the women in RMOs were also members of RUGs, which made up 36 percent of RUG membership. It was found that at least seven members of RMOs were women in the Hail *haor*.

Tangible Results

The nine-year program has demonstrated tangible positive results. According to an estimate, in MACH sites, fish yields were two to five times more than baseline (1998) yields; before intervention, it was 58–171 kg/ha, reaching 315 kg/ha in 2004–05, including the reestablishment of 8–10 threatened species. It is reported that fish consumption increased in the surrounding communities by 40 percent; MACH communities earned US$4.7 million more from local fish sales in 2004 than they did in 1999, mainly due to the revival of wetland habitats and fish stocks (Thompson et al. 2003).

There were ecological benefits as well. Wetland diversity expanded, as threatened fish species successfully reestablished, migrating birds returned, and aquatic plants recovered. The MACH program's success is rooted in community self-interest and ownership. In return for adopting conservation measures and sustainable fishing practices, community organizations (each representing several adjacent villages) receive 10-year leases to manage local waterways as well as grants to excavate silted *beels* and create wetland sanctuaries. To offset the hardships caused by fishing restrictions, poor households also received skills training and micro-loans to start new *enterprises*.

Conclusion

The MACH program and similar other projects have definitely provided a promising national road map for protecting NRs while enhancing livelihoods. By protecting wetlands from further destruction, overexploitation, and degradation, communities have also improved the environmental resilience of the resources on which their lives and livelihoods depend. The MACH program demonstrated the viability of an approach that empowered community members, including women and the poor, to support conservation efforts at an ecosystem level. The turnaround shows how CBNRM can nurture enterprise, generate income, and improve the state of local ecosystems. Through a CM approach, MACH increased linkages between community groups and local governments, leading to improved watershed management, increased fish yields, and improved biodiversity of wetlands. While the long-term sustainability of above benefits remains to be seen, community-led wetlands management has improved the ability of some of Bangladesh's poorest inhabitants to survive economic downturns, environmental disruption, and the potential impacts of climate change on the country's low-lying floodplains. It is now well recognized that local communities should have direct control over the management, utilization, and benefits of wetland resources so that these can be used on a sustainable basis. It demonstrates that the poor can manage complex landscape units while restoring biodiversity and improving their incomes; it certainly has practical application and management implications.

Recommendations

Our recommendation on the challenges of the sustainable management of freshwater wetlands in Bangladesh highlights the need for innovating new multilevel governance structures. This leads to NRM partnerships, harnessing the complementarity of the capacities and comparative advantages of different institutional actors, linking entitlements and responsibilities in the NRM context. Top-down and bottom-up approaches have their own relative advantages and disadvantages; each approach contributes to sustainable management from its own perspective, but none is capable of performing the management job alone. In confronting the situation, one can find a direction ahead—alternative governance through CM that is found to be the most favored strategy in bridging the gap between the two complementary yet contrasting approaches. To effectively fight global poverty, the CM idea of successful programs such as MACH in Bangladesh needs to be 'scaled up' geographically, economically, and politically. Expanding these efforts can provide the world's rural poor with ways to derive more sustainable income from nature-based enterprises while at the same time developing their resilience to emerging new environmental threats such as climate change.

References

Agarwal, A. 2001. "Common Property Institutions Sustainable Governance of Resources." *World Development* 29 (10): 1649–1672.

Agarwal, A. and C.C. Gibson. 1999. "Enhancement and Disenchantment: The Role of Community in Natural Resource Conservation." *World Development* 27 (10): 629–649.

Ahmed, I., B.J. Deaton, R. Sarkar, and T. Verani. 2008. "Wetland Ownership and Management in a Common Property Resource Setting: A Case Study of Hakaluki Haor in Bangladesh." *Land Economics* 68 (3): 249–262.

Banuri, T. and F. Amalric, eds. 1992. "Population, Environment and De-responsabilisation: Case Studies from the Rural Areas of Pakistan." Working Paper POP No. 1. Islamabad: Sustainable Development Policy Institute.

Barton, T., G. Borrini-Feyerabend, A. de Sherbinin, and P. Warren. 1998. *Our People, Our Resources.* Gland: IUCN (Available also in French and Spanish from the IUCN Social Policy Group and from http://www.iucn.org/themes/spg/opor/opor.htm).

Basnet, K. 1992. "Conservation Practices in Nepal: Past and Present." *Ambio* 21 (6): 390–393.

Berkes, F., ed. 1989. *Common Property Resources: Ecology and Community-based Sustainable Development.* London: Belhaven Press.

Borrini-Feyerabend, G. 1996. *Collaborative Management of Protected Areas: Tailoring the Approach to the Context.* Gland: IUCN (Available also in French, Spanish, and Portuguese from the IUCN Social Policy Group. http://www.iucn.org/themes/spg/Tailor/index.html).

Borrini-Feyerabbend, G., Farvar M. Taghi, Jean Claude Nauinguiri, and Vincent Awa Ndangang. 2007. *Co-management of Natural Resources.* Kasparek Verlag, Heidelberg.

Chowdhury, M.A.T. 2012. "Meeting Food Security Needs Through Sustainable Agriculture." *The Palawija* 29 (2): 3.

CNRS (Centre for Natural Resource Studies). 2009. *Climate Change and Wetlands.* Dhaka: CNRS.

Dugan, P.J., ed. 1990. *Wetland Conservation: A Review of Current Issues and Required Action.* Gland: IUCN.

FAO (Food and Agriculture Organization). 1988. *Land Resource Appraisal of Bangladesh for Agricultural Development.* Rome: Food and Agriculture Organization of the United Nations.

Freeman, C. 1996. "Local Government and Emerging Models of Participation in the Local Agenda 21 Process." *Journal of Environmental Planning and Management 39* (1): 65–78.

GoB (Government of Bangladesh). 2002. *Tanguar Haor Biodiversity Conservation Project,* Final Report. Dhaka: Ministry of Environment and Forests, Government of Bangladesh.

Halder, S. and P. Thompson. 2006. "Community-based Co-management: A Solution to Wetland Degradation in Bangladesh." MACH Technical Paper 1. Dhaka: Winrock International.

Fisher, R.J. 1995. *Collaborative Management of Forests for Conservation and Development.* Gland: IUCN.

IUCN (International Union for the Conservation of Nature). 1980. *World Conservation Strategy.* Gland: IUCN.

IUCN-Bangladesh. 2004. *Introduction to Community Based Haor and Flood Management.* Dhaka: IUCN Bangladesh Country Office.

Kabir, M.H. and M.N. Amin. 2007. *Tanguar Haor: A Diversified Freshwater Wetland.* Dhaka: Academic Press and Publishers Library.

Khan, M.H.H. 2011. "Participatory Wetland Resource Governance in Bangladesh: An Analysis of Participatory Community-based experiments in Hakaluki Haor." PhD Thesis, Institute of Natural Resources, University of Manitoba, Manitoba.

Khan, M.S., E. Haq, S. Huq, A.A. Rahman, S.M.A. Rashid, and H. Ahmed, eds. 1994. *Wetlands of Bangladesh.* Dhaka: Bangladesh Centre for Advanced Studies (BCAS) in association with Nature Conservation Movement.

Murphree, M.W. 1994. "The Role of institutions in Community-based Conservation." In *Natural Connections,* Chapter 18, edited by D. Western and R.M. Wright. Washington, DC: Island Press.

Nishat, A., Z. Hussain, M.K. Roy, and A. Karim, eds. 1993. *Freshwater Wetlands in Bangladesh: Issues and Approaches for Management.* Gland: IUCN—The World Conservation Union.

Rahman, A.K.A. 2005. *Freshwater Fish of Bangladesh.* Dhaka: Dhaka University Press.

Redclift, Michael. 1987. *Sustainable Development Exploring the Contradictions.* London: Methuen.

Roy, T.M.D. 2012. "Wetland Depletion and Its Impacts on Resource and Livelihood of Local People: A Case Study of the Khaliajura Haor Area under Nettrakona District, Bangladseh." Unpublished MS Thesis, Department of Geography and Environmental Studies, University of Chittagong, Bangladesh.

Sachs, Jeffrey D. 2012. "From Millennium Development Goals to Sustainable Development Goals." available at www.thelancet.com.

Scott, J.S. 1985. *Weapons of the Weak: Everyday Forms of Peasant Resistance.* New Haven and London: Yale University Press.

Thompson, P., P. Sultana, and N. Islam. 2003. "Lessons from Community-based Management of Floodplain Fisheries in Bangladesh." *Journal of Environmental Management 69* (3): 307–321.

UNGA (United Nations General Assembly). 2015. *Report of the Open Working Group of the General Assembly on Sustainable Development Goals A/68/970, August 12.* Available at:

http://www.un.org/ga/search/view_doc.asp?symbol=A/68/970&Lang=E (Accessed on November 2, 2015).

WECD. 1987. *Our Common Future, World Commission on Environment and Development (WECD)*. Oxford: Oxford University Press.

Western, D. and R.M. Wright, eds. 1994. *Natural Connections: Perspectives in Community-based Conservation*. Washington, DC: Island Press.

SECTION 2

Governance II: Experiments with 'Bottom-up' Approaches

7

Dynamics and Pay-offs in Community-based Water Resource Management: A Case Study from Indian Sunderbans

Satabdi Datta

Introduction

Water scarcity induced by lack of availability, access, and quality of water for consumption and production has emerged as one of the growing problems in the present century (Mehta 2006). Over the years, people's dependence on surface water is getting threatened by increasing water scarcity out of climate change, pollution, population pressure, overexploitation of resources, inequalities, power relations, conflicts among users, etc. Depleting ground-water table and groundwater quality are also restricting the water access in many parts of India. The available and utilizable water resources per capita per year are 2,384 billion cubic m (bcm) and 1,086 bcm, respectively, which declined from an estimated availability of 6,008 bcm in 1947 (Jain et al. 2007). Thus, natural as well as man-made phenomena constitute the growing water scarcity in India. The deterioration in availability, access, and quality of water resources severely affects the process of sustainable development through various channels such as impact on health, opportunity cost of fetching water, loss of employment opportunities, loss of agricultural productivity, impact on water intensive small- and large-scale industries, etc.

Water is not merely a commodity but a "source of life, dignity and equality of opportunity" (Watkins 2006). Thus, recognition and enforcement of human right to water is essential. Right to water is not directly identified by any international declaration of human rights (Bluemel 2005). It is only seen as 'subordinate and necessary' for achieving other recognized human rights. In Indian Constitution also, right to water is not included as a fundamental human right, although it is implicit in the right to life. In the world as well as in India, a large part of the population lacks access to basic water requirements and as a result suffers from several hazards. Therefore, along with proper recognition of right to water, it is required to have necessary institutional, economic, and management strategies (Gleick 1998) to deal with the crisis. Water as a major ecosystem resource provides a number of ecosystem services (ESS). These services include (a) Provisioning Services such as supply of drinking water, food, and fish; (b) Supporting Services such as productivity or *biodiversity* maintenance; (c) Regulating Services such as climate regulation, and groundwater recharge; and (d) Cultural Services such as tourism or spiritual and aesthetic appreciation. Since among the range of ESS provided by water, several are not visible or accountable in market transactions, those are often ignored severely (Bandyopadhay 2009). Among the visible ESS, water is utilized for both direct consumption and several economic activities, including agriculture, industry, energy, etc. According to the projections made by the National Commission for Integrated Water Resource Development Plan (NCIWRDP), the requirement of water for irrigation in India will grow by more than 50 percent in the next 50 years and the water requirements for household consumption and for industry would rise even faster (Rao 2002).

In order to cater to the survival necessities of the expanding population of India, there is a need for proper adaptation strategy to manage the shrinking water resources. While doing so, adopting a relevant approach to addressing the problem is a challenge. As realized by experts (Bandyopadhay 2009; Mehta 2006), given multifaceted failures of large-scale irrigation projects directed by expert-led knowledge since the late nineteenth century, use of surface water in irrigation and small-scale, locally based water harvesting measures based primarily on indigenous knowledge are the most viable alternatives. Thus, management mechanism combining participation of local communities[1] could be a possible way out for coping with degrading ecosystem resource. While taking the bottom-up approach that puts emphasis on local actors, that is, local communities, certain challenges might take place too. In this regard, understanding various institutional relationships along with identification and elimination of the constraints in community management and access to resource are required for ensuring a sustainable water resource management mechanism.

---

[1] By 'community', we mean a group of people who share a commonplace of residence and a set of institutions (Fabricius et al, 2007).

Objectives and Methodology

This chapter is based on a case study of a community-based water resource management program implemented by an international NGO in an agricultural rural region exposed to water crisis in Sundarbans area of West Bengal. The basic idea behind the NGO-initiated development project was to adopt bottom-up approach that is characterized by 'self-governance', 'grassroots-level action', 'self-sufficiency', 'inclusivity', etc. However, while implementing development programs by an international NGO at the grassroots level, program design and decisions made at various stages of program implementation are not necessarily always bottom-up. It is often a combination of top-down and bottom-up strategies where it is common that one overpowers the other.

In this context, the broad objective of this study was to examine community-based water resource management system in the Sundarbans area through a case study. In an attempt to do that, it was tried to find out the impact of the rainwater-harvesting (RWH) program implemented by the NGO on the village communities and to explore the dynamics within the communities in the entire process of resource management. Primarily, qualitative methodology was followed in the field study of village communities, and it was supported by some quantitative techniques. Qualitative methodology used to collect field data included individual interviews using structured and semistructured questionnaires, case studies, and focus group discussions. The field study was conducted during January–April 2012.

Water and Economic Scenario of the Study Site

Sundarbans, the densely forested wetlands of the delta of the rivers the Ganges, the Brahmaputra, and the Meghna, and belonging to a moist subhumid agro-ecological subregion, is shared between India and Bangladesh. Among the 25,500 sq. km spread of Sundarbans, approximately 9,630 sq. km constitute the Indian part (Vyas and Sengupta 2012). In the east of the Indian Sundarbans delta is the Ichamati–Haribhanga–Raimangal River; in the west, the Hugli River; in the south, the Bay of Bengal; and in the north, the Dampier–Hodges line[2] (Danda et al. 2011). The Indian Sundarbans

---

[2] The line, drawn in 1829–30, indicates the northern-most limits of estuarine zone affected by tidal fluctuations.

Figure 7.1:
Population expansion in Indian Sundarbans

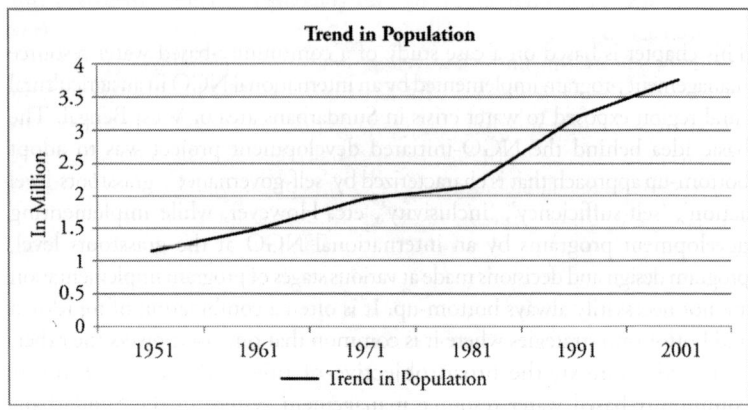

Source: Government of West Bengal (2010a, 2010b).

is located within the 24 Parganas districts of West Bengal. Of the 19 development blocks within the region, districts South and North 24 Paraganas share respectively 13 and 6 development blocks.

Most of the 4.5 million people (2001 Census) residing in Indian Sundarbans face considerable hardship in survival. For both the districts, in Sundarbans community development blocks of both the districts of Sundarbans, the poverty scenario is much worse than the non-Sundarbans blocks. Inhabitants in this region are the worst sufferers in terms of education, health, and basic infrastructural facilities and most of the villages under Sundarbans live without electricity and proper access to safe drinking water. The ever-expanding population here (Figure 7.1; estimated at 1.159 million in 1951, 2.133 million in 1981, and 3.756 million in 2001) has added to the vulnerability.

Water, be it for drinking and domestic use or for irrigation, is hardly available in plenty in Sundarbans. People are exposed to lack of availability, access, and quality of water. Restricted access to freshwater is one of the key threats to agriculture, the major source of income in Sundarbans. Lack of irrigation facilities and cropping is more or less common in the entire region, thereby leading to a high unemployment rate of 63 percent, an average poverty ratio of 43.32, and a large amount of seasonal migration (Danda et al. 2011). The degree of scarcity varies across the blocks. With a higher degree of water scarcity, overexploitation of the existing groundwater and surface water resources further intensifies the problem by denying sustainability too often.

Of the over 4.5 million people living in the Sundarbans villages, about 95 percent depend on agriculture, since there is no major industry in the Sundarbans region. In the Island blocks around forest boundary, 34 percent of the total working people cultivate either on their own land or on leased land and 48 percent are agricultural or other daily laborers, whereas the proportion of landless farmers is greater in other Sundarbans blocks, where 55 percent of the total workers are agricultural labor. Although the eco-region receives abundant rainfall annually (175 cm approximately), mostly during monsoon months of June through September, agricultural productivity is low (Danda 2007). Lack of water for irrigation in the pre-monsoon seasons, saline water incursion due to frequent breaches in embankments, exceeding salinity in soil, poor drainage, and submergence (90 percent of the land is medium or low) are some of the major underlying causes. Pressure on land and agriculture is on rise due to increasing population density in the Sundarbans blocks. A reduction in availability of agricultural land (it reduced from 2,149 sq. km to 1,691 sq. km in 2001–08) (Danda et al. 2011) along with a rise in settlement area is bound to pose the threat of inadequate food security to the Indian Sundarbans population.

Thus, the Sundarbans communities are more or less exposed to multiple survival threats arising out of the economic and ecological inequalities in allocation of resources, poverty, lack of women empowerment, climate change, degradation of ecosystems, etc. In the state of high dependency on natural resources supplemented with limited livelihood opportunities, adaptive management of resources by communities[3] in Sundarbans under institutional support, either government or nongovernment, is being increasingly felt over the decades.

The study site is located in Mathurapur I and II blocks among 19 community development blocks of Sundarbans. Mathurapur I and II share 17 percent (approx) and 21 percent (approx) of the Indian Sundarbans population, respectively. But the population density of Mathurapur I is higher than that of Mathurapur II (Table 7.1). In Mathurapur I, it is 1,110 per sq. km, which is more than the average population density of Indian Sundarbans, whereas that of Mathurapur II is less than average. Thus, the population pressure on natural resources such as surface water and groundwater is much higher in Mathurapur I than in Mathurapur II.

The operational villages of Mathurapur I and II are susceptible to both natural and ecological hazards and poor infrastructural facilities, thereby resulting in insufficient livelihood opportunities and low standard of living.

---

[3] One of the policy objectives of the State Planning Board was ensuring community participation in local governance.

Table 7.1:
Block profiles of Mathurapur I and II

Block	Area (in sq. km)	No. of Villages	Total Population	Population Density*
Mathurapur I	148.29	96	164,585	1,110
Mathurapur II	230.51	27	198,261	860
Indian Sundarbans	**1,074.25**	**334**	**961,463**	**925.17**

Source: http://sundarbanaffairs.in/, Department of Sundarban Affairs, Government of West Bengal.
Note: * Population per square kilometer area.

The entire region under these two blocks is in the low-lying delta of Sundarbans. Agriculture and allied activities are the major sources of income here. Thus, the communities in these villages are critically dependent on land and water resources for survival. The ecological, economic, and social vulnerabilities that they are exposed to place them in a disadvantaged position in getting availability of and access to water. This is the primary rationale behind selecting this region for the study.

The Case Study

An international NGO started the community-based resource management program of RWH involving the communities since 2005 in Mathurapur I and II blocks of South 24 Parganas. But unlike the existing studies, this study on community-based water resource management is under private property regime. It is because the entire Sundarbans, more specifically Mathurapur I and II, lack common property surface water resources utilizable for productive purposes.

The Land Shaping Project (LSP) encompassing RWH had been first developed by an NGO and philanthropic organization through a series of modification efforts and change of land use pattern in the coastal part of Sundarbans in the 1980s. This model was found to be effective in creating year-round crop production, aquaculture, income generation, preventing migration and displacement, ensuring food and water security to the communities, minimizing climate dependability (in complex, diverse, and risk-prone (CDR) agricultural system), balancing the monocropping risk, and ensuring income stability across the communities.[4]

---

[4] www.indiawaterportal.org

Table 7.2:
GPs under the two blocks where the program of rainwater harvesting has been implemented

Block	Mathurapur I	Mathurapur II
Gram Panchayats (GP)	Abad Bhagabanpur	Raidighi
	Shankarpur	Kumrapara
	Lakhi Narayanpur Dakshin	Nagendrapur
		Kankandighi

Source: Primary survey.

At present, a total of 170 village communities of 7 Gram Panchayats (GPs) from Mathurapur I and II blocks are incorporated in the program (Table 7.2). In the period of 2005–11, total 560 new ponds had been excavated for RWH and 125 ponds had been reexcavated.[5]

The NGO first approaches the villages for implementing the program. It organizes meetings and awareness campaigns in the selected villages in order to spread general consciousness about the LSP through RWH among the villagers. Since the program is designed in such a way that involves the participation of the community instead of a single individual, therefore, formation of Self-help Groups (SHGs) is a necessary condition for the program to take place. The size of the SHGs remains at 10–15 members. The SHGs are in most of the cases gender specific, that is, either male SHG or female SHG.

Marginal farmers with no more than 2.5 acres of landholding are considered eligible for the program. Willing farmers with a minimum possession of 1 *bigha* or 0.33 acre of land apply to an NGO through a community-based organization (CBO)/SHG (Figure 7.2). Then site investigation and feasibility checks (including income source, land size, land usage pattern, irrigation, socioeconomic status, etc.) are done by the NGO. After the primary selection procedure of farming households is over, the NGO makes a joint agreement with the concerned CBO or SHG and farmer. It provides partial financial support, that ranges from 60 percent to 80 percent, in the form of credit for land shaping to the farmer, who is entitled to payback that amount to the respective CBO/SHG with due rate of interest. Due to absence of common property surface water resource in the region except the canals,[6] almost all RWH cases are on private land. On an average, there are 18–20 beneficiaries

---

[5] NGO staff interview.
[6] In the pre-monsoon seasons, canal water remains saline and hence unutilizable for irrigation.

Figure 7.2:
The mechanism of land shaping through rainwater harvesting process involving community with institutional support

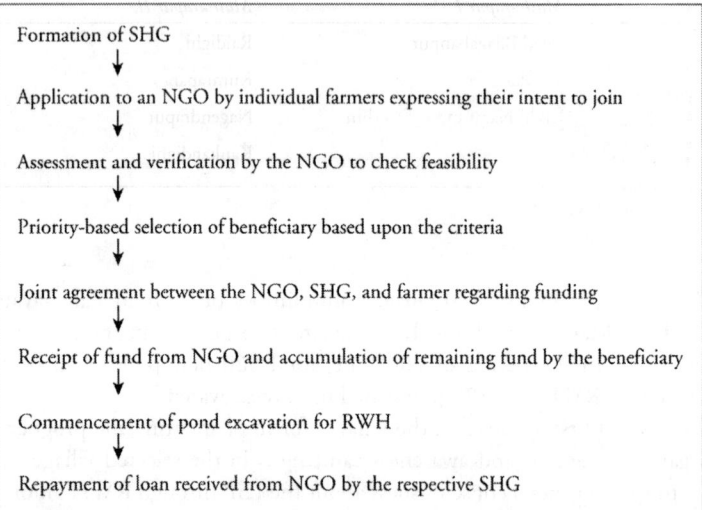

Formation of SHG
↓
Application to an NGO by individual farmers expressing their intent to join
↓
Assessment and verification by the NGO to check feasibility
↓
Priority-based selection of beneficiary based upon the criteria
↓
Joint agreement between the NGO, SHG, and farmer regarding funding
↓
Receipt of fund from NGO and accumulation of remaining fund by the beneficiary
↓
Commencement of pond excavation for RWH
↓
Repayment of loan received from NGO by the respective SHG

Source: Field study.

in the target villages (primary source). Several training and awareness programs are organized by the NGO (with joint collaboration of a biotechnology group) to promote organic farming and ensure sustainable agriculture.

The introduction of land shaping through RWH in low-lying areas of Mathurapur I and II was made to allow second cropping during winter (rabi) season by making water available for irrigation at critical stages and to reduce water logging during kharif. Excavation of pond and RWH are done to ensure year-round availability of water for irrigation and to improve surface drainage and reduce the top soil erosion of the land. It is also meant to check the loss of soil fertility through control of run-off by putting embankments surrounding the entire land, according to the environment impact assessment of the program by the NGO. Apart from these primary objectives, it also serves other commercial and noncommercial purposes, for example, vegetable and oil seed cultivation, fishery, duck rearing, water supply for paddy processing, water for cooking and washing, etc.

It was found that the overall impact of the ESS derived from the RWH pond on human well-being was significant. Some services benefited the individual beneficiary, whereas some benefited the community as a whole.

The present generations experienced well-being from both material and nonmaterial sources such as freedom, self-dependence, security, self-respect, improved knowledge base, empowerment, etc. But the intertemporal well-being is difficult to assess initially. Finding out whether the beneficiaries become able to sustain the resource and whether the stream of material and nonmaterial benefits flows to the future generations, at least to the extent as the present ones are accessing, would require wider time span analysis. The sustainability of the resource and their ESS would depend on proper management strategy.

In the villages where the program was implemented, there was not full coverage of the population. On average, 18–20 households from each targeted village were covered between 2005 and 2012 (primary source). In villages that were incorporated in the program much later, the percentage of coverage was less. The percentage of covered households or beneficiary households has a wide variation across village communities. It is primarily a result of differences in year of implementation. But even within a GP, disparity in coverage exists while the year of implementation is more or less same in village communities within a GP. Thus, there were other associated factors that defined the profile of nonbeneficiaries.

It came out that a number of economic and social determinants of nonbeneficiaries were present in the villages where the program was implemented. Although each case was different from the other in some dimensions, there existed a common pattern also in those. From the case studies, some of these generalized determining factors were identified, including both selection criteria of the NGO and factors that created economic exclusion. A household with either criterion was viable for exclusion and hence was tagged as nonbeneficiary. Those are briefly discussed as follows:

1. Size class of farmer: Marginal farmers with landholding of 0.33–2.5 acres are considered eligible for the program by the NGO. Thus, the target group is that whose one of the sources of income is agriculture. In Mathurapur I and II, the percentages of marginal farmers in total number of farmers are 81 percent and 74 percent respectively (estimated from Agricultural Census, Government of India). Thus, the size of the target group of population is much higher compared to the actual targeted group size. Exclusion occurred within the eligible group of people either due to unwillingness or else.

2. Small land size: One of the major causes behind nonparticipation within the eligible farmer size class, that is, marginal farmers was small land size. According to them, in their very small plot of cultivable

land if they excavate pond for RWH, then it will reduce their net cultivable area and hence would not be enough beneficial to them. It was observed that nonparticipations due to possession of small land size were for separate reasons. Although the primary factor of land size was same for some, some did not participate since they had some alternative sources of water, whereas some failed to participate because their dependence on the land they had was much higher for survival.

3. Inability to accumulate fund: There are cases where willing individuals could not take the scheme due to their failure to accumulate fund for pond excavation. The loan given by the NGO ranged between ₹4,000 and ₹12,000 depending on the year of implementation and pond size and depth. For the rest 20–30 percent of the fund, households are required to accumulate by their own. Also, the concerned households are liable to pay back the sum of money lent by the NGO to the respective SHG with due rate of interest. The more one is economically vulnerable, the more he/she is prone to risk of getting less than expected benefits from the pond and adjoining land. They feel it to be too costly to take the risk.

4. Inefficiency in SHG operations: Formation of an SHG is a necessary precondition for participation in the RWH. Successful operation of the SHG is beneficial to its members, since then they can borrow money from the SHG for pond excavation or for proper maintenance of the excavated pond and adjoining land at low interest rates. In a number of communities, there existed some people who were willing to excavate ponds and satisfied the eligibility criteria yet could not get included in the program due to inefficiency in their respective SHG or lack of incentive among people to form SHGs. In certain cases, willing individuals failed to borrow money from their SHG for pond excavation or maintenance, since nonrepayment of loans, that is, bad loans prevented the SHG to raise its fund and create additional lending facility.

5. Lack of social integration: Lack of social integration was found responsible for preventing people from community participation and accessing the program. On the one hand, it constrained people in forming SHGs, and on the other, it created a barrier against knowledge or information flow. Again, once a person is disjointed from the community life, it increasingly places him/her in a disadvantaged position by making the disparities larger.

6. Community with lesser number of successful cases: It was observed that in the villages that had smaller number of beneficiaries, people

are less likely there to participate. Furthermore, in communities where majority of the existing beneficiaries could not generate adequate benefits from it, additional people seldom joined.

There existed certain variations in the profiles of nonbeneficiaries across village communities. But several basic features remained the same. Among them, the most fundamental one was nonpossession of land and land size, since in these cases the water resources are created on private land, although community effort is required to manage it. Broadly, the nonbeneficiaries were those who do not fall in the category of marginal farmer, who do not depend on agriculture, who have access to sufficient source of water for cultivation, who depend on alternative sources of income in order to safeguard against uncertainty in agriculture, who are economically disadvantaged, etc.

Figure 7.3 shows the distribution of beneficiaries across the annual income from cultivation groups, before and after joining the community-based water resource management (RWH) program. It is seen that before joining, the households were distributed around a lower income group, whereas after joining the program, they were shifted to a high-income group. Accordingly,

Figure 7.3:
Distribution of beneficiary households across annual income from cultivation groups before and after joining the program

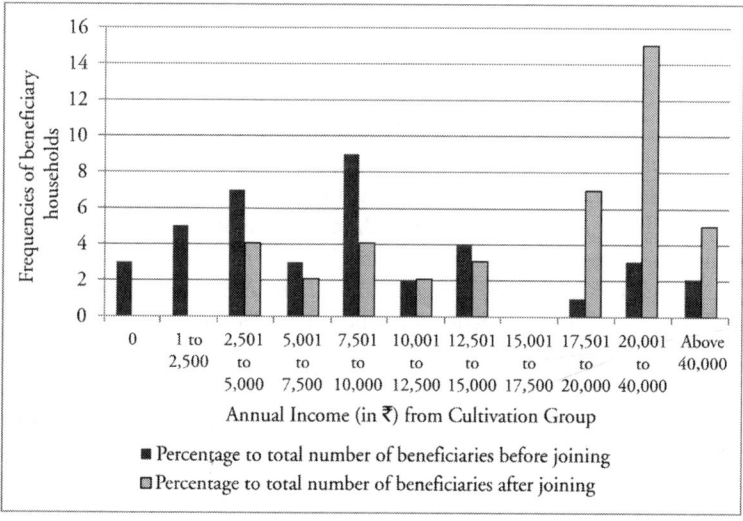

the average income from cultivation almost doubled after joining. Some households who had no income from cultivation earned positive income from it after they joined. But variation in income has risen after joining the program among the beneficiaries, as the standard deviations show. This is primarily due to the reason that some beneficiary households have been benefited to a larger extent compared to the other.

From a statistical analysis of standard of living it was observed that on an average, the households have low standard of living in the communities of Mathurapur I and II where the program of RWH was implemented. Furthermore, it was found that although the average standard of living of nonbeneficiaries is lower than that of the beneficiaries, the variation in living standard is much higher for them compared to the beneficiaries. It might be due to the fact that beneficiaries are only from a specific economic status of people, that is, marginal farmers, whereas the nonbeneficiary households include both the economically affluent and the distressed ones. But as the finding suggests, interhousehold inequalities in standard of living exist among the beneficiaries also. From the annual income of beneficiaries and nonbeneficiaries, it was observed that the average income of the beneficiaries is higher compared to the nonbeneficiaries.

Community Dynamics and Role of Institutions

In all the communities, the same project was implemented, but it had varied impacts. Communities were responding differently to livelihood challenges such as changes in livelihood options, ecological threats, economic backwardness, conflicts in access, etc. Some communities were found to be more efficient in adapting it than the others. Ideally, an adaptive community is one that has the capacity to cope with immediate threats and accordingly build up strategy for sustainable management of ecosystem resources by making necessary trade-off between short-term and long-term well-being (Fabricius et al. 2007). It depends on the knowledge base of the community and governing capacity.

In the present context of the study of RWH program under private property regime along with formation of community institution for management of the private resource, I categorized the studied village communities into three types, which are Type-I community, Type-II community, and Type-III community, which are to some extent based on the classification made by Fabricius et al. (2007).

Type-I community is the ideal one with both adaptive and governing capacities. In the context of the study, management of RWH system that involves management of both land and water (RWH pond) resources should follow a sustainable path. It is achieved when community becomes enabled to take adaptive strategies that involve and are associated with a planned decision based on traditional knowledge base of the community, intra and intergenerational knowledge exchange, optimal combination of short-term and longer term goals, formation and efficient operation of community institution (e.g., SHG), and access to minimum endowments of capitals, including natural, human, and social capitals. In the surveyed villages of Mathurapur I and II, existence of two categories of communities was found broadly. Ideal Type-I community with both adaptive and governing capacity was found in none of the villages. Some communities had good social networks and institutions such as full coverage of the village community by the SHGs but were not enabled enough to sustain the process due to the absence of any of the pre-stated attributes. As a result, strategies for sustainable resource management and long-term well-being of the community through proper management of water resources were missing there.

Most of the communities were of *Type-II*, that is, they had adaptive capacity but were not managing the systems properly due to lack of governing capacity. The principal characteristics either of which was found to be present in these Type-II communities in water resource management were as follows:

1. Economic and ecological vulnerability: Ecological threat and lack of access to and availability of water are common to these communities. Apart from that, some villages face inadequate employment and livelihood opportunities. Sometimes the communities lack access to natural and human capitals.

2. Absence of long-term vision: Formation of village institutions, that is, SHGs took place in order to cope with these multiple threats. Strategies for long-term sustainability as well as immediate action regarding water resource management depended on efficiency in the management of SHGs as SHG deposits would increase if RWH beneficiaries successfully repay their loan given by the NGO to the SHGs. Unless fund accumulation occurs for future, maintenance of the resource by beneficiaries will not be possible and hence constrain sustainability. Also, if there are enough deposits in the SHGs and they are handled properly, then the nonbeneficiaries can also utilize them for meeting their livelihood requirements instead of going to

the moneylender and borrow at high interest rates. But in some village communities, there existed only short-term responses.

3. Motivation to act and manage social institution: It was found in some villages that SHGs were operating successfully. There they had active participation by all members for decision-making and fund management. In case any member does not repay loans, they would arrange meetings at his/her home to pressurize him/her to repay. The strong social network makes it possible. Thus, it was the social capital that was required. But some communities lacked social capital. In a village, when one SHG was operating efficiently, others got motivated to follow their strategies. But in some other villages, lack of cooperation among members was found. In cases of nonrepayment, the SHG president decided to expel them from the group. It was found in big villages with considerable economic disparity. The social network was not strong enough there. As a result, lack of incentive to participate in SHG existed there. Also, awareness among SHG members about government schemes and programs, borrowing facilities from banks, and literacy status had an impact on sustenance of SHGs. In a village where female literacy was lacking, the female SHGs were not working properly.

4. Threat awareness and knowledge network: Threat awareness was present in Type-II communities. Both individuals and communities were concerned about the livelihood and ecological threats they were exposed to. In some cases, there existed good knowledge networks within the community.

Type-III community is characterized by weak adaptive and governing capacity and was found in one village in Mathurapur I block. The Muslim village community comprising of 86 households is in extreme ecological, economic, and socioeconomic vulnerability. The NGO incorporated this village under its project in 2008. Since its arrival, five SHGs have been formed among which two are women SHGs. Approximately, a little over 50 percent of the households participated in the SHGs initially, but the number has reduced over the years. Among them, until now seven have taken the scheme for RWH pond excavation.

The core problem areas of this village community are as follows:

1. Severe ecological threat: The soil remains saline in the pre and post monsoons. In each monsoon, a flood occurs that causes them a huge loss of crops and fish. Water intrusion from the saline water canal

flowing near the village ruins agricultural fields and ponds. Storm and cyclone are regular incidents.

2. Absence of consciousness to manage ESS: Adequate consciousness to manage ESS was not present in the community. Lack of knowledge network and lack of long-term vision might be the responsible factors.

3. Unproductive agriculture: Due to unavailability of irrigation water, agriculture is rain-fed mostly. Also, the nature of the soil makes agriculture unproductive and even fails to meet their consumption demand.

4. Lack of access to water: There is restricted access to both drinking and irrigation water. For drinking water, the entire village sustains on a single tube well, since the others are defunct. The women need to fetch water from the distant tube well three to four times a day for domestic use. Although several houses have an adjoining pond, such ponds dry up in non-monsoon seasons. Also, water becomes contaminated. So they have no option left but to use those ponds for bathing and washing. Conflicts arise too often when pond owners restrict people from using their ponds.

5. High population density: High population density compared to the neighboring villages creates population pressure on natural resources especially water and is responsible for low standard of living of the households.

6. Lack of alternative employment opportunity: Apart from cultivation, the major occupations in the village are tailoring on a contractual basis, wage labor, etc. But these are not regular. Many people seasonally migrate to Kolkata and Howrah to work as construction labor or something else. Even minor boys aged above 10 years migrate there seasonally to support their families. A woman told that her 14 years old son, who had dropped out in class six, works as a *zari*[7] worker for about 6 months a year in Howrah. She also pointed out that it is a common practice in many of her neighboring households also.

7. Lack of women empowerment: In the Type-II communities, women empowerment at varied degrees was found to be possible due to formation of women SHGs that enabled them for decision-making. But in the Type-III communities, that was not found to be present.

---

[7] *Zari* workers do hand embroidery on sarees and other types of clothes with a specific kind of thread.

8. Inefficient social institutions: Social institutions such as SHGs are working inefficiently in the village. A respondent shared her experience that when she first joined the SHG three years back, there were 13 women but it got reduced to 8 currently. Conflict of interest arises too often. Nobody wants to take care of others' problems. There were some problems with their SHG bank account. (The account was opened in a distant bank, since some of the SHG members had better access to the bank. But now some of those members have left the SHG and are finding it difficult to access that bank.) As a consequence, the amount that they deposited in the SHG account since their joining is left there unused. Due to the same reason, they cannot even take loans from the SHG and thus are deprived of the borrowing facility of the SHG. So, they have no option but to go to the local moneylender and borrow at high interest rates. Even if each member in the SHG is facing the same problem, nobody wants to take the responsibility, even the SHG president. Almost the similar fate is with the other SHGs too. In this way, inefficiency in the operation of SHG restricts sustainable management of water resources necessary for ensuring long-term well-being of the community.

9. Lack of cooperation: Lack of cooperation among villagers is prominent in the village and thereby involves high transaction cost in community-based resource management.

It was found that even after the implementation of the program, water scarcity was not entirely solved in some villages. Small land size, lack of resources, human and social capital, and weak social network were found to be responsible. From the surveyed village, it came out that apart from a single Type-III community, the community efforts of Mathurapur I communities are to some extent better than that of Mathurapur II communities. It can be said that even if the program has potential in water resource management for ensuring sustainable agriculture and livelihood opportunities, village-specific characteristics are important to consider for implementation, which were missing there. Also, involvement of larger number of people of the communities could have resulted in integrated development. In order to ensure sustainability in water resource management and improve access, strengthening of social institution is necessary so that even when external support is not there, community itself can manage it efficiently.

Conclusion and Policy Recommendations

In recent years, the significance of role of community has been realized in different spheres including natural resource management issues. But there are several instances where locally managed initiatives are taken that are not by the community itself, although the communities are needed to participate in management. Moreover, generally, community management of resources involves management of common property resources by the community. But in the absence of common property resources, the system works in a different manner. Thus, through this study it was attempted to explore the dynamics in such a system of community-based water resource management where there exists intervention by an external institution, rather than the community itself and where there is absence of common property land and water resources.

Observations suggested that the ESS derived from the RWH structure had potentials of both short-term and long-term benefits to the individual households and community. Since the short-term benefits were readily accessible, those were being valued highly by the resource users. The direct-use values of the resource were recognized by the communities rather than the option values in most cases. Accordingly, those services were found to have considerable impact on the well-being of the community, as perceived by the community members. But there existed discrepancies too due to the nature of property right assigned to the resource. Primarily due to lack of common property land and surface water resources in the selected region, the RWH structures were built in private lands of the individual households. As a result, the direct benefits of the program remained restricted within a small group of population with a minimum amount of landholding and the remaining uncovered population was categorized as nonbeneficiary.

It was found from the case studies of the village communities that inter-community variations were there regarding adaptiveness of communities after implementation of the program. Most of the communities (Type-II) had adaptive capacity but were lacking governing capacity. A single community (Type-III) was found to be lacking both adaptive and governing capacities. Thus, the efficacy of the communities in managing water resources depended on the degree of ecological vulnerability, economic opportunities, existence of minimum amount of natural, human and social capitals, knowledge network within the community, social integration, strengthening of community institutions, women empowerment, etc. The efficiency of community institutions

(SHG) depended on the transaction costs of enforcement of contract (e.g., loan repayment) and social capital. But in this particular study, no generalized pattern was found between community heterogeneity and collective action as stated by 'Olson Effect' (Bardhan and Johnston 2001). It might be due to the fact that 'Olson Effect' was applicable in the case of collective management of common property resources, which was not the case in the present study. It was further observed that in village communities where any SHG was operating efficiently with active participation by all the members, the other SHGs were more likely to work well. In heterogeneous communities (e.g., income or caste based), less incentive among people was found in joining collective action for managing resources, although there existed some exceptions also.

However, we could put forward the broad policy recommendations after the findings and observations of the present study. As we feel, these are as follows:

- Lack of common property water resources being the major constraint in the studied villages, there is a need for intervention of a local administrative body (e.g., GP) and/or nonprofit organizations, through financial support, in initiating community-based management of existing surface water resources (e.g., canal).
- During carrying out implementation programs, considerations of village-specific characteristics are always required for ensuring the success of the program.
- Formation of community institutions, such as the SHG, must ensure active participation by most of the members of the community. This could increase efficiency in contract enforcement by the community institution.
- In order to achieve women empowerment in managing resources, improvements in female education and female participation in the SHG are highly necessary.
- Government is needed to take steps in formulation and implementation of proper policy to ensure resource availability (human and physical) in the village communities and good governance from the local governments and stakeholders is also needed.
- In ensuring success and sustainability of locally managed water resource structures, it is required to empower the local community by recognizing its active role from the very initial stages of the process that includes decision-making, participation, and inclusivity.

References

Bandyopadhay, J. 2009. *Water, Ecosystems and Society: A Confluence of Disciplines.* New Delhi: SAGE Publications.

Bardhan, P. and J.-D. Johnson. 2001. "Unequal Irrigators: Heterogeneity and Commons Management in Large-scale Multivariate Research." In *Drama of the Commons,* edited by E. Osterm, T. Dietz, N. Doisak, P.C. Stren, S. Stonich, and E.V. Weber, 87–112. Washington, DC: The National Academic Press.

Bluemel, E.B. 2005. "The Implications of Formulating a Human Right to Water." *Ecology Law Quarterly* 31 (4): 957–1006.

Danda, A. 2007. "Surviving in the Sundarbans: Threats and Responses." PhD Thesis, University of Twente, the Netherlands. Available at: http://doc.utwente.nl/68915/1/thesis_A_Danda.pdf (Accessed on February 16, 2017).

Danda, A.A., G. Sriskanthan, A. Ghosh, J. Bandyopadhyay, and S. Hazra. 2011. *Indian Sundarbans Delta: A Vision.* New Delhi: World Wide Fund for Nature.

Fabricius, C., C. Folke, G. Cundill, and L. Schultz. 2007. "Powerless Spectators, Coping Actors, and Adaptive Co-managers: A Synthesis of the Role of Communities in Ecosystem Management." *Ecology and Society* 12 (1): 29.

Gleick, P.H. 1998. "The Human Right to Water." *Water Policy* 1 (5): 487–503.

Government of West Bengal. 2010a. *District Human Development Report: North 24 Parganas.* West Bengal, India: Development and Planning Department, Government of West Bengal.

———. 2010b. *District Human Development Report: South 24 Parganas.* West Bengal, India: Development and Planning Department, Government of West Bengal.

Jain, S.K., P.K. Agarwal, and V.P. Singh. 2007. *Hydrology and Water Resources of India.* Netherlands: Springer.

Mehta, L. 2006. "Water and Human Development: Capabilities, Entitlements and Power." Background Paper 2006/9, *Human Development Report.* UNDP. Institute for Development Studies, UK.

Rao, C.H.H. 2002, May 4. "Sustainable Use of Water for Irrigation in Indian Agriculture." *Economic and Political Weekly* 37 (18): 1742–1745.

Vyas, P. and K. Sengupta. 2012. "Mangrove Conservation and Restoration in the Indian Sundarbans. In *Sharing Lessons on Mangrove Restoration,* edited by Macintosh, D.J., Mahindapala, R., and Markopoulos, M., 93–101. Bangkok, Thailand: Mangroves for the Future and Gland, Switzerland: IUCN.

Watkins, K. 2006. "We Cannot Tolerate Children Dying for a Glass of Water." *The Guardian,* March 8. Available at: http://www.theguardian.com/environment/2006/mar/08/water.comment (Accessed on February 16, 2017).

8

Local Solutions to Local Disasters: Governance in Flood Management in Assam

Arpita Das and Partha Jyoti Das*

Of all the physical, geological, and natural phenomena that affect the earth, floods are the most widespread. There is no place in the world that has never experienced a single event of flooding. The consequences of floods can range from expected to floods being unexpected either in the nature of their intensity or that of their duration. Floods have an ever-present social gradient to them. This means that certain groups are almost always more affected than others in the same geography. When floods are viewed as integral parts of environmental and human systems, they become a formidable test of societal adaptation and sustainability. Coupled with erosions, floods result in the loss of land, livelihood, and identities. Floods in this chapter have been discussed as a sociopolitical phenomenon and as an issue that challenges the notions of governance for sustainable development.

Introduction

Assam is one of the most flood-affected states of India. Annual flooding is caused during the pre-monsoon and the monsoon seasons by its numerous rivers, rivulets, and streams that belong primarily to the two major river systems, namely the Brahmaputra and the Barak (Meghna). The state's flood-prone area amounts to 3.1 million ha, or some 40 percent of the total geographical area. This includes over 90 percent of the agricultural land and

* Partha Jyoti Das acknowledges the financial and technical support received from ICIMOD, Kathmandu, for carrying out a research on governance of flood mitigation structures in Assam, the results from which are liberally used in this chapter.

urban population centers, as well as its most valuable economic assets such as tea estates, oil fields, roads, and airports. On an average, an estimated US$47 million in annual crop production is lost due to floods, while damage to homesteads and livelihood affects some 3 million people (ADB 2006). Riverbank erosion is also a chronic problem caused by dynamic shifting of channels flowing through unconsolidated heavy sand or silt strata of the floodplain, with high sediment discharge. Since 1954, Assam's 17 riverine districts have lost 7 percent of their land area to erosion. Some 8,000 ha of land (valued at US$20 million) is lost annually. As a result of floods and erosion, about 10,000 families are displaced (ADB 2006), many of whom become landless each year causing significant social and economic disruptions. Taken together, the aggregate impact of annual flooding and round-the-year riverbank erosion is the single most important threat to the sustainability of development initiatives and their benefits. And the fact that the prevailing perception of development is only economic progress centric lacking in consideration of ecological integrity of landscapes, social justice and equity, and people's participation, the development projects themselves often prove to be misadventures and socio-ecologically counterproductive.

The conventional approach, primarily driven by top-down approach to flood management by the state government of Assam, is dominated by structural measures and embankments are the main structures that are constructed to protect selected reaches of riparian areas from floods. Assam has about 4,459 km of embankments and 851 km of drainage channels and about 681 town/village protection works (for protecting river banks and embankments from erosion) covering an area of 1.64 million ha, while 2.18 million ha of land remained to be protected until 2004 (MWR 2004). Almost 80 percent of these embankments were built during the period 1950s–70s, and the rest in the 1980s–90s.[1]

Formal flood mitigation strategies characterized by top-down approaches are constrained with serious limitations with severe consequences for human security and environmental sustainability. Flood continues to be viewed as a 'natural' hazard that can be 'dealt' with by engineers alone using only structural measures in vogue since colonial times. The knowledge base on the hydrological and geomorphological regimes of the Brahmaputra and the Barak (Meghna) river systems, which together constitute the dominant drainage system of the state, is also not adequately developed. Lack of

[1] As informed by Mr A.K. Mitra, former Secretary Water Resources, Government of Assam, interviewed on July 19, 2010 in Guwahati.

hydrometeorological database is also a constraint in taking appropriate decision in regard to flood management, be it in the designing of flood mitigation structures or in issuing reliable flood forecast and flood warning. Embankments can make or mar people's lives and livelihoods depending on whether they survive or give in to a flood wave. Yet, the maintenance of these structures is abysmally poor, resulting in frequent breaches, which then perpetrate the flood damage. Although the National Disaster Management Act (2005) and the National Policy on Disaster Management (2009) emphasize mainstreaming disaster management into development goals, the lopsided view of 'development' and the ecologically insensitive ways and means adopted to achieve development make disaster management ineffective and socially unacceptable. There is hardly any regular and comprehensive review or assessment of the performance of the embankments on the part of the Indian state and its agencies. On the academic front, there is hardly any critical discourse on mainstream or alternative approaches to flood management.

There is almost no informed debate that critically examines different aspects of flood management structures to result in any effective recommendation for governance and policy reform. Given the power equations and the prevailing state of (lack of) governance in the state, dialogue of this sort among the government agencies, academia, and civil society can hardly be expected. Given that 27 of the 29 districts in the state are ravaged by floods to different extents every year, the lack of any serious effort for constructive deliberations on the pros and cons of existing governance of flood management and consequent absence of advocacy for a paradigm change in flood management policy and practices can only push the state into the throes of further impoverishment, marginalization, and backwardness.

This chapter emerges from empirical research work carried out in the Lakhimpur and Dhemaji districts of Assam during the years 2008–11 and mainly from a case study on the Jiadhal river in Dhemaji district. It examines the governance mechanisms both formal and informal that conduct flood management at the local and state levels as its core issue. Moreover, the chapter explores how the existing formal governance system interacts with the political structures and the communities to formulate a politics that justifies the status quo of dominance of structural interventions imposed on rivers and local people by a monolithic technocracy sans local community's views and traditional knowledge.

National and Regional Paradigms of Flood Management

Flood and riverbank erosion emerged as a serious natural calamity for the people as well as the Government of Assam in the aftermath of the 1950 earthquake. Structural measures, especially embankments, popped up as a natural choice for the planners because people needed and the government was also eager to provide immediate protection and safety to the people with ways and means that were cheap and easy to develop in a short time and produced demonstrable results in almost real time. Lack of financial resources in those initial years of nation building was also a reason why embankments were preferred over other measures.

But embankments were not supposed to be the only solution prescribed to cure the problem of floods as seems to be the case today. The first formal policy statement on flood control enunciated by the Government of India on September 3, 1954 envisaged short- and medium-term measures mixed with long-term protection like storage reservoirs and other nonstructural measures to be carried out in phases to deal with the flood problem in the country. However, because of the shortsightedness of the policy-makers buoyed with the success of embankments in containing floods in many places, other options were not given due importance in Assam (Das and Bhuyan 2013).

The approach to flood management in the state has not changed much over the years. Even six decades after the first flood control policy of the country was adopted in 1954, the strategies being adopted are almost the same. Embankments have become the symbolic icon of flood management, although their limitations and negative consequences are well known (Dixit 2009; Mishra 2008a). The flood management regime of the state is controlled exclusively by the technocrats and bureaucrats comprising engineers and state-level administrative officers who propose, plan, and approve the technical interventions. It has become a monolithic structure in the subsequent decades, an entity that has remained static, rigid, and insensitive to environmental and social perspectives that matter as much, if not more, in flood management. However, in a pan-Indian context, the inherent weakness and side effects of the embankments were fully exposed time and again during and after every wave of floods triggered by breaching of these structures (Dixit 2009; Majumdar 1942; Mishra 2008b).

As in many other government sectors, the flood control system too was slowly transformed into a self-propagating cycle of mismanagement and corruption. Indigenous knowledge and opinions of local communities and views of civil society were isolated from the flood management process in a planned manner. A phenomenon like flood being experienced by people is often local and typical to a geographical context. While people far removed from this annual ritual may see it all as catastrophic, people experiencing these floods share a unique relationship with floods. The knowledge so obtained is a result of centuries of living with rivers and holds the key to offer insights while planning to manage floods. However, in a situation when planning happens in a context far removed, these precious learnings tend to be overlooked often in a deliberate and nonparticipatory manner. The 'solutions' thus formulated are rarely useful, thereby furthering the mismanagement. In a state like Assam, nonstructural measures, especially flood forecasting, early warning, and catchment treatment could have made a significant difference to reducing flood risks for the people, especially to those riparian populations residing on the banks of the tributaries of the Brahmaputra and the Barak rivers. Unfortunately, these activities were never taken up seriously. Embankments were promoted as something inevitable and unavoidable to counter floods in flood-prone areas.

There are many examples throughout the state where serious technical flaws have been observed in selecting locations of and designing the flood mitigation structures. Embankments and spurs have been built in such stretches of rivers where there was no need of such intervention. Apart from bad engineering and poor quality of construction, one reason of perpetuation of such mistakes and flaws over the years is that the local people are usually not consulted before or during the planning of the embankments and other structures. For example, in the case of the Jiadhal river of Dhemaji district, Assam water resource engineers ignored local community's traditional knowledge about the river and its characteristics and proactive suggestions in constructing embankments on many occasions and as a result those structures collapsed in the very next flood season because they were made in the wrong place and using inappropriate design (Das 2013).

As a result of dominance of the structural approach to flood management, people became so dependent on embankments that the traditional coping practices lost their importance and grew weak (Das et al. 2009). While some areas were given temporary protection by building embankments, many others in the downstream of those structures were rendered vulnerable to flood and erosion because of the hydraulic and hydrologic impact of embankments. Ill-maintained embankments often trigger floods and such embankment-induced floods are more devastating on most occasions than normal riverine flooding in open rivers. Yet the official regime has nothing better to offer to the people who are victims of a false sense of security that the embankments provide. It is astonishing to witness the incapability of the state government to ensure regular monitoring, repair, and maintenance of the very embankments that are considered so vital and touted as the 'only' remedy to floods. The result is that 27 of the 29 districts in Assam have become flood prone and with every passing year, floods have become a greater challenge.

Role of the State Government in Flood Management

The existing hierarchy of the State Water Resources Department (WRD; earlier known as the Flood Control Department) provides no space for the participation of local communities, civil societies, and experts from nonengineering disciplines in the planning and implementation of the structural flood management projects. The Technical Advisory Committee (TAC) that examines and approves the technical feasibility of the proposed projects is composed solely of experts and officials who represent various engineering disciplines by virtue of their association and employment with the government agencies. Owing to their primary orientation, which is technical, concerns about the social and environmental impacts of the structures built on the rivers remain unaddressed. Rivers and floods are officially governed and managed mainly by the engineers and technocrats with little or no space given to experts from social sciences and environmental disciplines. This results in irreversible and detrimental impacts on the river morphology, ecology, and settlements in downstream areas (Bandyopadhyay and Ghosh 2009).

The structural project proposals never incorporate the crucial issues of social and environmental impacts and the views of the community in the planning stage. Major projects with a long-term impact on river morphology such as river course diversion, laying of spurs and geotube-based embankments, or any embankment that covers a long stretch of a river are exempted from environmental impact assessment (EIA) or public hearing or community approval before being taken up. There is a complete absence of representation of stakeholder communities at any level—be it at the grassroots, district, or state during the planning and execution of the flood management structures. This is contrary to the spirit of the 73rd Amendment of the Constitution of India, which envisages planning and decision-making at the lowest levels possible.

The Role of Local Governance in Flood Management Practices: The Case of PRIs

The Panchayati Raj Institutions (PRIs) constituted under Article 243B of the Constitution of India and introduced as a country-wide uniform system of local governance by the Constitution (73rd Amendment) Act, 1992, provide for the lowest strata of democratic local self-governance in rural areas in India. In the case of Assam, the three-tier Panchayati Raj system consists of the Gaon Panchayat[2] (GP; equivalent to the Gram Panchayat in other states of India) at the village level, the Anchalik[3] Panchayat (AP) at the intermediate or block level, and the Zila Parishad (ZP)[4] at the district level (Ghosh 2008).

There is no direct reference to any activity concerning flood mitigation or construction of structures for flood protection in the Assam Panchayat Act, 1994. However, the Panchayats have been given power to take up some activities where flood-related tasks could be included within their respective jurisdictions. For example, the GP can incorporate tasks related to construction of structures such as canals, sluices, embankments, and maintenance thereof in the 'annual development plan' that they are required to prepare for their area of jurisdiction. It has to mobilize relief in the case of flood calamities and construct and maintain village roads that can qualify for small road-cum-embankments.

This is also true for the APs with the added advantage that the state government can assign any work related to rural development to these middle-level Panchayats under an existing scheme. This can well include flood protection work. In addition, it can implement irrigation work and construction and maintenance of rural roads, public ferries, and waterways, thus providing them a scope for inclusion of flood protection work, which are also useful as adaptation measures.

Similarly, the ZP can also construct and maintain roads; reclaim and develop land; and construct, renovate, and maintain minor irrigation work, all of which provide a scope for including flood protection work. It is the highest authority to approve of the schemes of the two lower agencies and get them implemented.

---

[2] Gaon means village in Assamese.
[3] Anchalik in Assamese means spread over a small spatial area,
[4] Zila in Assamese means a district; Parishad means a council.

As a result, PRI agencies can hardly take positive and effective roles in providing flood protection or facilitating any platform for the local people to organize public opinion or putting across public concerns to the higher agencies or incorporating people's opinions in flood protection and other developmental work. The control of the state government and the ruling class on the functioning of the PRIs is implicit in the way these institutions are constituted. According to the Assam Panchayat Act, 1994, some of the key functionaries of these agencies at high designations with considerable power are appointed by the state government or deputed from the serving bureaucracy.

The Block Development Officer (BDO) is appointed as the Executive Officer (also the Ex-officio Secretary) of the AP. The Chief Executive Officer of the ZP (Ex-officio Secretary) is a bureaucrat of not below the rank of an Additional Deputy Commissioner appointed by the state government. Moreover, the Chief Accounts Officer and the Chief Planning Officer, the other two key posts of the ZP, are also held by appointees of the government. Although such arrangements are made in principle to facilitate coordination of the Panchayats with the state government through the district administration, it has also paved the way for the bureaucrats acting on behalf of the government to exert political influence on the PRI agencies.

Furthermore, the capacity of performance of these agencies depends overwhelmingly on the financial grants from the state government. The ZPs can function only if their annual budget is approved by the state government, and therefore their functions are likely to be controlled at every step by the authorities in the state capital and this arrangement is far from the ideal of democratic decentralization (Bhattacharjee and Nayak 2001).

Since the Panchayat institutions are manned by democratically elected representatives affiliated to different political parties, political considerations play a major role in influencing distribution of funds and schemes to different areas depending on the political affiliation of the members and the people in those areas. The interests of a particular party that has majority in the three levels of Panchayats or the ruling party of the state call the shots in receiving projects. Moreover, the presence of the local Member of Parliament (MP) and the local Member of Legislative Assembly (MLA) as the members of both the AP and the ZP provides opportunities for politicians to influence the governance of the PRIs.

Thus, although PRIs are in a position to contribute substantially to flood management at the grassroots, the functioning of the PRIs is plagued with political biases, political affiliation to the ruling party, or loyalty to the government being the deciding factor for approval of schemes and funds.

People's representatives cannot perform at their full potential unless they are aware of the real power they have been given and gather the honest courage to exert that power rising beyond petty political and party considerations.

The Role of Local Governance in Flood Management Practices: The Case of the Mising Autonomous Council

Autonomous Councils (ACs) have been created by the state government in Assam to ensure welfare and development of certain tribal communities under self-rule in those areas that do not qualify under the Sixth Schedule of the Indian Constitution. These are local democratic institutions meant to facilitate decentralized governance by people's representatives dedicated to the welfare and development of particular communities. These councils are supposed to take proactive actions to address local problems such as flood and erosion in a participatory manner and engage intensively in the development of the community and the area for which they have been given the mandate, power, and funds. However, it is observed that the ACs have not been able to meet the aspirations and needs of the ethnic groups for which they were created because of some lacunae in the governance system. Rather these institutions have become instruments for fulfilling the state's political interest by patronizing financial corruption and ethnic hegemony.

The Misings are the largest tribal community living in these flood-prone areas that mostly come under the jurisdiction of the Mising Autonomous Council (MAC).[5] The MAC was established in 1995 under the Mising Autonomous Council Act, 1995, passed in the Assam State Assembly with the objective of social, economic, educational, ethnic, and cultural advancement of the people residing within its area of jurisdiction consisting of 'core' and 'satellite areas'[6] spreading over eight districts of Assam (MAC 1995). Unlike the PRIs, the ACs have been given specific responsibility to implement 'flood

---

[5] The performance of the present MAC, which was elected to power in 2014, is out of the purview of this article since this study was done during 2008–11.

[6] The 'core area' means the compact and contiguous areas predominantly inhabited by the Mising population having 50 percent and above as a whole in the area and not necessarily in the individual villages. The 'satellite area' means the area or areas consisting of noncontiguous cluster of villages predominantly inhabited by the Mising population having 50 percent and above as a whole in the cluster and not necessarily in the individual villages.

control schemes (not to be of highly technical nature) for protection of villages along with minor projects on irrigation, land development, and roads. However, the MAC has done precious little for flood protection in the study area. While lack of efficiency and financial irregularity on the part of different MAC authorities that have run the agency so far are considered a main cause for such nonperformance, there are administrative bottlenecks on the part of the state government that are also responsible for the failure of these bodies. Elections have never been held for this council since its inception. It has been run by interim committees appointed by the state government. The last ad hoc body that came into power in 2006 after prolonged demands of some Mising social and political organizations consisted of people having some acceptability among the community. However, it was suspended by the government on charges of financial misrule in 2009 and the same body is now facing enquiry by the Central Bureau of Investigation (CBI). The MAC was not audited nor was given any administrative and financial guideline for conducting its activities until 2006 for 10 years after its inception.

Given the fact that Misings are riparian communities preferring to live near rivers, most parts of the MAC area are critically flood affected, whereas the status of flood protection like embankments is in a bad shape all over. The MAC has not been given power to execute technical schemes of the right size that would help in protecting people from floods. Under the present Act, it can take up only small schemes of nontechnical nature. People think that elections should be held and responsibilities of flood management at the local level including the functions of the WRD should be entrusted to the MAC because since local candidates are themselves flood affected, they will understand the problem and perform better than the government officials who are not sincere in duty and have no feeling for the victims.

While allocating powers and functions to the MAC, it was not sorted out how it will coordinate with other agencies or line departments engaged in similar work or how the geographic and functional jurisdictions will be demarcated amongst the different agencies working for rural development. As a result, now one sees several agencies such as the PRIs, the District Rural Development Agency (DRDA), and the Integrated Tribal Development Project (ITDP), all doing the same kind of work under different schemes resulting in misuse of funds. Lack of coordination and transparency among these agencies is also a reason of their addressing people's problems effectively.

In the present arrangement, the Department of Welfare of Plain Tribes and Backward Classes (WPT & BC) of the state receives funds from the central government and distributes the same among the ACs according to the size of population under different councils. The state government uses this opportunity to exert control on the council's affairs. It creates situations where funds are allocated based on political subservience of the councils and kickbacks. Another hindrance in timely implementation of the projects is the time of release of funds. Funds are released generally toward the end of the financial year (December or January) that too in installments compelling the council to spend the amount hurriedly even resorting to false utilization certificates to get the next installment within the same financial year (April–March). This inappropriate funding cycle hinders proper planning and implementation of flood protection work for the next flood season that starts as early as April and continues until October.

In order to implement flood management locally, adequate resources and powers should be given to the agencies of local governance. This includes the authority to employ all means and methods, technical or nontechnical, required for managing floods along with minor irrigation, land development, and roads. However, the MAC has been given specific responsibility to implement "flood control schemes for protection of villages that should not be highly technical [in] nature" (MAC 1995). This would also involve measures of disaster risk reduction such as flood preparedness, a judicious mixture of structural and nonstructural methods of flood mitigation, and mainstreaming of adaptation practices to flood management and development programs, promoting traditional coping strategies and socioeconomic empowerment of vulnerable communities.

As a result of prolonged debate and varied discourse over flood management practices of the country, new ideas and concepts have found some acceptance in the governance system, but it will take a long time to bring about a complete change in institutions, policies, and practices to implement a holistic framework of flood management. This would mean the prevailing paradigm of controlling giving way to mitigating flood impacts with adaptive and participatory paradigms that will ensure application of only sensible and flexible river engineering along with empowering people to cope with floods. Mainstreaming of adaptation into the development process will be the hallmark of such a regime.

Community and Civil Society Initiatives in Flood Mitigation

The failure of the government's schemes to mitigate flood and erosion, mismanagement of funds, lack of transparency, and efficiency in implementing flood management projects for decades has made people disillusioned with the governance system. Years of suffering by the communities, without having any opportunity for meaningful participation in the flood mitigation process, have made them disgruntled and provoked them to spontaneous protests as well as organized campaigns against the faulty flood management regime leading sometimes to conflict situations with the WRD and other government agencies and contractors.

People affected by the floods of the Jiadhal river in Dhemaji district have been demanding proper repairing and strengthening of embankments. But over the last 30 years, when the floods have become more devastating, hardly any technically appropriate intervention has been made. The '*Jiadhal Nadi Baan Pratirodh Oikya Mancha*' (JOM),[7] a platform of villagers from eight different Panchayats affected by floods and erosion from the Jiadhal river, has been consistently engaging with the WRD and district administration exposing cases of misappropriation of project funds and sloppy work.

The JOM is an umbrella platform of various community organizations and individuals. It is leading a local movement demanding a permanent solution to the flood problem through properly implemented structural measures. It has resorted to nonviolent and peaceful agitations such as submission of memorandum, awareness meetings, rallies, and sit-in protests. It has resisted schemes that have sought only to repair embankments in piecemeal effort without treating the flood problem holistically. It thinks that instead of wasting money on such seasonal stitching work on the embankments, investment should be made in scientifically designed, technically sound robust bank protection with embankment that will contribute to long-term mitigation of the flood hazards.

The JOM has also developed a plan of embankments with schematic diagrams on its own, incorporating its learning and experience of structural measures and local knowledge of the river's behavior. It has highlighted the loopholes in the official schemes of embankments and related structures in the community plans. Suggestions have been offered from its side toward making more efficient and durable structures as well as containing the river in a defined channel. However, the WRD engineers are reluctant to recognize the element of knowledge and wisdom in the community plans and to give credence to the community demands.[8]

---

[7] Meaning United Platform for Resisting Floods of Jiadhal River, the leading movement protesting against bad governance of flood management specific to Jiadhal river.

[8] As told by Gandheswar Bora, a leader of the Jiadhal floods movement, in an interview at Barman Gaon on January 19, 2010.

While there is a plethora of examples of how community knowledge and participation have been ignored by the technocratic establishment leading to alienation of the local stakeholders from flood management projects, some technical experts want to put limits to such participation. They feel that although it is important to recognize community's knowledge in official flood protection work, one needs to be careful about implementing technical schemes wholly based on community recommendations, especially in the case of unstable and dynamic rivers that drain the landscape of the two districts under consideration.

People's initiatives are not limited to movements and resistance. Local civil society and media have brought to light numerous instances of illegal practices, inadequate and substandard work, and financial mismanagement over the years in embankments and related projects. They have demanded immediate compensation to and rehabilitation of all affected people, engagement of unemployed youth in Jiadhal projects, exemption of affected people from land revenue, etc. There are numerous occasions when the villagers repaired degraded parts of embankments and plugged breaches on their own without waiting for the government to respond.

During the rainy season when the possibility of floods increases due to heavy and continuous rain, the people in villages such as Holoudonga, Bokulbari, Maisa, Chaporigaon, Nagaon, Mangoloti, Podumoni, Kakoihal, Dihiri, etc. located on the banks of the Jiadhal river organize vigil on the embankment round the clock, day and night.[9] Groups are formed to keep watch on the vulnerable parts of the embankment so that the eroded sections can be plugged instantly or the villagers can be warned when flood level rises alarmingly or flood water starts entering the countryside. Mobile phones play an important role nowadays in informing people about the state of the river, flood, and the embankment.

The Asom Jatiyatabadi Yuba Chatra Parishad (AJYCP)[10] is strongly raising its voice demanding urgent and fruitful steps to control flood and loss of land due to bank erosion in the district of Dhemaji by organizing meetings, street protests with slogans, mass gatherings, road blockades, sit-in protests, etc. since 2008. It has asked the state government to ensure completion of flood protection work between November and February and rebuilding

[9] As told by the community in a focus group discussion at Holoudonga village, Barbam Panchayat, on January 19, 2009.

[10] AJYCP is a frontline organization of students and youth in the state having experience of leading movements on various social and political issues of the state such as on large river dams, flood and erosion, total autonomy for Assam, etc.

of all those embankments that have completed their lifetime (older than 30 years) preceded by inspection and scientific assessment of the structures. They, accompanied by the local press, inspect the embankments of the rivers in different parts of Dhemaji district on their own, identify the most vulnerable ones, and try to pursue authorities to ensure that proper action is taken. Similarly, the All Assam Students' Union (AASU)[11] is taking a proactive role in demanding the flood problem of Assam to be declared a national problem or national disaster. They are monitoring anti-erosion work in Majuli through video recording of the execution of the work by contractors. Ethnic Mising organizations like the Bane Kebang[12] and the Takam Mising Porin Kebang (TMPK)[13] have frequently voiced their concerns over the problems of floods and riverbank erosion. They have demanded rehabilitation of flood-affected people and appropriate flood protection measures.

Discussion and Conclusion

The cases of Lakhimpur and Dhemaji districts present a novel way in which floods and flood management can be looked at. Based on our work, we have argued that floods are not merely a hydrometeorological issue. Rather, they are essentially a sociopolitical issue. People in these two districts are not entirely made vulnerable due to floods. Rather, in many cases, their existing vulnerability is enhanced many times over because of floods. The technical intervention of floods has taken away the people's mandate from a natural process they have lived with for generations. This has resulted in one-sided actions with far-reaching consequences. An inclusive and integrated flood management plan along with comprehensive institutional reforms is the need of the hour.

At a global level, conventional flood management has been gradually giving way to emphasis on holistic understating of flooding and its impacts as a combination of hydrological, ecological, and social processes. There has

---

[11] AASU is an influential students' organization pursuing issues such as immigration, flood and erosion, large river dams, etc. that are critical for the survival of the indigenous people in Assam. It led the historic Assam Movement in the 1980s and is engaged in negotiations with the Government of India on matters related to the Assam Accord.

[12] The Bane Kebang is an umbrella organization of all Mising organizations. It enjoys support from all sections of the community and exercises considerable influence on the community in all matters related to culture, economy, and politics.

[13] TMPK is a students' organization of the Mising community.

been a definite shift in the way water, rivers, and floods have been understood so far. This is part of a much larger change with regard to expertise on rivers and floods. This alternate line of thought has begun from the South. Bangladesh has contributed to this change in a big way where living with floods has been recognized as a viable alternative.

At the national level, recent reforms in the policies and functioning of institutions in the context of disaster management in India have introduced many progressive ideas and practices. The mainstreaming of disaster management strategies in development projects and programs is a noteworthy example. The Disaster Management Act, 2005 (GoI 2005) that brought a paradigm change to the way disasters were dealt with in India before seeks to ensure the integration of measures for prevention of disasters and mitigation by ministries or departments of the Government of India into their development plans and projects. Section 5.1.6 of the National Policy on Disaster Management 2009 (National Disaster Management Authority 2009) upholds that "environmental considerations and developmental efforts need to go hand in hand for ensuring sustainability." It advocates for restoration of ecological balance in Himalayan regions in disaster management plans. It also stipulates that zonal regulations for disaster management must ensure the preservation of natural habitats (National Disaster Management Authority 2009: 18). This concept has opened the possibilities for linking disaster management to the goals of sustainable development.

However, like many other areas of governance, nothing worthwhile has been done on ground at the state and local levels to actually implement these ideas into practice in the context of flood management. As a result of prolonged debate and varied discourse over flood management practices of the country, while new ideas and concepts have found some acceptance in the governance system, it will take a long time to bring about a complete change in institutions, policies, and practices to implement a holistic framework of flood management. This would mean the prevailing paradigm of controlling giving way to mitigating flood impacts with adaptive and participatory paradigms that will ensure application of only sensible and flexible river engineering along with empowering people to cope with floods. Mainstreaming of adaptation into the development process will be the hallmark of such a regime. Although the emphasis has shifted from relief and rehabilitation to preparedness, disaster management in practices in India is still largely delinked from the risk-adaptation-resilience perspective. Therefore, mainstream disaster management as practiced by the government agencies has failed to make any significant contribution to the achievement of the goals of sustainable development in states like Assam.

To actualize the vision of sustainable development and integrated disaster risk reduction for flood preparedness in Assam, a judicious mixture of structural and nonstructural methods of flood mitigation and mainstreaming of adaptation practices to flood management and development programs, promoting traditional coping strategies and socioeconomic empowerment of vulnerable communities, is required. The relevant institutions such as the MAC, PRIs, and the WRD need to undergo reforms to ensue improvement in efficiency of their functioning empowered by adequate responsibility, power, and financial resources with a flexible and wider mandate to comanage water resources and water-induced disasters with a holistic and inclusive socio-ecological perspective.

The WRD must adopt a flood and erosion management policy and a set of technical guidelines without any further delay. If needed, the Government of Assam should consider formulating an Act for flood and erosion mitigation. The land acquisition policy and laws need to be much more sensitive in recognizing people's right of dissent to dispense with land for construction of embankments. It must be remembered that with growing population, human settlements have encroached upon the riverine spaces (like flood plains). At the same time, rivers are also eating up vast amounts of riverine land and causing degradation of farmland by depositing coarse sand year after year. Thus, land has become a scarce commodity and the most crucial asset that people strive to possess and protect in the valley region. If land is acquired by the government with due consent for common good (e.g., to make an embankment which is genuinely necessary), compensation must be given to the people expeditiously and adequately so that they can create additional resources and assets for their future survival and adaptation.

The functions of WRD and other related line departments should be decentralized to transfer more responsibility, power, and resources to PRIs and the MAC to urgently take care of construction and maintenance of flood mitigation infrastructure with participation of local communities in decision-making as well as in engagement of workforce. The government agencies work in isolation from the local communities and do not recognize the indigenous wisdom about rivers and water, but the ACs if run by local people's representatives can fare better in flood management if adequate powers and resources are provided to them.[14] The PRIs will have to embark on such activities with due consultation with the concerned departments for technical guidance, if needed.

---

[14] Interview with Mr P Chayengia, former Chief Executive Councilor, MAC, 2006–09, in Gogamukh, Dhemaji on April 20, 2010.

The argument about the paucity of funds is not a new one. Neither is it untrue that the paucity of funds is a major factor that leads to substandard work. However, it is convenient to cite this deficiency to camouflage rampant corruption. Making funds available to the WRD in time is a key to ensure good work from its officers. Similarly, a new financing policy should be followed by the central government in which release of funds directly to the MAC should be ensured much early in the financial year so that all important work related to flood protection and development can be planned and executed before the advent of the rainy season. To bring in the changes necessary, the MAC Act, 1995, should be further amended to give it more funds and more functional powers in flood mitigation.

To make the technical and financial estimates of the schemes proposed by the executive engineer's office more feasible and technically sound, a Research and Development (R&D) wing could be opened so that it can examine and help in improvement of the quality of the proposals in a preliminary scrutiny. The existing Planning wing working under a director, Planning (an officer of the level of a superintending engineer), can be reactivated to become such an R&D cell. Schemes that pass through rigorous scrutiny and review by this R&D cell may be taken up for approval by the TAC with adequate provision of funds. At the same time, monitoring and evaluation of the work done must be made very strict. Engaging community organizations as monitors can pay good dividends. Approval of projects and release of money should also be linked to the projects being technically sound and community friendly.

More transparency and coordination need to be assured among the PRIs, MAC, WRD, and line departments so that projects with multistakeholder involvements are completed smoothly. Guidelines should be provided to all relevant departments and agencies to ensure coordination and transparency so that the MAC and PRIs (in this particular case discussed) can work independently as well as in coordination with state agencies for flood protection work. The District Development Committee should incorporate the MAC and its views while preparing development plans for the rural areas of the districts to be implemented by the PRIs.

The Report of the Task Force formed by the Government of India after the catastrophic floods of 2004 in eastern India recommended establishment of a single authority for water management of the Brahmaputra river basin in northeast India and called it the North East Water Resources Authority. It was supposed to integrate all aspects of use and management of water so that water governance for the entire basin becomes unified and interrelated. It has not become a reality so far due to noncooperation of some stakeholders. Along with this proposed new institution, other ideas of erosion and flood mitigation such as confinement of the rivers in critical reaches by channel improvement, silt management, etc. are also gaining currency.

Meanwhile, we will have to live with both embankments and floods. There is no better way than to strengthen the existing network of structures so that they perform better and the investment already made in them remains useful. Utmost care must be taken before implementing further embankment projects on both technical and socio-environmental grounds. The government must be bold enough to admit the limitations of the flood management strategies it is adopting at present and encourage and empower people to enhance adaptation to floods and rivers. The state also needs to build the capacity of institutions for flood risk management and integrate their operations with the disaster management systems.

It is a fact that lack of sufficient information and database is a reason why our knowledge base about the state's rivers and the water regime in general is far incomplete than is desired. However, this gap would have been less had we given credence to the knowledge of people's discourses on floods. The fact remains that there is almost total neglect of the ethno discourses on floods over a more technical approach to floods. This results in a knowledge base that is lopsided, highly skewed toward structural interventions and macro-level solutions, and, hence, found wanting in multiple ways. Ethno discourses of people living in flood-prone areas for generations have hardly been considered while addressing the issue of floods. It is people and communities who have lived with floods.

With the introduction of the state's role in disasters, people began to lose control over their modes of adaptation. They began to live with flood control and flood policies, the second of which was top-down. A complete under-standing of floods in a holistic sense remained incomplete in both the intellectual and the popular domains because ethno discourses or people's voices never found their way into flood studies. This case study is a step in the direction highlighting the importance of inclusion of people's voices in flood management policies and programs. Lessons have been learnt from mistakes made in the past, new concepts are getting recognition, and support to implement new ideas is in the offing. Assam is now at the crossroads, and it is the right time to take the right course. Policies are made for people and unless people are included in the process, development can be neither sustainable nor inclusive.

References

ADB (Asian Development Bank). 2006. *India: Preparing the North Eastern Integrated Flood and Riverbank Erosion Management Project (Assam)*, Technical Assistance Report, Project No. 38412. Available at: http://www2.adb.org/Documents/TARs/IND/38412-IND-TAR. pdf (Accessed on October 12, 2010).

Bandyopadhyay, J. and N. Ghosh. 2009, November 7. "Holistic Engineering and Hydro-diplomacy in the Ganges–Brahmaputra–Meghna Basin." *Economic and Political Weekly* 44 (45): 50–60.

Bhattacharjee, P.R. and P. Nayak. 2001. Panchayati Raj in Assam. Presented at the National Seminar on "Local Governance and Rural Development: The Indian Experience," at Utkal University, Bhubaneswar, February 24–25, 2001.

Das, P.J. 2013. "Jiadhal River Catchment: Conflicts over Embankments." In *Water Conflicts in Northeast India: A Compendium of Case Studies*, edited by Partha J. Das, Chandan Mahanta, K.J. Joy, Suhas Paranjpe, and Shruti Vispute, 35–44. Pune: Forum for Policy Dialogue on Water Conflicts in India.

Das, P.J., D. Chutiya, and N. Hazarika. 2009. *Adjusting to Floods on the Brahmaputra Plains, Assam, India*. Kathmandu. Available at: http://lib.icimod.org/record/8025/files/attachment_669.pdf (Accessed on March 27, 2017).

Das, P.J. and H.K. Bhuyan. 2013. "Policy and Institutions in Adaptation to Climate Change: Case study on Flood Mitigation Infrastructure in India and Nepal." ICIMOD Working Paper No. 2013/4. Kathmandu: ICIMOD. Available at: http://lib.icimod.org/record/28382/files/WP_4_13.pdf (Accessed on March 27, 2017).

Dixit, A. 2009, February 7. "Koshi Embankment Breach in Nepal: Need for a Paradigm Shift in Responding to Floods." *Economic and Political Weekly* 44 (6): 70–78.

Ghosh, B.K. 2008. *The Assam Panchayat Act, 1994 with Rules*, 344. Guwahati: Assam Law House. Available at: http://atingl.nic.in/Downloads/THE%20DISASTER%20 MANAGEMENT%20ACT%202005.pdf (Accessed on February 17, 2017).

GoI. 2005. The Disaster Management Act, 2005. Government of India. Available at http:// ndma.gov.in/images/ndma-pdf/DM_act2005.pdf (Accessed on March 27, 2017).

MAC (Mising Autonomous Council). 1995. *The Mising Autonomous Council Act, 1995*. Available at: http://www.macgov.in/pdf/MAC_Act_as_amended_.pdf (Accessed on December 24, 2011).

Majumdar, S.C. 1942. *Rivers of Bengal Delta*, 9. Calcutta: University of Calcutta (as cited in Mishra 2008b).

Mishra, D.K. 2008a, September 6. "Bihar Floods: The Inevitable has Happened." *Economic and Political Weekly* 43 (36): 8–12.

———. 2008b. *Trapped Between the Devil and Deep Waters: The Story of Bihar's Koshi River*, 208. New Delhi: Peoples' Science Institute, Dehradun and South Asia Network on Dams, Rivers and People (SANDRP).

MWR (Ministry of Water Resources). 2004. *Report of the Task Force for Flood Management & Erosion Control*, 135. New Delhi: Government of India.

National Disaster Management Authority. 2009. *National Policy on Disaster Management*. National Disaster Management Authority, Ministry of Home Affairs, Government of India. Available at: http://www.ndma.gov.in/images/guidelines/national-dm-policy2009. pdf (Accessed on February 17, 2017).

9

The Role of Rural Local Bodies in Sustainable Development

James Rajanayagam

Introduction

According to Voltaire, "uncertainty is uncomfortable; but certainty is an absurd one." As the civilization moves toward more uncertainties—climate change phenomenon, population explosion, etc.—we also need to find a comfortable position in certainty. A certainty in ability to meet unforeseen challenges by people closer to uncertainties will be a key factor in sustaining the presence of human beings. Effective infrastructure such as roads, electricity, etc. plays a major role in combating uncertainties.

Meeting local needs through local solutions is a key to long-term sustainable development. In India, infrastructure in rural areas is poor and majority of people living in rural areas do not have access to good-quality roads, electricity, clean cooking solutions, etc. The governments at the central and state levels are responsible for providing funds to meet infrastructure needs. Very often, the funds are not provided in the budget due to lack of planning and awareness of local realities at these levels. The objective of this chapter is to present how elected rural local bodies (RLBs) can play a role in building infrastructure through entrepreneurial approaches through selected case studies. The chapter will also analyze the roles of central and state government actors and nongovernmental organizations (NGOs) in providing infrastructure and will then show how RLBs can play a better role in the provision of infrastructure. These case studies highlight the problems and challenges that these elected representatives faced while executing the entrepreneurial solutions and how they overcame them through innovative approaches. The chapter will also highlight how infrastructure development is positively correlated to sustainable development. Based on the success of their approaches, the author recommends strategies to policy-makers for implementing this entrepreneurial approach of RLBs at the national and regional levels.

Sustainable Development, Human Development Indices (HDI) in Rural Areas in India

The succession of mankind can be imagined to be a relay race in which each generation passes the baton to the next. A society or a family passes its knowledge, skills, and natural resources to the next generation. In the case of intangible assets like knowledge, society multiplies or maximizes its content before passing it on. In the case of tangible assets like forests, the group tries to minimize its depletion or maximize its replenishment.

Sustainability is defined as an ability to continue a behavior act indefinitely. It can be imagined as the stake that each generation receives and has, at its disposal, to carry on the race. It is the equity of the organization. Current global discussions and forums center on the preservation of equity. Discussions on climate change are about limiting the negative effects of pollutants on equity, which is natural resources in this case, that is, how natural resources such as land and water provide the yield of crop for the continuity of mankind. Discussions on trade talks focus on how one derives maximum leverage from the conversion of natural resources for utility purposes. India is at the lowest rungs in many of the global indicators—67 out of 122 countries in the Global Hunger Index, ranked by International Food Policy Research Institute on general malnourishment; 119 out of 169 countries in HDI, prepared by UN Development Programme, etc.

In other areas such as poverty, according to World Bank data, 22 percent of the population was below the national poverty line in 2012. The proportion of poor people living in villages was higher than in urban areas. According to the report, India is the place with the highest concentration of poor in the world. According to the Ministry of Statistics and Programme Implementation, the infant mortality rate stands at 44 per 1,000 births. The proportion of households with access to sanitation facilities was 40 percent in 2005–06, the lack of which is one of the prime causes for preventable diseases. Agricultural productivity is lagging behind those of other countries and it is estimated that 30 percent of the agricultural produce is wasted due to lack of storage facilities.

Sustainable development—the act of preserving or improving the economic equity of the society—is the key to good HDI. We may increase the finance capital (higher GDP), but if we decrease the natural capital (agricultural land), then, in the long run, the HDI will deteriorate. Hence, in the next section, we will see the role of infrastructure in sustainable development for higher HDI.

Relationship Between Sustainable Development and Infrastructure Development

Good infrastructure is a key ingredient of sustainable development. Infrastructure is a broad term with many components. These components are hard infrastructure such as roads and electricity; soft infrastructure such as health, education, drinking water, and sanitation; and human resources (HR) infrastructure such as skill sets. Good infrastructure enhances all types of capital—physical, human, finance, and natural.

Pani (2008) defends that it is difficult to establish a direct relationship between development interventions and removal of poverty. But, with development in infrastructure, it is possible. History of nations shows that improvement in infrastructure naturally improves quality of life index. There is a saying that America built roads and roads built America. Soft infrastructure improves the well-being of a person, HR infrastructure improves the knowledge capital, and hard infrastructure improves the returns on finance capital. Hence, infrastructure is the necessary condition for improving the HDI. The following section will explain this in detail.

Effectiveness of Agencies in Infrastructure Development

As seen earlier, the achievement of self-sufficiency in all the components of infrastructure, namely hard, soft, and HR, has not been up to the mark of the HDI, resulting in a low quality of life for the majority of Indian citizens. This section will take a look at the players involved in this, analyze their performances, estimate the output, and see if there is any outcome.

The agencies involved are state and central governments, NGOs, and private organizations. The Indian government adopted a welfare policy and formulated five-year plans to implement centralized, large-scale infrastructure projects. During the eighth five-year plan, the government identified eight sectors of intervention. They are primary education, health, water, electrification, roads, housing, environment improvement, and nutrition. Seventy years have passed since independence. There have been improvements in literacy, life expectancy, and mortality rates of children and mothers. However, hard infrastructure is not yet available to the majority of the people.

In the areas of skill development, major studies by industry associations show that a significant proportion of the workforce is unemployable, thereby diminishing the demographic dividend.

India has an estimated number of 3.3 million NGOs, that is, one for every 400 persons. The objectives of NGOs are capacity building, provision of health services, employment opportunities, policy advocacy, etc. They have been doing this through establishment of schools and hospitals, capacity building, access to finance for vocational opportunities, and activism for various causes such as gender rights, etc. They have reached places where arms of governments have not been able to penetrate. It is no gainsaying their role in sustainable development. We need to bear in mind the difference between outputs and outcomes. NGOs have provided very basic amenities to the most impoverished of the people in remote corners. They have created more dependencies than liberating people from the clutches of poverty. Microfinance institutions provided access to finance to microenterprises and individuals. They have nurtured and sustained the entrepreneurial spirits at the bottom of pyramid of the class structure but they seemed to have failed in unleashing the wild entrepreneurial spirits of people to catapult them to the next level. This is where a new class of entrepreneurs has stepped in to fill the need and release the spirit.

Social entrepreneurs, as they are called, are defined as those who solve the problems of society through market-based approach. The Skoll Foundation defines a social entrepreneur as "society's change agent: a pioneer of innovation that benefits humanity." Prahlad (2012) coined the term 'Bottom of the pyramid' and advocated the market-based approach for solving social problems. Corporate sector identified products and services that could be marketed affordably to the poor people in the developing countries. Spearheaded by the World Bank and other international foundations, social entrepreneurs started a hybrid of for-profit and nonprofit initiatives to provide products and services for the poor people and through these initiatives, they built predominantly soft and HR infrastructure. Hard infrastructure such as large-scale power projects and roads was out of focus and scope for most of these entrepreneurs and their stakeholders such as venture capitalists due to financial and other considerations. Hence, without hard infrastructure, soft and HR infrastructure proved less effective in the sustainable development of the region. For instance, efforts in soft infrastructure in improving the productivity of agricultural produce without cold storage solutions went without any outcomes—improvements in the incomes of farmers as the price increase were not realized without any cushions for seasonal fluctuations.

Growth in the welfare of the society or the removal of poverty has been happening. The rate at which it is happening, it may take a few decades before we become a developed nation and have removed poverty. Is this rate content enough and sufficient? It is estimated that US$500 billion will have to be pumped in by the Government of India during the next five-year period. What are the other alternatives? What can be done to accelerate the development pace?

Gandhiji had firmly believed that only Panchayati Raj (local government) will make villages self-sufficient (Prabhu et al. 1960). Many of the committees (Balwant Rai Mehta Committee 1957) formed to study the effectiveness of five-year plans have echoed the same view and have recommended the devolution of powers to RLBs. How can RLBs be involved in the planning, implementation, and management of the infrastructure projects?

The following sections will explore the current functions of the rural bodies, their limitations and challenges, and ways to overcome the challenges and will look into some of the case studies that advocate pioneering solutions. The case studies will show how these solutions of entrepreneurship can be replicated on a massive scale to accelerate the development process. Finally, the author proposes strategies for policy-makers and other stakeholders to engage local bodies effectively and recommend a road map to implement.

Role of Rural Local Bodies

Historically, local governance in India is referred to as Panchayat system. This is when rulers did not have reach to the remote places of their kingdoms and village elders came together to sort out their common problems. However, this is not an institutional set-up and with the advent of British, most of the administrative powers went to the executives and functionaries of the British government. For instance, management of water bodies came under the Public Works Department. After independence, the Constitution of India recognized the role of Panchayati Raj system and through the Directive Principles of States required the states to implement it. However, the system remained weak and two committees (1957 and 1977) were formed to make recommendations to strengthen local governance and effective implementation of five-year plans of the central government. It is important to note that one of the committees observed that in order to effectively implement development interventions, it is necessary for Panchayats to be involved in these initiatives.

The watershed moment came in 1993 when the Parliament amended Article 243 through the 73rd Amendment. The 73rd Amendment required that states hold periodic elections, devolve 29 subjects, and form a State Finance Commission to share resources. The roles of Panchayati Raj bodies are to assist in the planning process, implement state and centrally sponsored schemes (CSS), provide basic amenities of services, and finally raise own revenue through taxes and fees.

Since 1993, almost all of the states have enacted their own Panchayat Acts, devolved functions, and allocated funds through State Finance Commissions and some of the states have devolved functionaries. Panchayats have become the prime agents, although not the only one, as functions are still in the concurrent lists of all the forms of the governments, for most of the schemes. For instance, Panchayats oversee the works under the Mahatma Gandhi Rural Employment Guarantee Scheme. However, most of the works and assignments come from above—higher levels of government.

There are some schemes that give freedom to choose projects and implement according to certain fulfilled conditions. For instance, in Tamil Nadu, under the Self-sufficiency Scheme (SSS), Panchayats can choose projects of local relevance, get the approval of Panchayat committee, obtain the approval from the District Rural Development Agency, and finance internally up to one-third of the project cost; the state government will finance the rest.

To summarize, the process of decentralization of governance and implementation of welfare state has been started. The degree of decentralization is varying in different states. Panchayats have become the last-mile executing agencies for infrastructure projects. However, as with any fledgling institution, they also face challenges, which are described in the next section.

Challenges for RLBs as an Autonomous Institution

1. Lack of Awareness: At the Panchayat level, many of them suffer from lack of administrative, planning, and executing capabilities. This also leads to information asymmetry and lack of alignment with state and central governments on various information requirements. Many of the state and central government schemes remain unknown at the Panchayat level. Many of the Panchayat elected representatives are not aware of their roles and responsibilities. Furthermore, this is also affected by relationship patterns with the state government officials and the effectiveness of grassroots planning committee to develop the annual plans and budgets.

2. Lack of Access to Finance: According to the World Bank (2008), infrastructure development is based on a simple framework where need is the problem, supply is the solution, and civil service is the instrument through service provider. Civil service is an instrument handled by state and central governments and this is where bottleneck comes in. The last level of government, that is, the local bodies are not empowered to handle private sector. It is acknowledged by policy-makers that there are three (3F) challenges before Panchayats to fulfill their roles: function, funds, and functionaries.

The Constitution of India, as per the 73rd Amendment, has delegated 29 subjects that are under the state list to the local bodies. Refer Appendix 9.1 for a list of subjects/functions. Most of the states have fully/partially delegated these functions to Panchayats. Refer Appendix 9.2 for the status of delegation of functions by all the states to Panchayats. It can be seen that functions have been devolved.

Traditionally, Panchayats and other local bodies have relied on state and central governments for the execution of various projects. Funds have come in as grants-in-aid from state governments and as CSS from the central government. Here, there are two issues. One, the projects are decided by someone a few thousand kilometers away from the problem and very often, these projects are not the immediate priorities of the local constituencies. Second, funds may not be available or sanctioned by the state and central governments for the projects that require immediate attention. Hence, due to mismatch, funds are not properly allocated to the right project or the right project remains suspended for want of funds.

Central and state governments, through the institution of civil service, have the capabilities and resources to bring in the right kind of talent, both technological and managerial, for the functionary role, to execute major projects. For instance, they can bring in the best talent for the execution of major metro rail projects. Unlike the higher level of governments, local bodies, with the exception of major municipal corporations, do not have the resources to bring in the best talent for the functionary role.

3. Lack of Access to Technology: In the recent past, infrastructure projects in power storage facilities have remained within the domains of state and central governments due to the complexities of technologies. With advances in scientific progress, technologies have become affordable, decentralized, and smaller in sizes. However, these advances have not been effectively utilized at the grassroots levels. For

instance, instead of depending on large-scale mega power projects, local governments can implement decentralized and smaller solar, wind, or biomass-based projects for their electricity needs. The barrier is the lack of access to technologies due to their own lack of knowledge and other issues mentioned in point (1) above. The next section addresses these challenges.

Opportunities for RLBs as Entrepreneurs in Infrastructure Development with Case Studies

Opportunities have always existed for implementing new projects and this has depended on a number of factors such as initiatives by the local elected bodies, aspirations of the people in the constituencies, access to a private player with technology, and supportive government officials and other stakeholders such as NGOs.

The previous section addressed the challenges. Awareness of opportunities is improving and will improve with the advent of proactive and capable leaders at the grassroots, not necessarily very highly educated, who have the desire to bring change and motivate and convince people to undertake changes and follow. They have the capability to interact with officials at the higher levels and challenge the status quo. In essence, they are the social entrepreneurs addressed earlier, who want to see change through administrative and governance vehicles. Just like for-profit social enterprises, these leaders will address social problems through market approaches but without direct profit realizations.

At the same time, these leaders will be able to bring latest affordable technologies. Now, that leaves us with one of the major challenges—lack of access to finance. As mentioned earlier, the funds are primarily grants-in-aid and shared taxes and to a small extent own taxes. There is also a mismatch between project allocated and actual requirements (projected) in terms of size, local relevance, and timely need. Devolution of funds from state and central government schemes alone may not be able to solve the infrastructure issues. Increasingly, local governments may have to raise money on their own. The case studies, presented below, provide a model in which Panchayats determine the needs, estimate the project costs, bring in margin from internal resources, raise money from the market, provide guarantee for return without the intervention of the state government, show ability to charge fees from users, and repay the money raised with returns to the global and local investors.

The model will attract immediately many questions. Is the model feasible? Can the Panchayats raise money? Do they have the capability to execute the projects? What are the interferences from the officials of the state government? How can the Panchayats charge fees from users and how can they repay money?

The Constitution of India and individual state government Acts (Government of Tamil Nadu 1994) do not prevent local governments from raising money. But the practice has been that state governments have acted as the guarantor. Many of the large municipalities (Mohanty et al. 2007) such as Ahmedabad and Chennai have raised finances to complete various sanitation-related projects. Of course, it is not yet time for the rural bodies to raise finance on their own—the following sections will highlight a case study where it is really possible. There are few cases where Panchayats have worked with private players and NGOs on a public–private partnership model where projects are completed on build–operate–transfer (BOT) basis. These Panchayats have also demonstrated that they are able to raise margin money from the people and charge user fees. The model is certainly feasible where Panchayats have a larger role as entrepreneurs to undertake infrastructure projects and solve social problems. This is similar to cases where state and central governments have undertaken commercial activities like large public sector undertakings (PSUs) in which state and central governments are the promoters.

The next two sections will explore two case studies and arrive at factors responsible for their successes.

Case Study 1: Achieving Self-sufficiency in Electricity

This is a case study about self-sufficiency in electricity needs of a small village. Odanthurai is a Panchayat about 500 km southwest of Chennai, the state capital of Tamil Nadu. It is a cluster of 11 villages with a population of 5,000. As with any other typical village in India, it suffers from frequent power cuts during the day. Today, it is a power surplus village selling excess power to the state utility. It is the first local body in India to build its own 350 MW windmill. How did it happen? It started with the election of Shanmugam as the Panchayat president for the terms 1996–2006. Odanthurai spends about 60 percent of its budget on utilities. Hence, Shanmugam calculated that he could save this money if the village could generate its own electricity needs. He had observed windmills in the vicinity installed by private players working successfully. He wondered if private players could do it, then why not village Panchayat, which is the ultimate owner of natural resources in the village.

He soon found out the financial requirements for the wind farm. A 350 MW windmill would cost about ₹20.36 million. With margin money of ₹ 3.5 million and government subsidies, he could get a bank loan of ₹11.35 million. The Gram Sabha (a meeting of all eligible voters in the constituency) gave an overall approval for the project. Shanmugam approached the district government officials, who are under the state government, to sanction the project. District officials declined the proposal on the grounds that a Panchayat could not take a loan. The Panchayat went to the court and finally quashed the order and asked the state government to facilitate loan. The Panchayat obtained a soft loan from a public sector bank under a government scheme and installed the windmill. The windmill produces about 0.9 million units of electricity and as the village requires 0.45 million units only, the remainder is sold to the state government utility company. This generates an annual income, which is currently used to pay the loan and in couple of years when the loan is fully paid, the income would be surplus.

What are the factors for success? First, it clearly stands out that a proactive and visionary leadership is responsible, backed by unanimous support of the people. Second, the legal mechanism and the Constitutional provisions that enable the Panchayat government to borrow loan from a bank.

Case Study 2: Solar Energy Projects— Private-Panchayat Partnership

Minda NexGenTech Ltd is a private enterprise that has been established with the mission to provide electricity to off-grid villages through renewable sources of energy. It works directly with Panchayat leaders and in some cases through collaboration with local NGOs. It brings expertise and funds. Until now, it has executed dozens of projects all across India but primarily in Madhya Pradesh. In one of the case studies documented is a village called Indira Nagar in Rajasthan. Prior to the introduction, this village was using kerosene lamps. Minda implemented a 240 W solar power project with the objective to provide basic lighting to the village. After the commissioning of the project, households get basic lighting and each pays ₹150 per month for usage. There have been outcomes such as children are able to study after sunset and women are engaged in entrepreneurial activities such as stitching and grinding pulses.

There are other case studies where villages have collectively come together under the leadership of the Panchayat to solve their common problems such as all-weather road and transportation. For instance, Hander village in Reasi district in Jammu and Kashmir has laid 14 km of road at a project cost of ₹20 million, total contribution from the villagers themselves.

These case studies are a microcosm and may be considered exceptions as there are more than 600,000 villages in India. It is certainly possible to learn and replicate from these initiatives. The advantages are that the initiatives are from the bottom and can be taken parallel across villages, districts, and states.

Strategies for Policy-makers and Multilateral Institutions

As per the Ministry of Panchayati Raj, Government of India, there are 237,539 village panchayats in India. If we assume that each village will require basic infrastructure such as good roads, storage facilities, and decentralized power generations, an approximate guesstimate of ₹2.5 million would be required in the next five years per village. That would mean an investment of ₹600 billion. What are the ways in which policy-makers, multilateral institutions, financial institutions, NGOs, and private sector participate in this and accelerate the investment and execution of infrastructure projects?

There have been pilots and efforts going on to infuse administrative and management capabilities of large municipal corporations to avail of finance from the market. A few government undertakings such as the Tamil Nadu Urban Infrastructure Financial Services Limited have been established to facilitate this. They work with large municipalities to help develop vision documents, forecast plans for the next decade, and facilitate financing. It would be indeed very difficult to bring together villages under this. But, it has been suggested to work with cluster of villages and club together projects of small and medium sizes. Municipal financing is a large area and is out of scope for this chapter. However, policy-makers can look into avenues and policies to facilitate easier financing options and to make it reasonable for investors to invest in these ventures. The government may also consider instituting a fund, similar to the ones available for Micro, Small, and Medium Enterprises (MSMEs) and other innovation development funds. This fund could be an investor in other regional funds or it could be a corpus similar to Credit Guarantee Fund established for MSMEs.

Multilateral institutions and financial institutions have been engaged in promoting social enterprises. This type of investment into for-profit enterprises with the objective to improve the lives of disadvantaged sections of the society is called impact investing. Institutions can add local governments to their portfolio and engage them as they would engage social enterprises. A new class of vehicle, similar to social venture funds, could be promoted to identify regional priorities and work closely with local governments. Terms such as duration, returns, and structure of investment could be worked out to enable working with Panchayats.

There are many NGOs working with local leaders in the areas of capacity building, implementation of projects through grants, etc. One of the pain points in working with the local rural bodies is the absence of well-documented financial statements. One of the priority areas could be in the area of computerization of financial statements of rural bodies. NGOs can work with Panchayats in the areas of awareness creation, project identification, and proposal preparation, identify right technologies, and help develop market-based approaches for charging fees. Private sector can be engaged in a partnership mode to execute projects.

Conclusions and Recommendations

As we have seen, one dimension of sustainability is preserving the economic equity of a society. Infrastructure is the instrument through which equity is preserved. Infrastructure development through the hands of local government puts the responsibility directly into the hands of the people. Let us take a look at the outputs first and finally predict and visualize outcomes for a prosperous India.

At the hard infrastructure level, we foresee self-sufficiency in attaining the basic conditions of life. Energy needs are met through sufficient power generated from renewable sources of energy. All-weather roads will enable transportation to the nearby towns and markets for commodity exchanges.

The last two decades have seen huge influx of entrepreneurs in the social and microfinance space to achieve a triple bottom-line effect. This is also by a new kind of political entrepreneurs, changing the political map of India and thereby changing the way politics is conducted in India. Inclusive growth will be achieved with the participation of most of the citizens, particularly at the bottom. The chapter proposes three changes for the policy-makers to effectively engage local elected representatives: first, local governments and Panchayat leaders are to be seen as agents of change and they should bring vision to achieve a self-sufficient economy; second, Panchayat leaders may explore other finance options as described in this chapter for the planning, execution, and maintenance of projects; and third, financial investors and governments should enable an environment to engage directly with local governments for overall development of the society.

The changes mentioned above integrate bottom-up approaches with top-down approaches. The case studies show that entrepreneurship or market-based approaches by RLBs are effective instruments to provide sustainable development.

Appendix 9.1: List of 29 Subjects Under Panchayats as per Eleventh Schedule (Article 243G)

1. Agriculture including agricultural extension.
2. Land improvement, implementation of land reforms, land consolidation, and soil conservation.
3. Minor irrigation, water management, and watershed development.
4. Animal husbandry, dairying, and poultry.
5. Fisheries.
6. Social forestry and farm forestry.
7. Minor forest produce.
8. Small-scale industries including food-processing industries.
9. Khadi, village, and cottage industries.
10. Rural housing.
11. Drinking water.
12. Fuel and fodder.
13. Roads, culverts, bridges, ferries, waterways, and other means of communication.
14. Rural electrification including distribution of electricity.
15. Nonconventional energy sources.
16. Poverty alleviation programs.

17. Education including primary and secondary schools.
18. Technical training and vocational education.
19. Audit and nonformal education.
20. Libraries.
21. Cultural activities.
22. Markets and fairs.
23. Health and sanitation including hospitals, primary health centers, and dispensaries.
24. Family welfare.
25. Women and child development.
26. Social welfare including welfare of the handicapped and mentally retarded.
27. Welfare of the weaker sections and in particular of the SCs and STs.
28. Public distribution system.
29. Maintenance of community assets.

Source: www.planningcommission.nic.in/aboutus/taskforce/tsk_pri.pdf (Accessed on February 8, 2014).

Appendix 9.2: Status of Delegation of Functions to Panchayats by All the States

Functions Dimensional Index		Finances Dimensional Index		Functionaries Dimensional Index	
States		**States**		**States**	
Karnataka	58.0	Maharashtra	55.5	Maharashtra	75.4
Maharashtra	56.3	Karnataka	50.0	Kerala	68.6
Uttarakhand	53.9	Kerala	48.5	Karnataka	63.1
Rajasthan	53.0	Tamil Nadu	46.3	Tripura	53.3
Kerala	52.9	Haryana	36.9	Gujarat	53.2
Madhya Pradesh	52.6	Rajasthan	35.6	Haryana	50.2
Tamil Nadu	52.3	West Bengal	35.4	Goa	48.2

Functions Dimensional Index		Finances Dimensional Index		Functionaries Dimensional Index	
Odisha	51.5	Odisha	35.1	Rajasthan	40.9
West Bengal	50.6	Himachal Pradesh	34.9	Madhya Pradesh	39.5
Tripura	46.0	Madhya Pradesh	34.4	Tamil Nadu	39.2
Sikkim	45.1	Chhattisgarh	31.8	West Bengal	37.7
Assam	42.8	Sikkim	31.4	Himachal Pradesh	35.4
Uttar Pradesh	41.0	Tripura	28.4	Chhattisgarh	33.7
Bihar	39.4	Jammu and Kashmir	28.0	Uttarakhand	32.0
Gujarat	38.9	Uttarakhand	27.2	Sikkim	29.3
Chhattisgarh	37.5	Gujarat	26.6	Uttar Pradesh	28.6
Haryana	31.1	Uttar Pradesh	26.2	Odisha	28.6
Punjab	24.3	Arunachal Pradesh	25.2	Bihar	24.3
Himachal Pradesh	22.4	Manipur	24.0	Jammu and Kashmir	24.0
Jharkhand	19.0	Assam	23.1	Punjab	23.6
Goa	17.8	Bihar	19.4	Jharkhand	23.5
Arunachal Pradesh	17.2	Goa	18.7	Assam	21.7
Jammu and Kashmir	15.3	Punjab	17.4	Manipur	20.4
Manipur	12.2	Jharkhand	14.0	Arunachal Pradesh	10.1
Union Territories		**Union Territories**		**Union Territories**	
Lakshadweep	20.8	Chandigarh	25.9	Lakshadweep	39.8
Chandigarh	7.2	Daman and Diu	8.0	Dadra and Nagar Haveli	39.2
Daman and Diu	3.4	Lakshadweep	7.3	Daman and Diu	33.6
Dadra and Nagar Haveli	1.1	Dadra and Nagar Haveli	0.8	Chandigarh	18.8
National Average	**34.1**	**National Average**	**29.5**	**National Average**	**37.0**

Source: Alok (2013).

References

Alok, V.N. 2013. *Strengthening of Panchayats in India: Comparing Devolution across States—Empirical Assessment, 2012–13.* Indian Institute of Public Administration and Ministry of Panchayati Raj, Available at: http://www.iipa.org.in/upload/Panchayat_devolution_Index_Report_2012-13.pdf (Accessed on February 10, 2014).

Balwant Rai Mehta Committee. 1957. *Balwant Rai Mehta Committee Report.* Available at: www.importantindia.com (Accessed on February 11, 2014).

Government of Tamil Nadu. 1994. *The Tamil Nadu Panchayats Act, 1994,* Chapter IX, Clause 181-C. Available at: www.tnrd.gov.in (Accessed on February 11, 2014).

Lokraj Andolan. n.d. "Odanthurai Panchayat Shows the Way in Power Generation." Available at: http://www.lokrajandolan.org/successstories.html#1 (Accessed on January 20, 2017).

Mohanty, P.K., B.M. Mishra, R. Goyal, and P.D. Jeromi. 2007. *Municipal Finance in India.* Department of Economic Analysis and Policy, Reserve Bank of India. Available at: www.saiindia.gov.in (Accessed on February 10, 2014).

Prabhu, R.K. and U.R. Rao. 1960. *The Mind of Mahatma Gandhi.* Navajivan Trust. Available at: www.mkgandhi.org (Accessed on February 11, 2014).

Prahlad, C.K. 2012. *The Fortune at the Bottom of the Pyramid: Eradicating Poverty through Profits,* 5th edition. New Jersey: Wharton School Publisher.

Raghunandan, T.R. 2007. "Rural Infrastructure, Panchayati Raj, and Governance." In *India Infrastructure Report 2007.* Available at: www.iitk.ac.in/3inetwork/html/reports/IIR2007/iir2007.html (Accessed on February 7, 2014).

Pani, Subhas. 2008. "Building Rural Infrastructure." In *Infrastructure & Governance,* edited by Sameer Kochchar, Deepak B. Phatak, H. Krishnamurthy, and Gursharan Dhanjal. New Delhi: Academic Foundation.

10

Deployment of Solar Home Lighting Systems in Rural India

Kartikeya Singh

Introduction

The Challenge of Powering India

Access to energy is vital for economic development and the alleviation of poverty, which is why it was thought to be linked closely to meeting top-down approaches to addressing development such as the Millennium Development Goals (Urban et al. 2009). In the lead up to 2015, the International Energy Agency (IEA) stated that the "UN Millennium Development Goal of eradicating extreme poverty by 2015 will not be achieved unless substantial progress is made on improving energy access" (IEA 2010). Given that the goal of eradicating poverty was not met, and the world needed a revised post-2015 development agenda, access to energy has found its rightful place in the newly adopted United Nations Sustainable Development Goals (UN 2015). India continues to face an immense challenge of trying to upgrade its existing electricity grid and provide electricity to over 300 million of its citizens (IEA 2015). This population, summed up as the 'base of the pyramid' (BoP), spends less than US$75 a month on goods and services and represents 76 percent of the rural population, or 114 million households (Bairiganjan et al. 2010).

To meet the challenge, in 2006, the Indian government established a 'minimum supply' of 1 kW per household per day "as a necessity by 2012" (MNRE 2006). The policy additionally provided assurance of reliable power supply at reasonable rates and access to electricity for all households by 2009. Unfortunately, this goal was not achieved because of lack of adequate investment in the energy infrastructure coupled with subsidies that have a crippling effect on the State Electricity Boards. The well-meaning pro-poor policies designed to expand energy access resulted in a system of 'untargeted producer and consumer subsidies' that resulted in the power sector's inability to generate enough revenue for continued investment in infrastructure and management expertise. Through successive governments' efforts, the program to help expand energy access in the country through top-down approaches has not yielded the results of universal reliable access. A revised program under the previous government claimed to have managed to meet its target of electrifying 100,000 villages and providing free electricity to 175,000 households below the poverty line by 2012 through continued expansion of its national grid system. Finally, the new government led by Prime Minister Narendra Modi aims to achieve universal access by 2019.

But India's struggle to meet its electricity demand nationally has resulted in inadequate service of electricity to even those villages that have access to the grid. Perhaps this is because such top-down efforts at meeting targets are limited by their desire to achieve scale over political timelines. One of the interesting caveats in the government's electrification program, for example, is that only 10 percent of the households in a village need to be connected to the grid for the entire village to be considered electrified. This glaring case of conflicting political goals and realities of implementation of policy could, theoretically, render the entire country 'electrified', but 200–300 million people will still be without access to grid power.[1] Herein lies the opportunity for decentralized energy solutions for rural India, with solar home lighting systems (SHSs) being perhaps one of the best options for meeting lighting needs, at least until a grid arrives. Furthermore, as scholars debate the role and impact of adaptation in a climate-constrained world, such bottom-up solutions enhance the "adaptive capacity of communities who are facing increasing uncertainties as the climate changes and their environment is transformed" (Ireland and McKinnon 2013).

Indeed, as the climate changes, people will need energy access to adapt to a warming world, and energy for refrigeration, cooling, and communication, among many other services, will need to be provided if people at the very margins are to have a chance for survival. Thus, the focus of this chapter is on a technological innovation that can help bridge the divide between top-down and bottom-up approaches to sustainable development. As the editors of this book have highlighted, bottom-up approaches 'may develop promising social and technological innovations precisely because they are more spontaneous, less planned, and at their small scale far less risk-averse than the large institutions of sustainable development apparatus'. The two

---

[1] Definition of electrified village under the Ministry of Power's Memorandum No. 42/1/2001-D(RE), February 5, 2004.

cases below highlight how the same technology can help meet the desired development goals or fall short of the promise based on the approach used by the distributor.

Solar Home Lighting Systems

India's solar photovoltaic (PV) program was launched in the 1970s and though the country is able to manufacture its own cells and modules, the pace of that growth has been slow (Bhattacharya and Jana 2009). In recent years, however, the pace at which SHSs have been installed in the country has increased: the cumulative installation in 2004 was approximately 52,000, while recent figures indicate that approximately 600,000 systems have been installed. Given that the number of systems installed globally stands at 3.6 million, India's share of all off-grid rural applications of solar PV systems in the world is approximately 17 percent, which reveals the seriousness with which the government views the solution to help meet its rural electricity needs (IRENA 2012). The biggest boost to the solar industry in India has been the launch of India's 'Solar Mission'[2] by the government to have 100 GW of solar energy installed in the country (both on- and off-grid installations) by 2022 as a part of an effort to help provide energy and put India on a low-carbon pathway (Costa 2009).

This is good news for the growing solar industry and provides an opportunity to help scale up solar energy applications across the country. A number of energy service companies have cropped up taking advantage of government subsidies to help deploy their emerging technologies. While this may be good to help meet energy needs and alleviate poverty, the challenge lies in ensuring that all such providers are considering the barriers to successful and long-term implementation of the technology in the field. As with any decentralized energy technology intervention in villages, proper knowledge of local conditions is vital to the success of projects. Furthermore, Rouse (2002) argues that community involvement is "vital for ensuring the design of appropriate, effective and sustainable interventions."

One important factor in the diffusion of these technologies will be innovations in the business models that help diffuse this technology

---

[2] The National Solar Mission first launched in 2010 envisioned adding 20 GW of solar energy to the grid by 2022. In 2014, the government increased the goal to 100 GW of solar energy within the same time frame.

(Tawney et al. 2013). Indeed, the SHS market is "entering a new phase that is being led by entrepreneurs providing solar portable lights" and while the scale is currently small and costs present a barrier, "the technology is improving at a rapid rate and business models are maturing" (Birol 2011).

As far as the technology is concerned, solar PV (IEA-PVPS) serves as an attractive source of electric power for lower-income rural communities to provide basic services, such as lighting and drinking water, and have thus been championed by multilateral/bilateral financing agencies (Chaurey and Kandpal 2010). Chaurey and Kandpal (2010) have categorized the existing body of literature on energy access from various country-driven programs into the following: insights from system design and configurations, policies and programs, economics and financing, institutional and financing models for dissemination, and technical aspects and experiences of the users.

When evaluating this literature, institutional and financing models, as well as general economics and financing, emerge as the main challenges with which to experiment in order to help SHS technology diffuse in the field. These findings support the Tawney et al. (2013) argument that business and finance innovation (including not only the products but also the processes) is required to help address the energy access challenge.

An evaluation of cases from the Indian state of Karnataka reveals that "the viability of SHS market is critically dependent on the role that banks play as intermediaries between consumers and solar firms in rural areas" (Harish et al. 2013). Such studies have important implications for SHS providers who are attempting to establish an appropriate price point for their product and design effective systems for the adoption of their technology. Some limitations of SHS include power generation being dependent on intensity and duration of solar radiation (Kamalapur and Udaykumar 2011; Raman et al. 2012) and large losses attributed to the battery and other components (Kumar et al. 2009). Martinot et al. (2001) adds that the SHS industry as a whole could use market formation policies such as effective equipment standardization and certification procedures to ensure quality of service and affordability. Pode (2013) highlights that "the high upfront cost of SHS and the absence of payment flexibility is deterring the penetration [of the technology] into larger market of lower-income group rural population." Wong (2012) advocates easy access to credit for users as well as a robust complaint system to address some of the maintenance and supply chain failures associated with SHS. In sum, according to the literature, the success of an SHS program depends on sound technical performance, an appropriate financing infrastructure, responsive after-sales support, and the extent to which it can fulfill the needs of the end users of having grid quality energy. These factors suggest the need for both bottom-up and top-down approaches toward designing and implementing policies and businesses for the deployment of SHSs.

This chapter helps shed light on the barriers to effective management of SHS projects in the country through the examination of two unique cases in different parts of the country with divergently different socioeconomic demographics of the populations that have been exposed to SHSs. An analysis of the two cases will help further the understanding on how to successfully deploy and scale up such initiatives that will provide for the economic development of India's rural areas while ensuring environmental sustainability.

Background on Case Sites

Dabkan Village, Alwar District, Rajasthan

Dabkan village is located within the Alwar district of Rajasthan. It is accessible by a dirt road from the nearest rural nodal town, Tehla, 3 km away. Dabkan is 55 km southwest of the city of Alwar and 105 km northeast of Jaipur. In February 2006, each of the 52 homes in Dabkan village was fitted with an SHS by Grameen Surya Bijlee Foundation (GSBF), in cooperation with the Rajasthan Renewable Energy Corporation. Each home lighting system in Dabkan consists of one 10-watt solar PV panel, one 12-volt battery, and two 15- or 22-lights LED bulbs. Equipment and installation services were provided as a donation by GSBF.

Sugatur Village, Kolar District, Karnataka

Sugatur village is located within Kolar subdistrict of Kolar district of the southern state of Karnataka. It is the gram panchayat headquarters for local area villages and thus is easily accessible by road. SELCO Solar Pvt. Ltd, an energy service company (ESCO), was launched in 1995 as a social enterprise. A two-time winner of the prestigious Ashden Award for sustainable energy, SELCO has long had a presence in southern India specializing in providing 'sustainable energy solutions to the under-served households and businesses'. Furthermore, it aims to provide a complete package of product, service and consumer financing through grameen banks, cooperative societies, commercial banks and microfinance institutions.

SELCO's solar lighting systems evaluated in Sugatur village ranged from being four months old to five years old (the majority of them being recent installations). The two-bulb lighting systems were the oldest with the majority of recently installed systems being four-bulb or eight-bulb lighting systems. A few families reported having lighting systems as large as 10- or 18-bulbs. A typical SHS from SELCO has a 35 W PV module and a 90 Ah/12 V battery to power four 7 W DC compact fluorescent lights (CFIs) for about four hours a day. Some systems include a socket for charging of mobile phones or running televisions and fans.

Based on the sampling of homes in both the villages, the following can be deduced of the demographics of the two village areas (see Table 10.1):

Table 10.1:
Demographics of the two villages

Case	Median Household Income (US$/year)	Household Size	Housing Quality[3]	Technological Amenities	Grid Electricity Access
Dabkan	$300[4]	Seven[5]	60% formal 40% informal	Low access[6]	Not connected[7]
Sugatur	$856[8]	Eight[9]	100% formal	High access[10]	100%

Source: Author.

[3] Informal constructions are based on using earthen compounds and thatch, whereas formal constructions are based on stone and mortar. The latter is understandably more expensive.

[4] This does not include the effective income generated by subsistence agriculture production nor the collection of fuelwood and dung for home energy use. In Dabkan, monetary income is primarily derived from milk sales from animal husbandry. Many young males also pursue seasonal work outside of the village, typically in nearby mines, where they earn US$2.20–US$3.60 per day.

[5] Of whom, an average of three are under the age of 12. On average, two members are attending school.

[6] In Dabkan, it is estimated that there are a total of two mobile phones, three radios, and four bicycles.

[7] Because the village is situated within a government-designated forest area, grid electricity is prohibited from reaching the area.

[8] In Sugatur, the primary occupation is agriculture and sericulture (silkworm farming). Many people interviewed were operating small restaurants or provision stores.

[9] Several of the homes visited during this interview had small families of just four people and several did not have children.

[10] Residents of the village had access to various technological amenities, including mobile phones, televisions, fans, and refrigerators.

Research Design and Primary Questions

Between November 2007 and April 2008, field surveys were conducted in Dabkan village and Sugatur village in order to

1. Evaluate the impacts of rural solar lighting, by documenting the key outcomes of these installation projects with relation to education, health, income, and productivity and
2. Identify barriers to adoption and management of rural solar technologies, by studying the dynamics of such challenges in these two villages located in varying geographies of India with Dabkan being a remote tribal village in the northern state of Rajasthan and Sugatur being a semi-urbanized village in the southern state of Karnataka.

Specific hypotheses this study aims to test are as follows:

1. SHSs can generate significant monetary savings for a family.
2. Solar lighting benefits children's education.
3. Solar can provide adequate lighting needs for a family in rural India, but further renewable energy development is required to help families meet their full need for small-scale industrial activity.
4. A complete supply chain of parts available locally is vital to ensuring a solar lighting project's long-term success.

Methodology and Limitations

During November 2007, surveys were conducted in 18 households—or 35 percent of all households in Dabkan—with the assistance of the village schoolteacher. The large majority of respondents were decision-makers within their households, and 89 percent fell within the labor-contributing age range 18–49. Only 28 percent of the respondents were women. Similarly, during April 2008, surveys were conducted in a random sampling of 15 households in Sugatur with the assistance of a translator. The respondents were decision-makers within their households, with one-third of respondents being women. In both cases, aside from basic household data, respondents were asked about (a) primary uses of the lighting systems; (b) changes experienced in household status of education, health, income, and productivity; (c) maintenance procedures and challenges; (d) desired further electrical uses and attitudes toward grid electricity; (e) willingness to pay for such systems and/or improvements to it; and (f) the quantities, costs, and sources of energy consumed before and after the system installations. In Dabkan, additional four households were informally surveyed during a nighttime tour, regarding luminosity and maintenance concerns.

In total, there were 33 observations from the two distinct districts: Alwar in the northern state of Rajasthan and Kolar in the southern state of Karnataka.

Data Limitations

First, as the editors of this book note in the introductory chapter, 'results of bottom-up approaches are difficult to quantify' and can 'seem idiosyncratic and anecdotal'. Understanding that the same applies to the two cases observed in this study allows us to view them as one piece of a broader research endeavor seeking to understand the interplay between bottom-up and top-down approaches to sustainable development. While one major objective of the surveys was to quantify improvements in household status as a result of the solar lighting systems, no data is available from prior to the installations. Hence, there is only one time-interval data series available for analysis, which limits the effectiveness of the subsequent analysis. However, the purpose of this research is to verify whether theories around solar home lighting deployment match field realities and how the two may inform each other. For the purpose of verification, this type of survey should suffice.

Conducting the survey in the southern state of Karnataka, language (Kannada) was the first barrier that could have led to data limitations, as information could be lost in translation. Additionally, respondents were unable to quantify nonfinancial impacts, resulting in only basic numerical data being collected by this study. For example, while we are able to describe the percentage of households that feel education or health has significantly improved, we can hardly quantify the actual improvement in those areas. Repeated surveys of the same households on an annual basis would create the necessary time-series data that could support the argument for or against the ability of the SHSs to improve household indicators of productivity. It is important to note that there is limited raw data, especially annually collected data of impacts of micro-energy systems on rural households around the world at large. With renewed interest in energy access, there will be a growing need to fill this knowledge gap.

Results and Discussion

Energy Profile

This study is unique in that it shows the impact of solar in the backdrop of two vastly differing socioeconomic settings. These settings determine the type of technological amenities the local population have access to and impact the energy profile of the homes in the area. Dabkan did not have grid connectivity as it is close to a national forest area, whereas all the homes visited in Sugatur village had access to electricity from the grid. Furthermore, while most rural parts of India are plagued with a rather erratic electricity supply, respondents noted that because Sugatur is the gram panchayat headquarters, it has better grid connectivity than the surrounding villages.

Households in Dabkan cited two major household energy needs: (a) lighting and (b) cooking. Households in Sugatur reported three major household energy needs: (a) domestic lighting, (b) cooking, and (c) energy for business. In both cases, the use of SHS helped reduce the consumption of either kerosene- or grid-based electricity—both used for lighting purposes.

In Dabkan, kerosene was previously used for lighting in every household. Post-installation, kerosene was still used in 78 percent of the households— primarily for care of livestock at night—though at considerably lower levels. On an average, households previously consumed 6.8 liters of kerosene per month, while post-installation of SHS, they consumed an average of 2 liters per month, representing a decrease of 4.8 liters (see Figure 10.1).

In Sugatur, purchasing SHS has allowed nearly all the respondents to save on their electricity bills. All respondents were asked the difference in their electricity bills pre- and post-installation of the SHSs. A few households were not able to answer due to having only recently purchased the system.[11] One respondent was already paying a very small monthly amount of US$1.30/month under the Karnataka government's 'Bhagya Jyothi' scheme, which provides a minimal amount of electricity at a set low cost for low-income families. The resulting difference in the monthly electricity bills of all the families between the pre- and post-installation is shown in Figure 10.2.

---

[11] They may also not have been visited by a utility agent to collect their dues, an indicator of the low rates of collection efficiency in many parts of rural India.

Figure 10.1:
Changes in kerosene consumption as a result of solar lighting intervention, Dabkan village

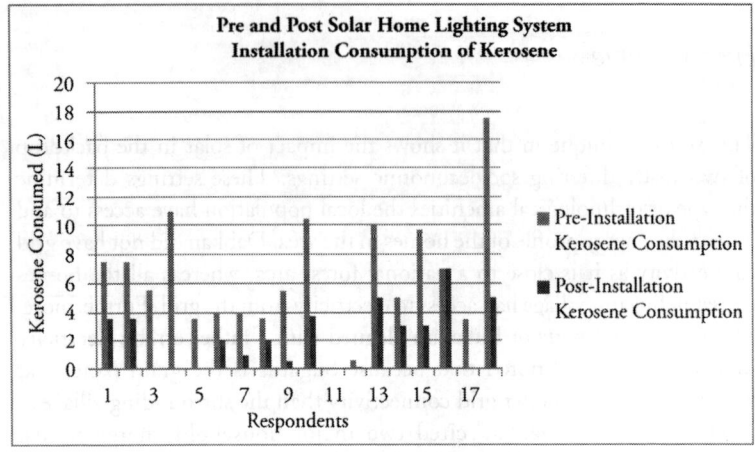

Source: Author.

Only 66 percent of the respondents in Sugatur claimed to prefer solar lighting to lighting provided through grid-based electricity.[12] One respondent stated that the grid supply used to be more erratic but is now more reliable. Still another family said that they would have had to bribe local officials in order to get their house connected to the grid. In order to avoid this situation, they resorted to buying an SHS. The family reported managing to get grid connectivity after purchasing the solar lighting system without any problem. Politicians have often meddled in electrification policies at the state level in India, which has led to uneven progress of electrification across the country. This government behavior violates one of Douglas Barnes' (2007) five principles[13] of electrification, which specifically includes keeping politics separate from electrification projects.

[12] The consumption of candles and kerosene prior to the installation of the solar lighting system was hard to quantify as it was negligible due to the existence of the grid. The presence of the grid allowed people to store electricity in back-up inverters.

[13] Barnes' five principles of electrification include sustained government commitment, effective prioritization and planning (includes establishing a rural electrification body, coordinating with other rural development goals, limiting political interference, and reducing construction and operating costs), sustainable financing (includes developing effective subsidies), and having a customer focus by effective distribution companies.

Figure 10.2:
Changes in monthly electricity bills as a result of solar lighting intervention, Sugatur village

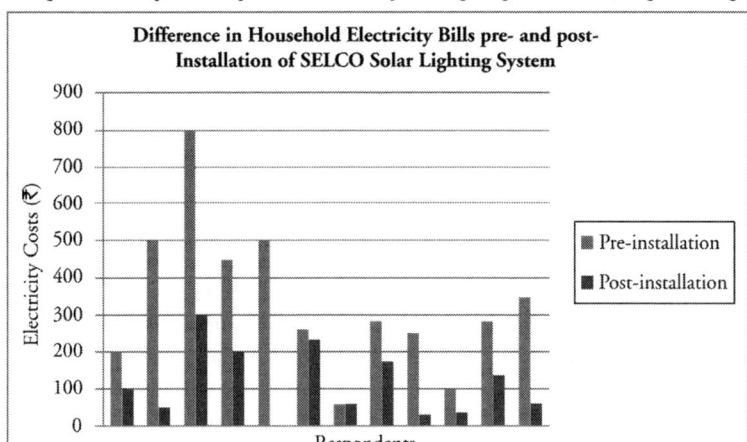

Source: Author.

In Dabkan, the situation was quite different with 79 percent of respondents stating they prefer power from solar lighting systems to access to grid electricity. The most commonly cited comparative disadvantages of grid electricity were (a) inconsistency of power; (b) safety hazards for children from wiring; and (c) the monthly cost—as they had not paid for their existing systems.

Changes in Household Status

All households in both cases reported high satisfaction with their systems and a major improvement in the quality of their lives from the availability of solar light. Observed improvements in the status of households specifically relate to (a) education; (b) health; and (c) income and productivity (see Table 10.2).

Education

By and large, the most significant improvement experienced from the solar lighting systems has been the enhancement of educational opportunities for

Table 10.2:
Changes in household status as a result of solar lighting intervention

Case	Education Improvements[14] (%)	Health Improvements (%)	Income and Productivity Improvements (%)
Dabkan	100	72	100[15] and 100
Sugatur	87	40	60 and 73

Source: Author.

children. All households with school-going children cite a strong increase both in the amount of time children study each night and in the children's educational achievement. In the remote village of Dabkan, post-installation of the SHS, children studied for an average of 1.5 hours each night, from not being able to study at home at all previously. The village schoolteacher verified that the effect on children's education as a result of the lamps was so dramatic that his entire curriculum had become significantly more advanced as a result. Currently, scores of 60 percent on exams earn a passing grade. According to the schoolteacher, if he had administered these same exams previously, then not a single student would have passed. At that time, he said, few students would retain much knowledge from one day to the next. He now estimates that on a daily basis, students retain an average of 70 percent of what was taught the day before, because of the present ability to do homework exercises in the evening.

Similarly, in Sugatur, though it is hard to quantify the level of achievement, approximately 50 percent of the respondents with children cited improvement in their children's grades. One mother said that her daughter was the top student in her Kannada-medium school and was able to transition to an English-medium school.

Health

The difference in the number of respondents who cited an improvement in their health as a result of the technology transition may be due to the fact that kerosene was predominantly used in Dabkan for lighting, whereas those

---

[14] Of households with children.
[15] Largely through savings in Dabkan: an average of ₹50 per month—or 5 percent of household income—from the reduced use of kerosene.

in Sugatur were mostly dependent on candles or the grid for lighting needs prior to having the solar lighting system. Kerosene being a dirtier fuel, its substitution is bound to have a greater health impact for the residents of Dabkan. Respondents cited better breathing capacity and less eye and nose irritation as a result of no longer being exposed to the smoke from kerosene. One interviewee described waking up each morning with black soot around his nostrils; post-installation of the SHS, he claimed, that is no longer the case. Meanwhile, some respondents stated that the lights allow them to avoid snake and scorpion bites, hence eliminating a considerable health challenge.

Income and Productivity

Total 73 percent of the households surveyed in Sugatur and 100 percent of those surveyed in Dabkan reported a change in their daily activity schedule as a result of having access to solar lighting systems. While no one stated that they were able to have more leisure time as a result, several stated that their number of working hours increased. For those practicing sericulture in Sugatur, it meant that longer hours could be spent in the previously dark sericulture houses or that more people could participate in the processing of silk worms. In Sugatur, the access to solar lights has allowed 60 percent of the respondents to increase their daily income. This is true especially of those running small restaurants and provisions stores where on an average solar lighting increased income by US$2.20–US$4.40 daily. Others were able to start businesses including sericulture and a license to manufacture firecrackers.

In Dabkan village, not a single family earns from skilled trades (such as weaving or artistry) or through the operation of small businesses—activities which are typically augmented by the availability of nighttime light. Thus, any perceived improvements in income are largely through the savings accrued from the reduction of kerosene purchases for lighting.

Expanded Uses

One of the interesting results of the surveys was the discovery of how solar home lighting products might be better designed for use by people in villages. In Dabkan, a remote village compared to Sugatur, with few technological

Table 10.3:
Percentage of households citing their desired uses for electricity

Use	Percent
Fan	78
TV	56
Lantern	17
Radio	11
Irrigation Pump	6
Air Cooler	6

Source: Author.

amenities and little hope of ever getting electricity, respondents were asked (a) whether their current systems are sufficient in the brightness and quantity of bulbs and (b) which additional uses of electricity beyond lighting they would most prefer to avail (see Table 10.3).

Total 78 percent of respondents stated that when properly functioning, the luminosity of their SHSs was sufficient. In addition, an equal number of respondents stated that they would like more bulbs (a larger SHS) than their current two-bulb system. The most popular responses to which other uses respondents would like electricity for were overwhelmingly fan and television. This suggests a technological challenge of providing enough energy so that households can meet not only their lighting needs but also enough so that some other technological amenities may be afforded. Three respondents mentioned that the lighting systems would be more useful if they were mobile. The latter suggests that perhaps solar lanterns may be more beneficial or certainly fill a need that fixed lighting systems cannot.

The situation in Sugatur was somewhat different, where 93 percent of respondents said that the luminosity of their systems was sufficient and 100 percent conveyed that they found the SELCO SHS useful; its uses were meant to augment the electricity being supplied through an erratic grid. The most popular use for the lighting system was having access to light in the kitchen while cooking. The second most popular use of the lighting system was to use it for extending work hours for businesses such as provision stores, small restaurants, and sericulture. In some cases, the primary motivation for purchasing the lighting system was so the individual could start one of those businesses. Some unique uses of the solar lighting system include (a) lighting for special functions including weddings; (b) starting a firecracker manufacturing business; and (c) doing household chores at nighttime.

Maintenance Procedures and Challenges

Over the years, different maintenance issues have arisen with solar lighting systems in the field. Careful attention has been given by some companies to ensure that after-sales support and care exist but in the case of Dabkan, an experimental trial of LED-based SHSs, we can see what happens to technology in the field without appropriate maintenance systems in place. In Dabkan, following installation, households must perform regular maintenance of their systems and must also be able to source and pay for replacement equipment.

System Upkeep

In both cases, all households report performing regular upkeep of their systems, which primarily consists of wiping dust from the PV panels regularly—reportedly done an average of once per week in Dabkan—or ensuring that the entire battery is not discharged and using distilled water in the battery to ensure smooth operation as in the case of Sugatur. All households were trained in upkeep procedures at the time of installation, and 83 percent of respondents in Dabkan feel that they have adequate knowledge to perform the regular maintenance tasks needed. Remaining respondents believe that they do not have adequate maintenance knowledge for the reason that the initial training was administered to the head of household only. In Sugatur, 100 percent of the respondents state that they have proper knowledge to operate the SHS. Only one respondent says that system maintenance or operation is difficult. One respondent claims to have enhancing the efficiency of his battery by using distilled water. Finally, 40 percent of the respondents in Sugatur state that the system performance is compromised during cloudy days and especially during the monsoons. In Dabkan, the figure was much higher with 67 percent of respondents stating that cloudy and monsoonal weather conditions decrease the performance of their systems.

Equipment Damage and Degradation

Given that the majority of the surveys conducted in Sugatur were of systems installed within the previous year, there was limited data available on system

Figure 10.3:

Percentage of total households experiencing equipment disrepair

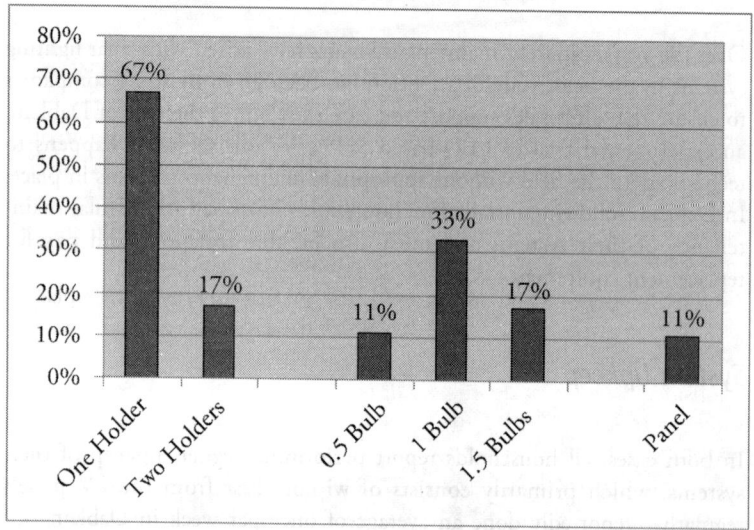

Source: Author.

damage and degradation. In Dabkan, the situation was much different (see Figure 10.3). In 78 percent of surveyed households, one or more pieces of equipment are damaged or broken. In two-thirds of households, bulb holders have cracked. In these households, an average of 1.25 holders are now broken and cannot be used. To compensate, households have used string and cloth to wire the bulbs directly, in a haphazard fashion. In 61 percent of households, some of the LED lights within each bulb have fused and are thus no longer functional. Among these systems, an average of half of one full bulb has fused, thus reducing overall light output. In two households, the PV panels are cracked and thus no longer work. According to these households, the cracks were the result of tampering by monkeys or peacocks. Additionally, in some households, rats have begun to chew through wiring.

Four of the systems in Dabkan were received during a second round of equipment delivery, six months after the initial installation. Recipients claim that this equipment is of higher quality, from their comparisons to other systems; none of it is currently in disrepair.

Servicing and Replacement Opportunities

Among the 78 percent households in Dabkan possessing equipment in disrepair, all are uncertain about how to resolve their issues. According to them, there are no technical specialists or servicing outlets in the nearby area, nor are there relevant retail outlets for purchasing replacement components. The challenge is magnified by the fact that LED light bulbs are a relatively new technology in India, and the fact that they are largely imported makes it difficult to have them available locally in remote corners. All households expressed the desire to pursue repair or replacement.

Additionally, no household conveyed a plan to finance impending replacement needs—such as system batteries, which have a working life of three to five years (same as SELCO's). Without such a plan, it is likely that there will be considerable friction at these maintenance checkpoints, as households struggle to pool funds to afford the considerable sums needed.

According to SELCO, each of its customers is provided with a system that is appropriate for their needs. SELCO provides not only installation but also training on proper use of the system and an after-sales maintenance and support guarantee for up to eight years. The warranty is eight years for solar panels, three years for the battery, and one year for the electronics. SELCO provides one-year free service to every system. That includes two scheduled services—one every six months and one emergency call if required. After the free-service period, customers can sign up for an annual maintenance contract or use the pay per service option as and when they require.

Response time to queries is quick, thanks to the distribution system being designed with the ability of a technician to be able to reach the customers within 5 km from the local field office using a motorbike. Customers can call the technician if they experience any problem with the system that they are not able to fix. All the replacement parts are provided through SELCO's field offices with the help of technicians. CFLs are available in local shops.

Finances

Although it is difficult to ascertain the true 'willingness to pay' for such technologies, questions regarding the capability of people to finance the purchase of such products might indicate whether or not the technologies are considered affordable. Different questions were asked depending on which was contextually appropriate—in the case of Dabkan, respondents were asked to consider the following questions within the context of taking bank loans between US$67 and US$112, should they be available:

1. Given the realized benefits of the present solar lighting system, would you be willing to invest in a similar system or replacement parts?
2. If yes, would you be willing to pay for it using a loan from the bank?

Eighty-one percent of Dabkan households said yes—they would be willing to pay for the system or improvements through a bank loan. In the remaining households, the inability to pay back the loan—because of exceptionally low income—was cited as the reason for unwillingness to pay. In several homes, respondents were asked to give the maximum amount of a loan that they would be willing to take to pay for the system or improvements. Typically, the answer was US$112.

In Sugatur, all respondents said that they would be willing to buy another system as needed, given the experienced benefits. Approximately 73 percent of the respondents said that they would be willing to pay for it using a loan. The remaining did not wish to use a loan and many had paid for the first system using cash. Some people reported that they were about to purchase their second system during the time of the field visit. Some respondents claimed getting a small discount on a third system if they purchased two systems. The cost of an SHS from SELCO currently starts at US$168 and increases based on the need of the customer. Most of its products are customized based on the need and affordability of the customer. The price list keeps on changing but a Manager of Innovations at SELCO states the following on affordability of their systems:

> [A]nybody who can afford to pay $2.20–$3.36 per month for five years can afford a single light system which lasts more than 15 years with one battery replacement. Similarly, somebody who can pay $5.60 per month for five years can afford a 4 light system lasting for a similar time period.

Customers are able to access small credit loans from a variety of banks with which SELCO has formed partnerships. According to SELCO, the interest rates vary depending on the source, but they were between 10 percent and 13 percent at the time the survey was conducted. Customers make a down payment of between 10 percent and 25 percent and pay the remaining balance over a period of three to five years. For those who were able to make an income profit of US$2.20–US$4.40 daily as a result of the new SHSs, the payback period for a four-bulb SHS comes to 1.2–2.5 months. This suggests that for small business owners, solar home lighting is a valuable and considerably easier investment decision.

Points for Further Analysis

Based on the findings of the two cases, the following points need further evaluation to give a complete picture of rural SHSs in India:

Pre-evaluation and Customization

Prior to installation, a bottom-up approach such as an effective pre-evaluation process is necessary in order to match the power and maintenance needs of a community with available technology options. A study that interviewed 400 professionals working in the field of energy for rural development revealed that 41 percent of the respondents believed that 'appropriateness of technology' was extremely important as a potential barrier affecting the promotion of solar lighting systems (Chandrasekar and Kandpal 2007). The two unique cases examined here offer insight into how solar home lighting projects might be carried out in rural India. While the GSBF case was a test pilot and SELCO has been lauded as a successful end-use energy provider model for rural India, looking at the time scale and the choice in technology can help us evaluate how rural energy planning should proceed. Hiremath et al. (2010) have suggested that more attention needs to be given to the factors that determine the "implementation, acceptance and spread" of technological improvements. They draw upon previous studies that champion a relatively recent concept of decentralized energy planning that can help meet rural India's energy needs in a reliable, affordable, and environmentally sustainable way (Ravindranath and Hall 1995; Reddy and Subramanian 1980).

The solar lighting systems installed in Dabkan may not have been the most appropriate renewable energy option for its needs, nor even the best use of solar technology in that context. While integrated energy planning can be beneficial to help a community meet its needs, the most common misconception is that it focuses on how different types of technologies can be integrated into a community rather than how technologies can be integrated into an existing need in a community to help meet and advance its economic development (Sinha et al. 1994). In a classic example of the limits of a top-down approach to development, Dabkan's residents were not prepared to handle impending maintenance needs, nor was there infrastructure in place to facilitate repair and replacement.

Customized Financing

GSBF made a technology intervention in an extremely below poverty line (BPL) village. Unfortunately, its products were given away free of charge as these customers were not prepared for purchase. Rao et al. (2009) state that "a lack of energy-finance options is hampering the 'quality of life' of the BPL community." This affects between 27.1 percent and 23.6 percent of India's population in both urban and rural areas. The researchers argue that the lack of appropriate finance locks this segment of population into the poverty cycle.

SELCO is very clear about its target customer segment, the 'underserved households'. Since the initial investment cost on solar is high, SELCO has worked closely with various financial and microfinance institutions to bring appropriate financing for solar end users. SELCO claims that it is the first company that used bank financing for solar lights in India and because of its efforts, a solar loan portfolio began to be offered to end users. Now most banks in India finance solar just like any other product. The rate of interest on SELCO's products varies between 10 percent and 13 percent and the loan term between three and five years. In some cases, SELCO has used some of its profits to facilitate the process of financing for its end users, as many of them cannot afford to pay the cost of financing, though they can afford the equated monthly installments (EMIs). Additionally, SELCO has partnerships with many banks offering a subsidy on interest rates for solar loans. Some of these banks facilitate margin money requirements and some provide a standing guarantee for loans.

But the efforts of firms such as SELCO in bridging the gap on financing for the rural poor cannot be sustainable. As long as kerosene continues to be subsidized by the government, solar lighting will continue to face a disadvantage (Blake 2009). Thus, perhaps policy interventions (in the form of subsidy swapping) from the government can be used to incentivize[16] the

---

[16] The Direct Benefit Transfer scheme is slowly being rolled out by the central government, which provides a direct cash transfer to customers to use for things like energy. Evaluation of this new program will be essential to understanding its impacts on the solar home lighting market.

purchase of SHSs instead of subsidizing kerosene or providing schemes for low-income families such as Bhagya Jyothi that provide a minimal and erratic electricity supply from the grid. Furthermore, advancing financial inclusion, as the central government has begun to do, could unlock the ability of millions of households to purchase SHSs and take the burden of expanding financial inclusion off of the firms who should focus on creating and supporting quality solar products. This is an area where smart top-down policies can work to support bottom-up solutions to sustainable development.

Maintenance Infrastructure

A community must have nearby means of soliciting replacement and repair services. The opinion survey-based assessment conducted by Chandrasekar and Kandpal (2007) reveals that 50 percent of the respondents believed that the "availability of after sales and services" was extremely important to the success of solar lighting system interventions in rural India. In Dabkan, households claimed that there were no nearby options to repair broken equipment, or to purchase replacements. Despite the low maintenance needs of solar lighting systems, this study clearly indicates that maintenance availability is a crucial component of their success rate. Ideally, a top-down approach would support the creation of a network of servicing centers within a reasonable distance of any village with SHS installations. These facilities would be supported with trained technicians who could perform home visits when needed for maintenance and repairs. The center's primary functions could be (a) to facilitate the delivery of replacement parts from suppliers, when needed and (b) to perform installation work according to uniform standards, to avoid the accidental damage that a user may cause while tampering with her or his system. The network of 6,000 government-run industrial training institutions across the country could serve as the backbone of such a network to support these bottom-up interventions to address sustainable development.

The Sugatur case makes a strong argument for these top-down and bottom-up approaches. Given that none of the houses in Sugatur reported having any damages to their SHSs during the time of the field visit and many had full and effective knowledge on how to properly maintain and operate their systems, it is clear that having a strong maintenance and after-sales support service is crucial for customer satisfaction. This satisfaction results in the desire for people to purchase more systems as made evident by the results of this case. Having field technicians available for quick response to deal with customer needs and provide replacement parts ensures that customers are indeed energy independent.

Conclusion

If a country like India is to meet its targets of providing universal access to reliable electricity and doing so with low-carbon technologies, then SHSs have a role to play. As stated before, the government has outlined its top-down approach of targets and timelines under which it hopes to deploy these technologies across diverse geographies. In order to genuinely achieve these well-intentioned targets, supporting top-down policies that are not directly related to the development objective, in this case expanding energy access, is critical. For example, government policies to advance financial inclusion, removing subsidy distortions, creating uniform specifications and standards for technologies, removing trade barriers to help companies gain access to new innovations, and increasing support for technical training of entrepreneurs would create an ecosystem in which SHS technologies can diffuse rapidly. But as this study demonstrates, the technologies will only scale if the firms responsible for their diffusion undertake bottom-up approaches to understanding their market of customers and providing them with the kind of support they need to adopt these technologies. This interplay sets the stage for how sustainable development must be undertaken in a climate-constrained world if its objectives are to succeed.

References

Bairiganjan, S., Ray Cheung, Ella Aglipay Delio, David Fuente, Saurabh Lall, and Santosh Singh. 2010. *Power to the People: Investing in Clean Energy for the Base of the Pyramid in India*, 74. Washington, DC: World Resources Institute and IFMR.

Barnes, Douglas. F. 2007. *Meeting the Challenge of Rural Electrification: The Challenge of Rural Electrification*. Washington, DC: World Bank.

Bhattacharya, S.C. and Chinmoy Jana. 2009. "Renewable Energy in India: Historical Developments and Prospects." *Energy* 34 (8): 981–991.

Birol, F. 2011. *Energy for All: Financing Access for the Poor*. Paris: International Energy Agency.

Blake, L. 2009. "Solar Lamps Face Subsidy Shadow in Rural India." *The Wall Street Journal*. Available online at https://www.wsj.com/articles/SB125991486832876383 (Accessed on January 11, 2017).

Chandrasekar, B. and Tara C. Kandpal. 2007. "An Opinion Survey-based Assessment of Renewable Energy Technology Development in India." *Renewable and Sustainable Energy Reviews* 11 (4): 688–701.

Chaurey, A. and Tara Chandra Kandpal. 2010. "Assessment and Evaluation of PV-based Decentralized Rural Electrification." *Renewable and Sustainable Energy Reviews* 14 (8): 2266–2278.

Costa, A. da. 2009. "India Launches Solar Mission, Seeks International Support." *Eye on Earth.* Available at http://www.worldwatch.org/node/6325 (Accessed on January 11, 2017).

Harish, S.M., Kaveri K. lychettira, Shuba V. Raghavan, and Milind Kandlikar. 2013. "Adoption of Solar Home Lighting Systems in India: What Might We Learn from Karnataka." *Energy Policy* 62: 697–706.

Hiremath, R.B., Bimlesh Kumar, P. Balachandra, and N.H. Ravindranath. 2010. "Bottom-up Approach for Decentralized Energy Planning: Case Study of Tumkur District in India." *Energy Policy* 38 (2): 862–874.

IEA (International Energy Agency). 2010. "Energy Poverty: How to Make Modern Energy Access Universal?" *World Energy Outlook 2010: Special Early Excerpt for UNGA Assembly on the MDGs.* Paris: IEA.

———. 2015. *World Energy Outlook 2015.* Paris: OECD and IEA.

Ireland, P. and K. McKinnon. 2013. "Strategic Localism for an Uncertain World: A Postdevelopment Approach to Climate Change Adaptation." *Geoforum* 47: 158–166.

IRENA (International Renewable Energy Agency). 2012. *Renewable Energy Jobs and Access,* 1–80. Masdar City: IRENA.

Kamalapur, G.D. and R.Y. Udaykumar. 2011. "Rural Electrification in India and Feasibility of Photovoltaic Solar Home Systems." *International Journal of Electrical Power and Energy Systems* 33 (3): 594–599.

Kumar, A. P. Mohanty, D. Palit, and A. Chaurey 2009. "Approach for Standardization of Off-grid Electrification Projects." *Renewable and Sustainable Energy Reviews* 13 (8): 1946–1956.

Martinot, E., A. Cabraal, and S. Mathur. 2001. "World Bank/GEF Solar Home System Projects: Experiences and Lessons Learned 1993–2000." *Renewable and Sustainable Energy Reviews* 5 (1): 39–57.

MNRE (Ministry of New and Renewable Energy). 2006. *12th Plan Proposals for New and Renewable Energy,* 1–66. M. o. N. R. Energy. New Delhi: MNRE, Planning Commission, Government of India, 1–66.

Pode, R. 2013. "Financing LED Solar Home Systems in Developing Countries." *Renewable and Sustainable Energy Reviews* 25: 596–629.

Raman, P., J. Murali, D. Sakthivadivel, and V.S. Vigneswaran. 2012. "Opportunities and Challenges in Setting Up Solar Photovoltaic-based Micro Grids for Electrification in Rural Areas of India." *Renewable and Sustainable Energy Reviews* 16 (5): 3320–3325.

Rao, P.S.C., J.B. Miller, Y.D. Wang, and J.B. Byrne. 2009. "Energy–Microfinance Intervention for Below Poverty Line Households in India." *Energy Policy* 37 (5): 1694–1712.

Ravindranath, N.H. and D.O. Hall. 1995. *Biomass, Energy and Environment: A Developing Country Perspective from India.* Oxford: Oxford University Press.

Reddy, A.K.N. and D.K. Subramanian. 1980. "The Design of Rural Energy Centers." *Indian Academy of Sciences* 2 (3): 109–130.

Rouse, J. 2002. "Community Participation in Household Energy Programmes: A Case-study from India." *Energy for Sustainable Development* 6 (2): 28–36.

Sinha, C.S., Ramana P. Venkata, and Veena Joshi. 1994. "Rural Energy Planning in India: Designing Effective Intervention Strategies." *Energy Policy* 22 (5): 403–414.

Tawney, L., Mackay Miller, and Morgan Bazilian. 2013. "Innovation for Sustainable Energy from a Pro-poor Perspective." *Climate Policy* 15 (1): 1–17.

UN (United Nations). 2015. *Transforming Our World: The 2030 Agenda for Sustainable Development*, 1–29. New York, NY: UN.

Urban, F., Rene M.J. Benders, and Henri C. Moll. 2009. "Energy for Rural India." *Applied Energy* 86: S47–S57.

Wong, S. 2012. "Overcoming Obstacles Against Effective Solar Lighting Interventions in South Asia." *Energy Policy* 40: 110–120.

11

From Participation to Empowerment: Community-based Ecotourism in Goa

Rohini Fadte

Introduction

One of the defining features of true ecotourism is the participation and involvement of the local communities and people residing at and in close proximity to a site. An essential criterion of sustainability and development in any 'new' tourism scheme is community participation, which is envisaged in the principles of sustainable development. Participation, once considered a peripheral activity, forms a part of the core of the work of several national and international nongovernmental organizations (NGOs) and multilateral and bilateral agencies. Rightly so, the 1990s has been referred to as the decade of participatory development.

Through the evolution and development of Local Agenda 21 (LA21)—an action plan for achieving sustainability based on the involvement of local communities using a bottom-up approach—participation of local communities has become an integral part of the mechanism of development. Reflections of participatory development can be seen in Agenda 21 that emerged from the 1992 Earth Summit at Rio de Janeiro in a document entitled 'Agenda 21 for the Travel and Tourism Industry: Towards Environmentally Sustainable Development' (WTTC 1995). Agenda 21 incorporates the philosophy of community empowerment and proactive 'grassroots' development. It also systematically outlines the formal structures of planning, legislation, and governance in which participation ought to take place. Agenda 21 has been referred to as the 'sustainable development bible' (Doyle 1998) and has indeed been successful to some extent in bridging the gap between green ideology and an environmental policy that is politically feasible.

Tourism is identified as a resource-intensive industry; therefore, it needs to incorporate elements of sustainability at both local and global levels. The growing debate on the prospects and challenges of the tourism industry has led to the search for better approaches to tourism that are environmentally and socially responsible. Since the emergence of the concept of sustainable development, attempts have been made to link it with virtually every aspect of development. Tourism is no exception, hence the term 'sustainable tourism' (De Kadt 1990; Hunter 1997). The World Tourism Organization defines this sustainable form as one that improves the quality of life of host communities, provides a high quality experience for the guest, and maintains the quality of environment on which they both depend (WTO 1993). This goal of sustainable tourism development in destination areas is sought through the promotion of economic developments that conserve local natural, cultural, and built resources (Hunter and Green 1995). The three basic principles of sustainable development are its emphasis on ecological, social, and economic issues. Although sustainable tourism is a major focus in the debate on environmentally responsible tourism development, existing research shows that sustainability is a complex concept and one that requires more critical and comprehensive analysis. Ecotourism as a subset of sustainable tourism also faces the same theoretical and practical issues.

Participation and empowerment are key factors in sustainable development. Local 'ownership' is absolutely needed to support capacity but depends on widespread participation, which in turn requires empowerment. It is often seen that the local communities are neglected and left out of the planning and decision-making process of tourism activities. Mitchell and Reid (2001: 114) observe, "local people and their communities have become the objects of development but not the subjects." Therefore, the community approach to tourism has been proposed as a way of empowering communities and providing them opportunities to break free from the damaging influences of mass tourism. With economics as its basic motive, the tourism industry has been expanded and blindly promoted by forces outside the communities' control, a case in point being tourism development in Goa.

The case of ecotourism in Goa is used to illustrate that the potential for ecotourism as a tool for sustainable development can be realized only when community development takes place. In this context, the specific objectives of this chapter are: first, to analyze the concepts of participation and empowerment, and second, to examine the challenges and obstacles hindering the processes of empowerment of local communities in Goa and by extension realizing the goals of community-based ecotourism. Ecotourism development is in its nascent stage in the state, and its full impact is yet to be seen. The final objective of the study is to make recommendations that would help the tourism planners and policy-makers to adopt a participatory and integrated management approach toward development of community-based ecotourism in Goa.

This chapter is divided into six sections. The first section discusses the relationship between participation and empowerment; the second section studies essential features of community-based ecotourism; the third section analyzes the issues and concerns relating to ecotourism in Goa; the fourth section chalks out the status of mining in Goa; the fifth section examines community rights; while the last section discusses the essential features of the Western Ghats Ecology Expert Panel (WGEEP) report and follows this discussion with recommendations.

Participation and Empowerment: Interdependence and Interaction

A sustainable approach to tourism development is delineated by two features: first, by its identification and contribution of a full range of stakeholders and second, by the participation of local communities in the planning and decision-making process. Within development practice, there has been a movement toward 'participation' and 'empowerment'. In particular, 'the local' has emerged as the site of empowerment and hence as a center of knowledge generation and development intervention. The shift within the neoliberal development strategy from its sole emphasis on market deregulation to an additional emphasis on institutional reforms and social development (World Bank 1997) has brought the civil society from the periphery to the center of development activities.

Community participation, a Western paradigm in natural resource management and utilization, is currently a much-discussed issue relating to research in sustainable tourism development. Somarriba-Chang and Gunnarsdotter (2012) observe that community participation not only depends on the management system but also on governmental support to infrastructure and local entrepreneurship. Nault and Stapleton (2011) insist that the viability of community-based ecotourism in the long term requires close collaboration and consistent support from trusted community leaders and knowledgeable and committed stakeholders. Reed (1997) emphasizes the relevance of power relations to tourism settings, while Scheyvens (1999) distinguishes between four types of empowerment, namely economic, psychological, social, and political empowerment in the context of ecotourism development. Power relations are a part and parcel of emergent tourism settings, and hence, the need for participation is critical. Lai and Nepal (2006) suggest that while local residents may support ecotourism development based on international guidelines, their level of participation will be determined by local environmental, social, and politico-economic conditions. With a central focus on 'local', Palmer (2006) links natural resources and cultural environments, and reiterates the role of ecotourism as a means of promoting and conserving diversity in globalization debates.

Participation is not a new concept. Unlike the early development era, which was dominated by a global, top-down perspective, there has been a clear shift of focus to more locally sensitive approaches. As Henkel and Stirrat (2001: 168) argue, "It is now difficult to find a development project that does not ... claim to adopt a 'participatory' approach involving 'bottom-up' planning, acknowledging the importance of 'indigenous' knowledge and claiming to 'empower' local people." Arnstein (1969) has emphasized that citizen participation has to be accompanied by questioning of the power dynamics. She introduced a 'ladder of citizen participation' to explain the necessary steps, categorized into three levels of gradual evolution: 'non-participation', 'degrees of tokenism', and aiming toward 'citizen power', wherein local residents are given full control and power for policy and management. Haywood (1988) and Reid (2003) note that the ladder provides an insight into the situation at the tourist destination, as well as the current state of local involvement in tourism development.

Empowerment is a new name for development with a human face. It is a movement from power of the elite to the power of the people. Parpart et al. (2002) consider empowerment as a means and an end, while Rowlands (1997: 14) emphasizes the "processes that lead people to perceive themselves as able and entitled to make decisions." Community empowerment is a gradual process of achieving complete power, up to the top end of Arnstein's ladder. In applying this concept to tourism, such empowerment would mean that communities at the tourist destination rather than the agencies from the 'outside' and 'above', such as governments or the multinational corporations, have the authority and resources to make decisions, take action, and control tourism development (Timothy 2007). Thus, the empowerment of communities affected by tourism development will serve the purpose of political and socioeconomic justice (Sofield 2003) and help in fulfilling the objectives of sustainable tourism. Empowerment according to Reid (2003) should focus on enhancing the level of awareness within communities and transformative learning processes so as to enable communities to confront problems themselves (cited in Okazaki 2008).

Participation is often associated with 'empowerment' and sustainability, and hence, the varied direct and indirect impacts arising from it have tended to place it on a pedestal. However, despite its positive characteristics and as a powerful discourse, a critical examination of participation is necessary as participation has the potential for undue abuse of power. Participation has been criticized at various levels. Cooke and Kothari (2001) refer to participation as the 'new tyranny', while Henkel and Stirrat (2001) have commented on the 'evangelical promises of salvation' and argue that there is an element of spiritualism in participatory practices, which involves discussions on travel. Cleaver (1999) argues that participation has become an act of faith in development: something we believe in but rarely question. This faith in participation has its roots in three key tenets: participation is inherently good, especially for the participants; a sound technique will ensure success; and considerations of power structures should be avoided because of their latent divisive and obstructive traits.

Community participation may be negotiated at the household and community levels and at the same time may be shaped by prevailing norms and structures of the society. This raises critical questions about whether and to what extent participation can be empowering to individuals involved. Community participation ranges from passive participation, where people are told what a development project is, proceeding to self-mobilization, characterized by independent initiatives where local people are strengthened socially and economically by their involvement (Pretty 1995). The process of empowerment or decentralization involves participatory democracy and decision-making on issues that affect the communities. However, this is easier said than done. While the term empowerment has different connotations in different social and political contexts, powerlessness is embedded in the very nature of institutional relations. Sofield (2003) highlights the role of the state in establishing an environment (e.g., through legal mechanisms, structures as well as formal plans and policies), which supports communities in their efforts to engage with and benefit from the tourism industry through the exercise of their own power. Thus, traditional empowerment of communities must be transformed into legal empowerment if sustainable tourism is to be achieved. This means that the significance of the state in supporting and sanctioning community empowerment cannot be denied.

What Is Community-based Ecotourism?

A community-based and participatory approach to local development has been considered as the core essence of sustainable development. In discussions on community participation, it is important to recognize the fact that communities are not homogeneous but in fact heterogeneous and are influenced by power relations that are inherent in age, class, caste, ethnicity, religion, and gender (Cooke and Kothari 2001). Community-based tourism is considered a more sustainable form of development than mass tourism as it allows the host communities to break away from the hegemonic grasp at the international level and the oligopoly of wealthy elites at the national level. Community tourism is about grassroots empowerment as it aims to develop the industry in equilibrium with the economic and social needs and aspirations of host communities in a way that is acceptable to them (Fitton 1996).

It is often observed that the ecotourism label is used as a smart marketing ploy, and attempts have been made to 'greenwash' the tourism industry. Some writers have suggested that the term 'community-based ecotourism ventures' should be used to distinguish those endeavors that are not only environmentally sensitive but also aim to ensure that local communities have a high degree of control over the activities taking place, and a significant proportion of the benefits are plowed back to them (Ceballos-Lascurain 1996; Liu 1994), a strain that is also reflected in official documents such as the WWF-International (2001). Sometimes, the tourism ventures are controlled by exogenous forces, while in other cases government agencies are the sole beneficiaries (Akama 1996). Scheyvens (1999) observes that while the slogan for East Africa of 'wildlife pays so wildlife stays' is apt (Ziffer 1989: 2), to date it has mainly 'paid' for governments, foreign tourism enterprises, and local entrepreneurs, rather than the benefits trickling down to the local communities.

A community-based approach to ecotourism emphasizes a two-pronged approach: the need to promote both the quality of life of people and the conservation of resources. It is now recognized in parts of Africa, for example, that local people should be compensated for the loss of livelihood they suffer when wildlife parks are created. For instance, the Narok Country Council, which has jurisdiction over the Masai Mara Park, puts money into a trust fund that is used to fund school, cattle dips, and health services, which benefit the entire community (Sindiga 1995). Thus, community-based ecotourism specifically refers to tourism activities that involve local communities, operate in their lands, and are based on their cultural and natural assets and attractions (Nelson 2004). Community-based ecotourism can also be used as a strategic tool for poverty alleviation. Issac and Kuuder (2012) observe that the evolution of community-based ecotourism in Ghana has triggered enormous interest among people and has set the stage for development of ecotourism as well. It has received much attention from the policy-makers and tourists alike for its positive results it has delivered in terms of visitor numbers and revenue generation. Butcher (2011) argues that ecotourism has the potential to tackle poverty by creating a 'symbiosis' between conservation and development, a situation often associated with sustainable development and attuned to the aspirations of the MDGs.

An underlying feature of community-based ecotourism is the empowerment of people that enables host communities to experience various forms of benefits. Ecotourism generates economic benefits that can be equitably distributed to the communities for their upliftment. Ecotourism can contribute to the psychological empowerment of the local people by enhancing their sense of self-esteem by creating pride for their cultural and natural heritage. Additionally, ecotourism may foster cooperation and coherence among members, thereby strengthening social bonds within the community. Finally, ecotourism brings about political empowerment, since it provides a platform for expression of people's voices relating to issues of local development (Scheyvens 1999). However, it is possible that communities will develop negative attitude toward ecotourism development if they are denied benefits, due to nonparticipation, that accrue from ecotourism development. This might occur, for example, when indigenous people perceive tourism as threatening their livelihood by competing with others over land and resource (Ross and Wall 1999). In such situations, community-based ecotourism is likely to fail or not succeed to a great extent, thus defeating the purpose of sustainability (McCool and Moisey 2001).

It is imperative to involve local communities so as to empower them to tackle basic causes of underdevelopment and enable them to influence decisions that affect their lives. Several studies indicate the importance of incorporating the perceptions, values, and interests of the local people in the region where the ecotourism site is found (Vincent and Thompson 2002). To make the process of participatory development more effective, what is needed is a better understanding of local norms of decision-making and representation, and how people negotiate local power structures and are influenced by them. In this context, it would be pertinent to see how the interaction between participation and empowerment on the one hand and sociopolitical processes on the other plays itself in the context of Goa. The case study of ecotourism in Goa involves a detailed contextual analysis of a limited number of events and their relationships. Therefore, the study does not allow for generalization of findings. Nevertheless, this method is used to examine developments in Goa and is an excellent tool to understand a complex issue as well as to add strength to what is already known through previous research.

Ecotourism in Goa: Issues and Concerns

Of recent, the concept of governance has begun to receive international attention. While the concept of governance is not new, its application to tourism and protected areas is a new development. Governance is defined as the ability to coordinate the aggregation of diverging interests to promote policies, projects, and programs that credibly represent the public interests. Public involvement, transparency of decision-making procedures, conflict resolution, and leadership accountability are some of the issues of governance (Frischtak 1994 cited in Trousdale 1999). The size and scale of the global tourism industry seems to suggest that tourism is a hot political issue. The growth of tourism has to be seen against global monetary and political structures, such as the General Agreement on Tariffs and Trade (GATT). This is so because there is an interrelation between the global structures and tourism in India, of which Goa is an integral part. The promotion of ecotourism by many peripheral and lesser developed communities has often been presented as a rational, conscious response toward management of problems associated with mass tourism. Tourism in the Third World, where Goa finds itself, is understood, without doubt, in these terms.

After relentlessly tackling problems due to mass tourism, Goa has opened its green corridors to the onslaught of tourism, dubbed by policy-makers as 'ecotourism'. In a bid to divert attention from beach tourism and its impacts (see Naronha 1997), Goa, a small state and a former union territory of India, has added ecotourism as a development option. Two wildlife sanctuaries have been selected as ecotourism destinations. Much ecotourism development has taken place with scant regards to the needs of the community. The two ecotourism sites, namely the Bhagwan Mahaveer Wildlife Sanctuary and the Cotigao Wildlife Sanctuary, have varied historical backgrounds.

A key consideration in promoting sustainable tourism through effective governance is the development context. Governance needs to be understood in the context of varied factors such as historical experiences, economic forces, politics of the region, sociocultural influences, and legislation. The present status of ecotourism in Goa therefore needs to be understood in the context of the tribal struggle for inclusion and the ban on mining activity by the Supreme Court. Policy documents such as the Western Ghats Ecology Expert Panel Report and the Forest Rights Act (FRA 2006) would also be examined.

The Western Ghats[1] portion within the state of Goa is recognized as wildlife sanctuaries, national parks, and a bird sanctuary. Two wildlife sanctuaries, namely the Bhagwan Mahaveer Wildlife Sanctuary and National Park and Cotigao Wildlife Sanctuary, have been centers of experiments in ecotourism. While the exact details of initiating ecotourism activities are not known, ecotourism, according to the Department of Wildlife and Ecotourism's official records, generally means activities such as social forestry, joint forest management, afforestation, and the like, and is an extension of the ecodevelopment programs implemented by the state. Because of the historical, geographical, and sociopolitical factors, these places have become sites of contestation of claims and rights of people living in and around the vicinity of the sanctuaries.

The Bhagwan Mahaveer Wildlife Sanctuary and National Park (situated in Mollem village) forms part of the Western Ghats, with Dudhsagar Waterfalls as its primary tourist attraction. The Wildlife (Protection) Act, 1972, came into being in the union territories of Goa and Daman and Diu along with majority of the states and union territories. The Mollem Sanctuary was named after the ardent advocate of nonviolence and 24th *tirthankar*[2] observing Jainism, Bhagwan Mahaveer. The second sanctuary lies to the south of Goa and is known as the Cotigao Wildlife Sanctuary. The name Cotigao has been derived from the words *Khathe* meaning estate and *Gaon* meaning village. Apart from these, the Netravali Wildlife Sanctuary, Bondla Wildlife Sanctuary, and Mhadei Wildlife Sanctuary constitute the 'green corridor' of Goa.

Mining in Goa

Of all industrial activities in Goa, the most destructive one is mining, which has done more damage to the ecosystems than high-profile culprits such as

---

[1] The Western Ghats mountain range spreads across seven states from north to south on eastern side of the state of Goa and is renowned for its anthropological and natural resources. Recently, UNESCO's World Heritage Committee inscribed the Western Ghats of India as a World Heritage Site, bringing international attention and support for its conservation.

[2] In Jainism, a *tirthankar* is a savior and spiritual teacher of the dharma (righteous path). A *tirthankar* is an individual who destroys attachment with all earthly things and relations and frees himself from ignorance.

tourism and chemicals factories. The story of mining industry in Goa is replete with gross violations of the law and abuse of the land.

A good number of mining leases in Goa lie within the wildlife sanctuaries and are privately owned by a few influential families. The Ministry of Environment and Forests (MoEF) has approved opencast mining within 1–3 km of Goa's wildlife sanctuaries. Mining leases were granted 'environment clearance' even when the areas were notified as 'reserve forests" under the Indian Forest Act, 1927. The estimated area of forests affected due to mining in Goa is about 2,000 ha. Since mining is a non-forest activity, approval of the central government is required under the Forest (Conservation) Act, 1980 (Alvares and Saha 2008). The Goa government allowed mining to continue in these leases despite the Supreme Court's orders until they were stopped by the Central Empowered Committee, thus exposing the state–center nexus.

The legal cases related to mining involved central and state government institutions, the mining industry, and the civil society, all battling for diametrically opposite aims. On the stage fighting for Goa's destiny were the state of Goa, the MoEF, the Ministry of Mines (which commissioned the Justice MB Shah Commission of Inquiry on mining), the miners (who are challenging the state and central government orders and the Shah Commission findings at the Bombay High Court), the NGO Goa Foundation (whose writ petition has taken the mining industry to task), the NGO Paryavaran Sangarsh Samiti (which has filed a petition before the Green Tribunal against 100 mining companies for environmental damages), and the Central Empowered Committee (which has been appointed by the Supreme Court to study the findings of the Shah Commission and other documentation). The third and final Shah Commission report dwells on the financial transactions and losses through illegal mining, and it will probably be the final blow to the mining industry. Affidavits and counter-affidavits pile up high, with public interest litigations, writ petitions from all sides creating a formidable task for the Supreme Court. After the mining industry came to a sudden halt as a result of government order in September 2012, followed by a suspension of all environmental clearances by the MoEF, the process of unraveling the monumental damage has just begun (Miranda 2013). It was the lack of respect for nature's life-providing mechanisms, the breakdown of the state administration, widespread corruption, and illegalities in the mining business that led to the ban.

While mining is projected as the backbone of the economy of the state, what are not reflected are the environmental and social impacts of mining. An exploratory study to value some of the impacts of mining in Goa using 1996–97 data, for example, suggested that even if partial accounting of the environmental and social impacts is netted out of the value created by mining activity in terms of value added to the gross state domestic product (GSDP), the 'true income' would be only 15 percent of the reported income (Naronha 2001). More recent papers in response to a National Council of Applied Economic Research (NCAER) report (2010) suggest that the benefit cost ratios no longer favor mining in Goa (Mukhopadhyay and Kadekodi 2011). This proves that mining in Goa has crossed the social and environmental carrying capacity of this small state.

A second blow was delivered to the people of Goa in 2006, when the Government of Goa notified the highly ambitious Regional Plan for Goa 2011 (RPG-2011), which is a statutory land-use plan accompanied by a surface utilization map. The plan indicated a change in land use, public exclusion, destruction of biodiversity hotspots, especially at Mollem, Bondla, and Cotigao, abuse of the Land Acquisitions Act, and reducing the forest cover up to 70 percent, thereby making Goa a concrete jungle.[3] Due to major anomalies discovered by the public and an ensuing well-documented outcry, RPG-2011 was withdrawn with retrospective effect in October 2006. Interestingly, whereas a surface utilization map requires that all zones of impact be shown on it, the mining leases were not disclosed. Mining would not only cover 8.5 percent of Goa's land mass but was also proposed to cut right into the Bhagwan Mahaveer and Netravali Sanctuaries (Alvares and Saha 2008).

Forest Rights Act and the Recognition of Community Rights

The Scheduled Tribes and Other Traditional Forest Dwellers (Recognition of Forest Rights) Act, 2006, commonly referred to as the FRA, 2006,[4] is considered a pioneering legislation that acknowledges the historical injustice meted out to India's forest dwellers, particularly tribals. The legislation proclaims to "recognise and vest forest rights and occupation in forest land in forest dwelling and who have been residing in forests for generations but whose rights could not be recorded" (Samarthan 2011).

The FRA, 2006, recognizes and vests secure community tenure on 'community forest resources', which are defined as common forest land

---

[3] savegoa.com
[4] tribal.nic.in

within the traditional or customary boundaries of the village or seasonal use of landscape in case of pastoral communities, including reserved forests, protected forests, and protected areas such as sanctuaries and national parks to which the community had traditional access. A report by NGO Samarthan (2011) has evaluated the FRA and its implementation by states of Madhya Pradesh and Chhattisgarh. It has emphasized on the need to identify the range of forest assets and resources used by the communities and map claims that could be made for community and individual user rights to these resources. Places of worship, minor forest produce collection, water bodies, quarries, cremation or burial grounds, community hall, and other governmental infrastructure are potential resources that can be claimed. The report has also specified the status of user rights for forest resources claimed under the FRA.

Closely linked to the FRA is the issue of tribal struggles in Goa. The tribal communities—Gauda, Kunbi, Velip, and Dhangars—are the original settlers of Goa. The colonial administration launched census enumeration in Goa way back in 1850 and labeled the Gauda, Kunbi, Velip, and Dhangar communities as primitive tribes. In the post-Liberation Goa, the tribal communities and other traditional forest dwellers were not accorded ownership rights over their ancestral land in the hilly areas of Goa (Maske 2011).

While these tribes inhabited Goa for thousands of years, their political assertion began only in the second half of the twentieth century. The Commissionerate of Scheduled Castes and Tribes had made out a case for the inclusion of Gawadas and Kunbis in the schedule as early as 1965–66 (Munshi 1989). The rapid developmental activities carried out by the Government of India soon after the independence of Goa in 1961 alienated the original settlers from the mainstreaming efforts of the state due to their illiteracy, poverty, and overall backwardness. The idea of according constitutional status to tribes did not gain much momentum during the merger and language controversy (1961–67), which engaged the attention of the political leaders. The socioeconomic agenda was temporarily shelved.

Soon after the independence of Goa, in order to help overcome the disadvantaged situation, the then conscious Gaudas formed an organization called the Gomantak Goud Maratha Samaj. This effort took shape in 1962 with the sole aim of creating cohesion and awareness among the tribes of the region and was followed by several other organizations such as the Gauda Vikas Mandal and the Goans Organiser's Association (GOA). The GOA also tried to assert community rights before the state, demanding the scheduling of all the tribes in the region. In 1980, GAKUVED Federation, the federation of all tribes—the Gaudas, Kunbis, Velips, and Dhangars—was formed. The sole objective of GAKUVED Federation was to fight for inclusion of these four communities into the Schedule List of the Indian Constitution by giving voice and visibility to all the tribes in Goa under one banner. The inclusion of three tribes—the Kunbis, Velips, and Gaudas—in the Scheduled List for the state of Goa was notified by the government in 2003, but the Tribal Sub Plan[5] has not yet been implemented. As a result, the *Mand* or community lands considered as sacred sites (Devaraee) originally belonging to the tribes of Goa were acquired and sold in the name of development for the corporate sectors in mining, industries, and real estate. The mining companies have encroached on the 'commons', which were for generations reserved for religious and agricultural activities in the village (Datta 2014).

The Western Ghats Ecology Expert Panel (WGEEP) Report

The need to protect the environment as a sustainable development strategy finds reflection in the Fourth Five-year Plan of the country, as early as in the 1970s. The Constitution of the WGEEP by the MoEF (2012) of the Government of India is a step in that direction.

The chairperson of the WGEEP, Dr Madhav Gadgil, appointed by the MoEF, observed that the 'blueprint approach' reflects current practices of 'development by imposition' along with 'conservation by imposition'. This results in reckless development, destructive for nature as well as livelihoods: point in case is the mining scam of Goa and thoughtless conservation elsewhere (Gadgil 2013). The WGEEP proposed a 'greenhouse approach' for synthesizing conservation to development, moving away from the 'Develop recklessly—conserve thoughtlessly' pattern to one of 'Develop sustainably—conserve thoughtfully'. This requires the full involvement of local communities in fine-tuning development–conservation practices and keeping in mind locality-specific contexts (Gadgil 2013). Toward this end, the WGEEP advocates a layered, nuanced, and participatory approach. It is now widely accepted that development plans should not be cast in a rigid

---

[5] The Tribal Sub Plan is a program funded by the Government of India, especially meant for the development of tribals in scheduled areas through sectoral allocation made via respective departments, depending on the concentration of tribals in the specific region.

framework, but they ought to be tailored to prevalent locality and time-specific conditions with full participation of local communities, a process that has been termed as *adaptive comanagement*. Such a system would synthesize conservation and development, and not treat them as separate, incompatible objectives (MoEF 2012).

However, to date, there has been a total failure to implement the community forest resources provision of the FRA in Goa.[6] On the contrary, the administration has attempted to evict people from all wildlife sanctuaries—even before implementing the FRA—by declaring the regions as Critical Wildlife Habitats, causing immense hardships to the tribals.[7] The panel has observed a serious deficit in environmental governance all over the Western Ghats tract. However, it was impressed by the levels of environmental awareness and commitment of citizens toward the cause of the environment, and it also recognized their helplessness in the face of their marginalization in the current system of governance. The panel has urged the MoEF to take a number of critical steps to involve citizens and most importantly the proactive and sympathetic implementation of the provisions of the Community Forest Resources of the FRA.

The panel report observes that the revival of the scheme of *Paryavaran Vahinis* (committees of concerned citizens to serve as environmental watchdogs) and establishing a lean bureaucratic apparatus could play a coordinating, facilitative role to ensure that local communities can effectively enforce a desired system of protection and management of the natural resource base. This system would also channel rewards for conservation action to relatively poorer communities, thereby serving the ends of social justice and creating in the long run a situation far more favorable to the maintenance of biodiversity in the region (MoEF 2012). WGEEP has proposed decentralization as a means to help power percolate to the lowest level, namely gram panchayats, taluka panchayats, zilla parishads, and nagar palikas, as was done for RPG-2021.[8]

---

[6] To take a specific case, the Devapon Dongar mine of Caurem village in Quepem taluka of Goa is located on a hill sacred to the Velips, a scheduled tribe group, and a mine has been sanctioned on this hill against serious local opposition and without completing the implementation of the FRA.

[7] www.mandgoa.blogspot.in

[8] RPG2021 involved a compilation of a comprehensive, spatially referenced database on land, water, and other natural resources of the state of Goa; this information was then shared with all gram sabhas and their suggestions as to the desired pattern of land use obtained, consolidated, and used as a foundation for the preparation of the final plan.

Conclusion and Recommendations

The development of Goa as a tourist site has produced serious consequences for the state's people, ecology, and political economy, and has transformed the coastal sites of Goa into a dispensable space. Structural adjustments that have occurred within Goa's economy to facilitate tourist development have been made without the participation of those who will be most affected by the outcomes, for example, the migrants, marginalized, and indigenous populations. Various sociopolitical events in Goa, however sporadic, have shown that assertion of rights by locals have helped shape specific pro-people policies. If the recommendations of the pioneering documents—the WGEEP report and FRA—are implemented in letter and spirit, then it will provide evidence to support community-based ecotourism activities that are truly sustainable, supporting the rights of indigenous people to benefit from their traditional lands and wildlife and thereby providing a robust model for participatory government.

Both the policy documents show reflections of local participation and effective governance—key elements of sustainable development. These also provide opportunities for implementing the bottom-up approach with the support of top-down sustainable development actors. In the case of Goa, direct community support, participation, decentralized and multisectoral planning, management, and governance will provide higher chances for a project's sustainability. It is imperative to link the top-down and bottom-up approaches for capitalizing on the comparative advantages of each approach. Thus, "ecotourism can be a promising strategy, if and only if good institutional capacity exists, especially at the local level" (Keons et al. 2009: 1235).

When tourism is introduced into nature reserves where local inhabitants are excluded from decisions, there is the obvious risk that local people will not benefit. Strong political will and effective community leadership must come together to incorporate the concept of governance into systematic management of the ecotourism sites in Goa, as better governance is crucial to addressing the several challenges of sustainable tourism. Stakeholder involvement may occur in the tourism development process when local communities are considered a key group. Therefore, specific answers must come from the stakeholders vis-à-vis development of ecotourism, as stakeholders have their own interests: latent and manifest capabilities, strategies, and traditions. The extent of stakeholder involvement in community-based ecotourism will impact the outcome of tourism development at a specific site.

At this point, five contextual issues related to the recommendations need attention.

First, it has to be understood that it is not just external influences that affect tourism policy and development but internal factors as well. This includes the conflicts between governmental and nongovernmental interests, and conflicting interests and priorities within governments themselves. The latter are particularly prevalent where responsibilities over tourism resources overlap more than one government development. In Uganda, for example, there were three separate government departments with responsibilities for national parks. A similar administrative problem is reflected in Goa, as the lack of coordination and cooperation between the Department of Wildlife and Ecotourism, Department of Environment and Forests, the Department of Tourism, and the Department of Social Welfare seems obvious.

Second, involvement of local NGOs, especially those focusing on ecological and environmental issues, needs to be considered. The conservation measures that are designed to maintain ecological biodiversity undertaken by the local NGOs need to be in harmony with the priorities and aspirations of local communities attempting to secure their livelihoods.

Third, the local community should be more directly involved in the decision-making process, take greater responsibility for their own community affairs, become more politically active, and demand higher standards of governance from their politicians.

Fourth, it is necessary to recognize and negotiate the sociocultural constraints and kinship bonds that may inhibit strong and full participation of stakeholders.

Fifth, a broader decision context should be considered. It is often seen that most tourism development strategies focus on the limited short-term goals that will benefit an elite minority rather than the long-term interest of the whole community (Brohman 1996). Trousdale (1999) suggests longer planning horizons to incorporate intergenerational equity so as to enable stakeholders to move past short-term political agendas and business obligations that stress immediate economic benefits.

The case of Goa shows that tourism would cease to be destructive and exogenous if elements of equity, social justice, and empowerment are incorporated in its development. While no single institutional model for empowerment exists, certain defining elements are always present. Some of the empowering mechanisms are inclusion and participation of local community and treating them as co-producers with authority and control over decision-making, accountability, and gender equity (cited in Puthenkalam 2004). It is argued that while community participation needs to be seen as a means to an end and not an end in itself, empowerment despite its limitations should be encouraged, thus fulfilling the objectives of sustainable tourism development.

References

Akama, J. 1996. "Western Environmental Values and Nature-based Tourism in Kenya." *Tourism Management* 17 (8): 567–574.

Alvares, C. and R. Saha. 2008. *Goa, Sweet Land of Mine.* Goa: Goa Foundation.

Arnstein, S.R. 1969. "A Ladder of Citizen Participation." *Journal of the American Institute of Planners* 35 (4): 216–224.

Brohman, J. 1996. "New Directions in Tourism for Third World Development." *Annals of Tourism Research* 23 (1): 48–70.

Butcher, J. 2011. "Can Ecotourism Contribute to Tackling Poverty? The Importance of 'Symbiosis'." *Current Issues in Tourism* 14 (3): 295–307.

Ceballos-Lascurain, H. 1996. *Tourism, Ecotourism and Protected Areas.* Gland: IUCN (World Conservation Union).

Cleaver, F. 1999. "Paradoxes of Participation: Questioning Participatory Approaches to Development." *Journal of International Development* 11 (4): 597–612.

Cooke, B. and U. Kothari, eds. 2001. *Participation: The New Tyranny.* London: Zed Books.

Datta, S. 2014. "Insights into Ecological Struggles in Goa: Field Perspectives." Field perspectives and Notes. Available at: www.saded.in (Accessed on February 9, 2014).

De Kadt, E. 1990. *Making the Alternative Sustainable: Lessons from Development for Tourism.* Sussex: Institute of Development Studies, University of Sussex.

Doyle, T. 1998. "Sustainable Development and Agenda 21: A Secular Bible of Global Free Market and Pluralist Democracy." *Third World Quarterly* 19 (4): 771–786.

Fitton, M. 1996. "Does Our Community Want Tourism? Examples from South Wales." In *People and Tourism in Fragile Environments,* edited by M.F. Price, 159–174. Chichester: John Wiley and Sons.

Frischtak, L.L. 1994. "Governance Capacity and Economic Reform in Developing Countries." World Bank Technical Paper No. 254. Washington, DC: World Bank.

Gadgil, M. 2013. "Of Blueprints and Greenhouses." *The Times of India,* May 20, Goa.

Haywood, K.M. 1988. "Responsible and Responsive Tourism Planning in the Community." *Tourism Management* 9 (2): 105–118.

Henkel, H. and R. Stirrat. 2001. "Participation as Spiritual Duty: Empowerment as Secular Subjection." In *Participation: The New Tyranny,* edited by B. Cooke and U. Kothari, 168–184. London: Zed Books.

Hunter, C. 1997. "Sustainable Tourism as an Adaptive Paradigm." *Annals of Tourism Research* 24 (4): 850–867.

Hunter, C. and H. Green. 1995. *Tourism and the Environment: A Sustainable Relationship.* London: Routledge.

Issac, M. and Conrad-J. Wuleka Kuuder. 2012. "Community-Based Ecotourism and Livelihood Enhancement in Sirigu, Ghana." Available at: www.ijhssnet.com/journals/Vol_2_No_18_October_2012/12.pdf (Accessed on February 21, 2017).

Keons, J.F., C. Dieperink, and M. Miranda. 2009. "Ecotourism as a Development Strategy: Experiences from Costa Rica." *Environment, Development and Sustainability* 11 (6): 1225–1237.

Lai, P. and S. Nepal. 2006. "Local Perspectives of Ecotourism Development in Tawushan Nature Reserve, Taiwan." *Tourism Management* 27 (6): 1117–1129.

McCool, S.F. and R.N. Moisey. 2001. "Introduction: Pathways and Pitfalls in the Search of Sustainable Tourism." In *Tourism, Recreation and Sustainability: Linking Culture and the Environment*, edited by S.F. McCool and R.N. Moisey, 1–15. New York: CABI.

Liu, J. 1994. *Pacific Islands Ecotourism: A Public Policy and Planning Guide.* Pacific Business Centre Program, University of Hawaii, Manoa.

Maske, R. 2011, September 1. "Tribal Welfare in Goa: Uncertain Future under Mafia-Minister-Police Raj in Goa." *Atharva* 6 (9). Available at: http://atharvagoa.blogspot.in (Accessed on September 17, 2013).

Ministry of Tribal Affairs. 2006. *Forest Rights Act 2006.* Ministry of Tribal Affairs. Available at: tribal.nic.in (Accessed on March 1, 2017).

Miranda, C. 2013. "After Destruction, It's Time for Healing." *Hindustan Times,* September 27.

Mitchell, R.E. and D.G. Reid. 2001. "Community Integration: Island Tourism in Peru." *Annals of Tourism Research* 28 (1): 113–139.

MoEF (Ministry of Environment and Forests). 2012. *WGEEP: Part I.* MoEF, Government of India. Available at: moef.nic.in (Accessed on January 2, 2014).

Mukhopadhyay, P. and G.K. Kadekodi. 2011. "Missing the Wood for the Ore: Goa's Development Myopia." *Economic and Political Weekly* 46 (46): 61–67.

Munshi, D. 1989. "Goa's Backward Class Neglected." *The Times of India,* January 20.

Naronha, F. 1997. "Fighting the Bane of Tourism." *Economic and Political Weekly* 32 (51): 3253–3256.

Naronha, L. 2001. "Designing Tools to Track Health and Well-being in Mining Regions of India." *Natural Resources Forum* 25 (1): 53–65.

Nault, S. and P. Stapleton. 2011. "The Community Participation Process in Ecotourism Development: A Case of the Community of Sogoog, Bayan-Uligii, Mongolia." *Journal of Sustainable Tourism* 19 (6): 695–712.

Nelson, J. 2004. "The Evolution and Impacts of Community-based Ecotourism in Northern Tanzania." Issue Paper no. 131. Available at: http//:www.sandcounty.net (Accessed on June 15, 2012).

Okazaki, E. 2008. "A Community-based Tourism Model: Its Conception and Use." *Journal of Sustainable Development* 16 (5): 511–529.

Palmer, N. 2006. "Economic Transition and the Struggle for Local Control in Ecotourism Development: The Case of Kyrgzstan." *Journal of Ecotourism* 5 (1–2): 40–61.

Parpart, J.L., S.M. Rai, and K. Staudt. 2002. "Rethinking Em(power)ment, Gender and Development: An Introduction." In *Rethinking Empowerment: Gender and Development in a Global/Local World*, edited by J.L. Parpart, S.M. Rai, and K. Staudt, 3–21. London and New York: Routledge.

Pretty, J. 1995. "The Many Interpretations of Participation." *Focus* 16 (4): 4–5.

Puthenkalam, J. 2004. *Empowerment, Sustainable Human Development Strategy for Poverty Alleviation.* New Delhi: Rawat Publication.

Reed, M. 1997. "Power Relations and Community-based Tourism Planning. *Annals of Tourism Research* 24 (3): 566–591.

Reid, D.G. 2003. *Tourism, Globalization and Development: Responsible Tourism Planning.* London: Pluto Press.

Ross, S. and G. Wall. 1999. "Ecotourism: Towards Congruence Between Theory and Practice." *Tourism Management* 20 (1): 123–132.

Rowlands, J. 1997. *Questioning Empowerment: Working with Women in Honduras.* Oxford: Oxfam Publications.

Samarthan. 2011. "Forest Rights Act in Madhya Pradesh and Chhattisgarh—Challenges and Way Forward." Samarthan—Centre for Development Support. Available at: www.undp. org (Accessed on February 2, 2014); savegoa.com (Accessed on June 30, 2012).

Scheyvens, R. 1999. "Ecotourism and the Empowerment of Local Communities." *Tourism Management* 20 (2): 245–249.

Sindiga, I. 1995. "Wildlife-based Tourism in Kenya: Land Use Conflicts and Government Compensation Policies over Protected Areas." *Journal of Tourism Studies* 6 (2): 45–55.

Sofield, T. 2003. *Empowerment for Sustainable Tourism Development.* Oxford: Pergamon.

Somarriba-Chang, M. and Y. Gunnarsdotter. 2012. "Local Community Participation in Ecotourism and Conservation Issues in Two Nature Reserves in Nicaragua." *Journal of Sustainable Tourism* 20 (8): 1025–1043.

Timothy, D.J. 2007. "Empowerment and Stakeholder Participation in Tourism Destination Communities." In *Tourism, Power and Space*, edited by A. Church and T. Coles, 199–216. London and New York: Routledge.

Trousdale, W.J. 1999. "Governance in Context: Boracay Island, Philippines." *Annals of Tourism Research* 26 (4): 840–867.

Vincent, V. and W. Thompson. 2002. "Assessing Community Support and Sustainability for Ecotourism Development." *Journal of Travel Research* 41 (2): 153–160.

World Bank. 1997. *World Development Report (1997): The State in a Changing World.* Oxford: Oxford University Press.

WTO (World Tourism Organization). 1993. *Sustainable Tourism Development: Guide for Local Planners.* Madrid: WTO. Available at: www.mandgoa.blogspot.com (Accessed on February 21, 2017).

WTTC (World Travel and Tourism Council). 1995. *Agenda 21 for the Travel and Tourism Industry: Towards Environmentally Sustainable Development.* London: WTTC.

WWF-International. 2001. Guidelines for Community-based Ecotourism Development, from http://www.zeitzfoundation.org/userfile/guidelinesforcommunitybasedecotourism development.pdf (Accessed on May 5, 2012).

Ziffer, K. 1989. *Ecotourism: The Uneasy Alliance.* Washington, DC: Conservation International.

SECTION 3

Climate Change Adaptation: A 'Bottom-up' Challenge to 'Top-down' Sustainable Development

12

Downscaling Climate Change: Perceptions and Adaptive Behaviors of Rural Farmers in West Bengal[*]

Farhat Naz, Marie-Charlotte Buisson, and Archisman Mitra

Introduction

Technically, climate change could be defined as a change in the state of the climate that can be identified statistically and that persists for an extended period, typically decades or longer. This can be due to natural or anthropogenic (human) causes (Livelihoods and Forestry Programme 2010). Main human causes are the emission of greenhouse gases such as carbon dioxide and methane along with the changes in land use (Wheeler and von Braun 2013).

Moving from the causes to the consequences, climate change is expected to bring warmer temperatures, changes to rainfall patterns, and increased frequency and severity of extreme weather. At 4°C global warming, the sea level is projected to rise more than 100 cm by the 2090s, and the monsoon to become more variable with a greater frequency of devastating floods and droughts (Balasubramanian and Kumar 2014). The recent World Bank (2013) report focuses on how the effects of climate change on agriculture, water resources, and coastal fisheries are likely to increase, often significantly. Any country's vulnerability to climate change is decided by the presence of appropriate mitigation and adaptation options. Thus, changing weather

* The chapter is based on data collected for the project 'Ecosystems for Life: A Bangladesh–India Initiative' led and funded by the International Union for Conservation of Nature (IUCN). The International Water Management Institute (IWMI) works in this project through a focus on climate change, food security, and water productivity from the West Bengal perspective. An earlier version of this chapter was presented at the 'International Conference on Environment, Technology and Sustainable Development: Promises and Challenges in the 21st Century' held at ABV-Indian Institute of Information Technology and Management, Gwalior, India, in March 2014.

patterns in terms of less predictable seasons, increasing events of erratic rainfall, or prolonged droughts are the most imperative factors threatening the sustainability of agriculture and food security (IPCC 2001). The circumstances of each country in terms of its climatic conditions, socioeconomic setting, and growth prospects will also partly determine the scale of the social, economic, and environmental impacts of climate change (Stern 2007).

Climate change will dramatically alter the natural balance of local and global ecosystems and will infringe on human settlements. Consequently, the most vulnerable groups will face food insecurity, loss of livelihood, and hardships due to environmental degradation and extreme events such as droughts, floods, storms, and cyclones. These extreme events will also lead to displacement and a whole host of potentially devastating economic and social consequences. Various researches have shown that environmental factors display a role in human mobility (Afifi and Jager 2010). Agriculture is also intrinsically vulnerable to climate variability and change. It is expected that climate change would directly influence crop production systems, affect livestock health, and alter the pattern and balance of trade of food and food products (Wheeler and von Braun 2013).

The question that arises is how to have sustainable development with a bottom up-approach in terms of adaptation strategies in contexts of climate change. Various studies have shown that the links between climate change and sustainable development are several and varied (Banuri and Gupta 2000; Cohen et al. 1998; Robinson and Herbert 2001), and they both interact in a circular fashion. The international discourse on climate change and sustainable development represents different cultures; it is necessary to understand that while climate change has been primarily science-driven, sustainable development is more human-behavior-centered.

It has also been widely acknowledged that the top-down approach to climate change interactions in terms of adaptation starts with climate change scenarios and estimates impacts through scenario analysis, based on which possible adaptation practices are identified (Gbetibouo 2009), whereas the bottom-up approach explores the actual adaptation behavior based on the analysis of farmers' decisions in regard to how farmers perceive climatic change and variability and the types of adaptive responses they will adopt (Bryant et al. 2000). Therefore, the bottom-up approach plays a crucial role as it typically involves the local communities and the stakeholders of the systems in identifying climate change stresses, impacts, and adaptive strategies. The local knowledge, which comes from the farmers, reflects the farmers-initiated bottom-up approach, whereas the institution-led top-down approaches could facilitate the flexible and widely accepted adaptation to climate change. Consequently, top-down and bottom-up approaches are not mutually exclusive in nature and could complement each other (Fujisawa et al. 2015). In this larger context, the main purpose of this research is to downscale the climate change concept from the experts and policy level to the affected population and especially to the farmers level. The objective of this chapter is thus to examine the perception local communities have about climate change and the long- and short-term strategies they use in order to cope with climate change. Moreover, the chapter elaborates further that different groups within the rural communities have diverse coping strategies in order to adapt to climate change, which is largely determined by their socioeconomic conditions.

This chapter is divided into six sections. The section 'Climate Change in India' briefly discusses the understanding of climate change and how it will impact the natural balance of the local and global ecosystems. Moreover, it highlights the impacts of climate change on Indian agriculture. The section 'Methodology' elaborates the methodology used for data collection and presents the study area in brief. The section 'Perception About Climate Change in the Study Area' illustrates empirically the perception local communities have about climate change. Then in the section 'Adaptation Strategies', we elucidate the long- and short-term adaptation strategies employed by the local communities. Finally, the chapter concludes by summarizing the main findings and highlighting the policy recommendations.

Climate Change in India

Climate change in India is expected to be accompanied by high average temperature, change in rainfall patterns, increased severity and frequency of floods, droughts and cyclones, and oceanic acidification. The Indian monsoon period from 1871 to 2009 shows a well-defined epochal unevenness with each epoch of approximately three decades (Murali and Afifi 2014). Some of these events are already occurring; for example, according to the data of National Climate Centre in Pune, rainfall has decreased in July and greater rainfall has been recorded in August in key crop growing areas of the country. Himalayan glaciers are retreating at rates of 12–24 m per annum and about 28 percent of the geographical area of India is vulnerable to droughts, 12 percent to floods, and 8 percent to cyclones (Ray and Ray 2012). In India, the mean flow of the Indus may increase by about 65 percent. The Ganges may have a 20 percent increase in run-off by 2040 and a 50 percent increase in run-off by the 2080s. The late spring and summer flows of the Brahmaputra may substantially decline. The gross per capita water availability is projected to decline due to population growth (Balasubramanian and Kumar 2014: 66).

With about 1.24 billion people, India has the greatest population in the tropical–equatorial regions of the world (United Nations 2011), but India is also home to a third of the world's poor—the population that lives on less than one dollar a day—which constitutes over 32.7 percent of India's population (World Bank 2010). Poor households are the most vulnerable to the effects of climate change, having the least recourse from the status quo and minimal physical protection from environmental shifts.

The economic costs estimates of climate change for India are tremendous. This may affect transport, industries, agriculture, natural resources, pastures, fertility of land, and flora and fauna (Ray and Ray 2012). Considering only agriculture, this sector contributes 14 percent of India's gross domestic product (GDP) (United Nations Development Programme 2011) and is reliant on seasonal rains, a fertile and nonsaline coastline, and river-based irrigation–agricultural requirements that could be harmed by rising heat, rising sea levels, or river depletion, respectively. The Indian Agricultural Research Institute study estimates that with every 1°C rise in global temperature, India will lose four to five million tonnes in wheat production (Sharma 2008). Climate change would substantially affect food availability and supply systems by direct and indirect effects on crops, livestock and fisheries, and on their interrelationships.

Some other recent studies (Ray and Ray 2012) indicate a possibility of 10–40 percent loss in crop production by 2080–2100 in India and other South Asian countries due to increase in temperature, rainfall variability, and decreases in irrigation water. Aggarwal et al. (2000) have also shown that in northern India, rice yields during last three decades are showing a declining trend and this is possibly related to increasing temperatures. Nonetheless, the direct impact of climate change on agriculture includes shortage in grain production resulting in less availability of food items, especially to the economically poor people, changes in agricultural inputs such as fertilizers and pesticides, shift in planting dates of agricultural crops, preference for new crop genotypes resistant to changing climate, soil erosions, soil drainage, and lower fertility level.

Methodology

The research methodology for this chapter is based on the analysis from two sources of data. First, qualitative data was collected in a limited number of villages in March 2013. This step gave insights to better understand and contextualize poverty, food security, and water productivity questions in the study area. The qualitative data collection consisted of focus group discussions (FGDs) and key informant interviews (KII). This phase focused on the perception and on the long-term trends, two aspects that cannot be easily understood using quantitative questionnaires. Communities described their perception about climate change. In addition, the constraints and coping strategies related to droughts and floods as well as the adaptation strategies and the changes related to their livelihoods and agriculture were described by the participants.

Second, quantitative survey was conducted in May 2013. The household survey helped us to describe, measure, and compare the findings identified in the first phase. A sample of 600 households has been surveyed in 40 villages for this project in West Bengal, India. One learning from the qualitative phase was that farmers with different status in terms of land and water access have different perception about climate change and implement different adapting strategies. The sampling is therefore a proportional random sampling; the ultimate purpose of the household selection was to sample different categories of farmers. The geographic information system (GIS) location of each surveyed village has been recorded and is used to produce the map below showing the location of all the villages and their classification in terms of groundwater availability.

Study Area: In Brief

The study area is three border districts of West Bengal: Malda, Murshidabad, and Nadia (Figure 12.1). These three districts are localized in the Ganges basin, more specifically in the lower part of the basin for Murshidabad and Nadia. In the three cases, the climate is hot and humid, and these districts usually receive adequate rainfalls from southwest monsoon from June to October.

In our sample study, all the respondents were farmers with different landholding capacities. But then in spite of agriculture being the main activity, for most of the households, agriculture is practiced in small areas of land. Indeed, land fragmentation is important in West Bengal as seen in Table 12.1: from our sample, almost 60 percent of the households hold less than 1.5 acres (around 4 *bighas*).

Figure 12.1:
Study area and groundwater availability categorization

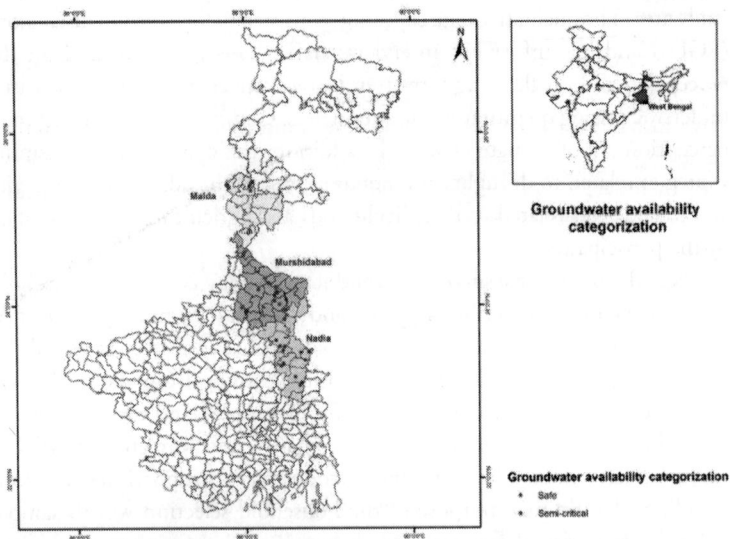

Source: International Water Management Institute (IWMI). Data from the Central Ground
Water Board, Ministry of Water Resources, Government of India, 2009.

Note: This figure is not to scale. It does not represent any authentic national or international
boundaries and is used for illustrative purposes only.

Table 12.1:
Landholding by district

Landholding (Agri-aqua-orchard)	Malda	Murshidabad	Nadia	Total
Marginal farmer (<0.5 acres)	17.58	17.19	8.33	14.64
Small farmer (0.5–1.49 acres)	43.64	44.92	46.11	44.93
Medium farmer (1.5–2.49 acres)	15.76	21.48	21.11	19.8
Large farmer (>2.49 acres)	23.03	16.41	24.44	20.63

Source: IWMI. Quantitative survey under project "Ecosystems for Life: A Bangladesh-India
Initiative," 2013.

Perception About Climate Change in the Study Area

To consider the climate change patterns at the level of our study areas, we have to rely on the perception from the community members on the one hand and on some secondary data on the other hand. Indeed, farmers do not base their decision on technical reports on climate change but on their own perception. This perception can be biased and limited by the memory; for example, recent events will be given higher importance. However, this study on climate change is essential to understand the climate consequences from the micro-level perspective of farmers. Ultimately, the adapting behaviors will be based on these biased perceptions.

We found that the local communities are aware of the phenomena of climate change. Farmers have their own understanding and are able to give definition of climate change.

> Climate change is the change in the pattern of nature's usual occurring in different seasons—something that should happen in a particular season is not happening while something unusual is taking place. We do not know exactly what is meant by climate change. We never heard about the word climate change as such, but we feel that seasons have changed; we feel the change in the weather. We are witnessing such changes taking place over a long time. (Men FGDs, Silinda village)

When considering the consequences of climate change, several phenomena are pointed out: erratic and lack of rainfall, rise in temperature, loss of fertility of land, decline of groundwater table, and rise of pest due to climate change. Figure 12.2 shows the perception of climate change as collected by the household questionnaire. As for the climate, households perceive a clear degradation in terms of temperature (increasing trends) and rains (decreasing trends). As for natural disasters, drought is becoming more frequent. As a consequence, they also note a decline in the groundwater tables. For the agriculture, households consider that the land fertility is degrading and that pest attacks have become more frequent.

The transcripts from the FGDs and KIIs also give some insight into what is occurring for farmers and how they are affected.

First, several of them mentioned a continuous rise in the temperatures in summer. Earlier, people used to work in the field from morning until

Figure 12.2:
Perception about the evolution of climate, disasters, and climate-related events

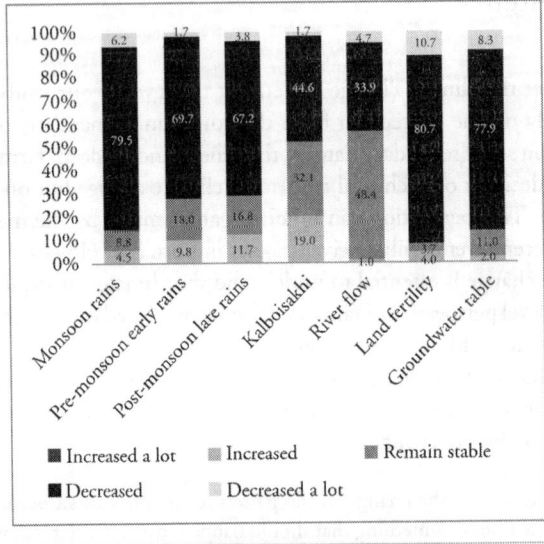

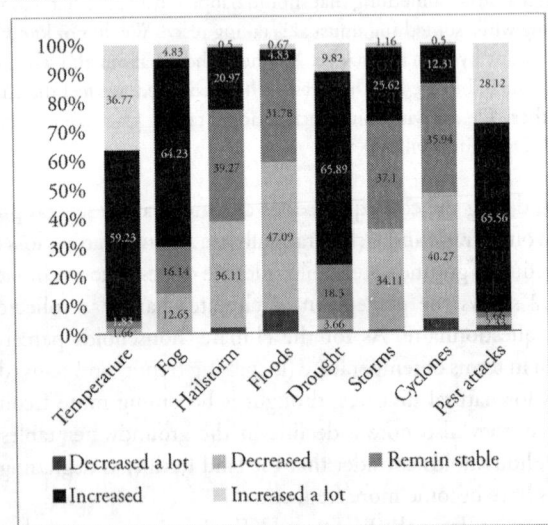

Source: IWMI. Quantitative survey under project "Ecosystems for Life: A Bangladesh-India Initiative," 2013.

1 o' clock in the afternoon, but now they cannot continue to work after 12 o'clock.

> The dry season seems to have increased with monsoon arriving late. Temperature is increasing; even the air is getting dry and hot during summer season. This was not the case 20–25 years ago. (Women FGD, Silinda village)

On the contrary, in the winter season, cold seems to have increased. They consequently consider that the temperatures are becoming more extreme. Farmers also largely discussed about the change in the rains. First, they noticed that the thunderous rains, known as *kalboishakhi*,[1] have completely disappeared during the last few years. The sudden rain in the *Falgun mas* (March) has almost vanished in comparison to 20–25 years ago. Similarly, the continuous rainfall for 5–6 days that marked the monsoon season has become a rare phenomenon. Monsoon has become so irregular that even in some villages, in July, the ponds and canals are not getting filled. Thus, monsoon rains are delayed and reduced over time. In addition, as per their perception, the rain that used to happen in seasons other than monsoon has almost completely stopped.

> We are facing mainly the problem of decrease of rainfall for the last few years. In fact, since the flood in the year 2000, rains and storms both have reduced a lot. And with time these seem to be further reducing. (Women FGD, Gajna village)

In the discussions, depletion of groundwater was also often mentioned by the farmers. Indeed, most of the villages selected for the qualitative data collection are localized in semicritical blocks; so, for these villages, the perceptions of the farmers have been confirmed by the secondary data.

> Water table has gone down 20 feet in the last 10 years. 20–25 years ago the depth of groundwater in the post-monsoon period was just 10–12 feet. Now we get water at a depth of 35 feet in the post-monsoon period. (KII, Poldanda village)

Some declining trends are also felt in villages localized in the safe blocks; nevertheless, the post-monsoon level remains constant, thanks to the monsoon recharge of the aquifer.

---

[1] This summer storm (early rains) brings some relief from the scorching heat in April–June; it also helps watering boro paddy to some extent, reducing the dependence on extensive pumping of groundwater.

Post-monsoon the groundwater table comes up to 15–18 feet below the surface and there is hardly any change in the post-monsoon water level. (KII, Silinda village)

Adaptation Strategies

The continuous climatic and societal changes require human beings to take up different adaptation strategies. This is even truer for the population that our sample survey deals with. Farmers in the Gangetic basin are experiencing many changes in the recent decades. As previously noticed, the climatic changes happening in this region include the decrease in rainfall, increase in temperature, and more uncertainty on the timing of rains. Along with this, the natural resources on which their livelihood depends so crucially are also undergoing major changes. For example, groundwater for irrigation is going down in some areas, soil fertility is decreasing, and sometimes the entire landscape is changing due to a change in the course of rivers. We define adaptation strategies as all the initiatives and measures aiming to reduce vulnerability of natural and human systems to actual or expected climate change effects. Adaptation strategies are intended to mitigate a situation that brings vulnerability in the farmers' livelihoods.

In terms of explaining the different strategies adopted by the farmers, we use the concept of actor-oriented approach to explain different responses to similar structural circumstances, even if the conditions appear relatively homogenous (Long 2001). The approach is useful for our study, as it places actors at the center of the natural resource management discourse, with the recognition that there are diverse actors. People involved vary from those who are present in direct encounters to those who influence the situation from 'behind closed doors', thus affecting the actions and outcomes in a given situation (Long 2001). This was apparent in the selection of different adapting strategies. Thence, the approach establishes why farmers have multiple rationalities, desires, capacities, and practices. The approach stresses the dynamic interaction between social agents and institutions.

These strategies could be categorized into different types depending on which aspect we put our focus on. One way would be to categorize as individual strategies versus collective strategies. For example, the decision regarding what crop to grow would be an individual decision, while all villagers can collectively decide to contribute land and dig up ponds to do rainwater harvesting. Similarly, some strategies like taking loans are short term in their scope, while some decisions like investing in education will require a long-term outlook from the farmers. There could be other different ways to classify these strategies; for example, some strategies will be conscious, whereas some will not be. Similarly, in the decision process, some farmers will feel constrained to change, whereas some will be more proactive and will voluntarily choose the change.

In the following section, we analyze the main adaptation strategies implemented in our sample. For this purpose, we have classified the strategies into two main groups: agricultural strategies and strategies beyond agriculture. Indeed, the primary adaptation strategies are related to agricultural practices; these involve decisions regarding what to grow and how to grow. But, in addition, farmers also mitigate the climate change effects by adapting beyond agriculture strategies.

Agricultural Adaptation

Cropping Choices and Calendars

One of the most important adaptation strategies is to change the cropping pattern and introduce new types of crops. From our survey, we see that farmers are shifting from crops like boro or jute, which require a lot of water to less water-intensive crops such as oilseeds and sesame. In the last five years, 37 percent of the farmers had to change their crops, while 42 percent of them introduced more diversity. Total 51 percent of the households reduced or stopped boro cultivation in the last five years, while for jute cultivation the corresponding figure is 45 percent.

This change in cropping patterns is largely related to the water requirement of the crops. As seen from Figure 12.3, boro and jute with high requirement of water are being reduced, while less water intensive crops such as sesame and oilseeds have become more important.

Several households have decided to stop or reduce their boro cultivation. Numerous reasons explain this evolution. Farmers first underline the cost of irrigation, which is increasing, especially in areas where water levels are falling. Second, the labor cost (boro requires quite intensive labor) is sharply increasing as well as the prices of the fertilizers. Farmers then see a decline in their boro yield or productivity. And finally, the selling price of the output is not improving over the recent years. In lieu of boro, diverse crops are cultivated. Some farmers shift from boro to sesame in summer season, some

Figure 12.3:

Percentage of households that changed their cropping patterns in the last five years

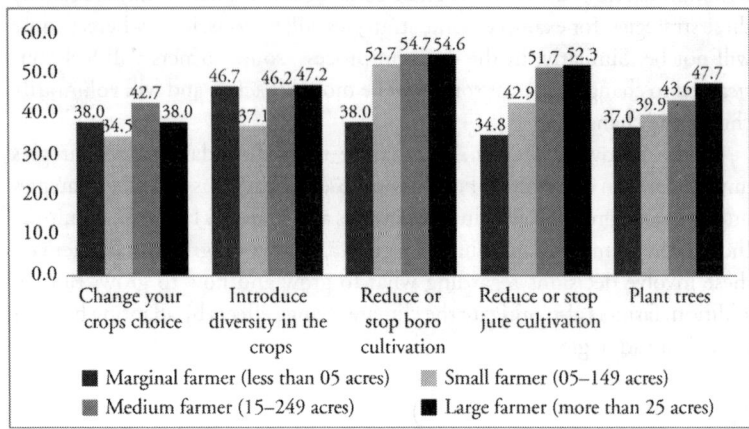

Marginal farmer (less than 05 acres)   Small farmer (05–149 acres)
Medium farmer (15–249 acres)   Large farmer (more than 25 acres)

Source: IWMI. Quantitative survey under project "Ecosystems for Life: A Bangladesh-India Initiative," 2013.

cultivate mustard and wheat, and some other convert their plot into orchard for fruits (banana and mango trees).

> The boro paddy has been replaced by til (sesame) this year in our family land. If we consider the whole village, now farmers are cultivating crops that have a good market and require lesser inputs, i.e. lesser cost of production. As for example, many farmers are now cultivating different varieties of flowers that fetch good price from the market. Many are now turning their paddy fields into fruit orchards. (KII, Silinda village)
> As rainfall is reducing and extracting groundwater to compensate rainfall is becoming costly, many farmers are opting for less water-consuming crops and leaving boro cultivation. (KII, Gajna village)

Jute cultivation is also suffering due to high water requirement, especially at the processing stage and from stagnant prices in the market.

> Both the quantity and quality of jute ha[d] suffered last year due to lack of water to retain the jute stems before extracting jute fibres. Farmers now prefer low water-consuming crops that have good demand in the market and fetch remunerative price. (Men FGD, Gajna village)

Through the questionnaire survey, we also analyzed which households are most likely to adopt these new cropping strategies. We establish that

Figure 12.4:
Percentage of households cultivating each crop, by land size

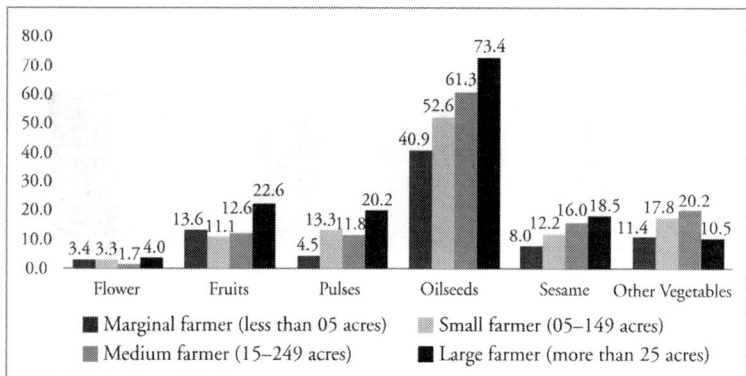

Source: IWMI. Quantitative survey under project "Ecosystems for Life: A Bangladesh-India Initiative," 2013.

these 'new' crops with relatively good market prices and higher value added are mostly cultivated by larger farmers (Figure 12.4). The reason for this is most likely that larger farmers have larger capacity to take risk and they are not doing subsistence farming, which allows them to shift from paddy.

Then apart from changing the crops they cultivate, some farmers bring flexibility in their cropping calendar. This is mainly to answer the changes in the climate and in the rain arrival in monsoon.

> Due to late arrival of monsoon, we could not sow the seeds of Aman crops in the month of Ashar (June) that we were used to do for generations. The sowing period in the kharif season has been deferred in the recent period. (Men FGD, Silinda village)

Adoption of New Technologies (HYV)

To adapt to changing climatic conditions it is also important to adopt new technologies. Most of the farmers (74 percent) increased their use of fertilizers in the last five years and introduced high-yielding varieties (HYV). These actions become necessary to ensure that agricultural output is not adversely affected by the harsher conditions brought about by climate change such as decreasing soil fertility or pest attacks. Again, the adoption of these technologies is easier for larger farmers who have better investment capacities and better access to the information (Figure 12.5).

Figure 12.5:

Percentage of households that introduced new agricultural technologies in the last five years, by land size

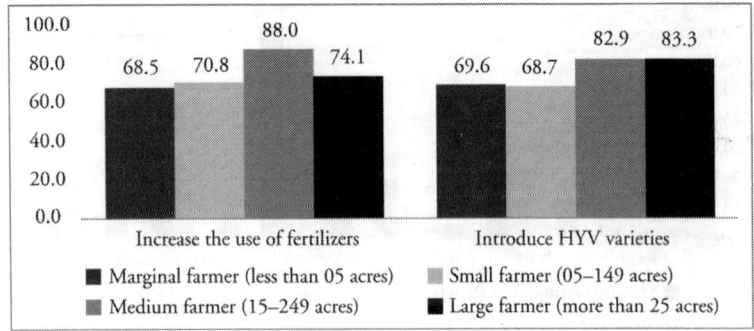

Source: IWMI. Quantitative survey under project "Ecosystems for Life: A Bangladesh-India Initiative," 2013.

Nevertheless, several farmers remain careful of the use of fertilizers and highlight that the land fertility is decreasing. The huge increase in the prices of fertilizers and questions on the quality of the inputs are also clear drawbacks.

Fertility is decreasing due to the usage of huge amount of chemical fertilizer and overuse of groundwater. This is also happening due to lack of floods that used to bring alluvial soil with it. (KII, Silinda village)

Land fertility is decreasing due to intensive cropping (three crops a year and no rest for the land) and excessive use of chemical fertilizers. Pests have also increased that are not eliminated by applying pesticides, which seem to be adulterated. (KII, Polsanda village)

IRRIGATION AND ALTERNATIVE TO GROUNDWATER

The primary source of irrigation is groundwater; however, in the study area, where groundwater is declining, farmers are realizing that this system of irrigation is unsustainable in the long term and becomes costly.

Earlier [...], one could irrigate one bigha (0.33 acre) of land having boro paddy by running the pump 2–3 hours. Even 3–4 years back such was the situation. Now one bigha of boro paddy requires 4–5 hours running of the pump set. [...] The command area of one submersible pump has reduced by more than 3 acres. The expenses of boro cultivation have increased almost two-fold in the last ten years. (KII, Silinda village)

Figure 12.6:
Percentage of households that stored rainwater in the last five years, by district

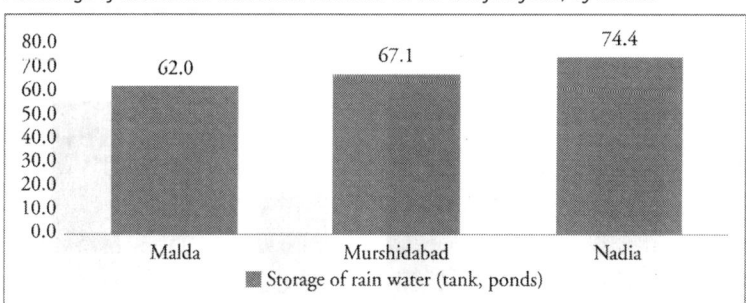

Source: IWMI. Quantitative survey under project "Ecosystems for Life: A Bangladesh-India Initiative," 2013.

This situation encourages the farmers to think about a possible alternative to groundwater. From our interviews with the farmers, we found that alternative sources of rainwater harvesting and storage through tanks or ponds are being re-introduced by some farmers. In fact, in Nadia, 74.4 percent of the households implemented that strategy in the last five years (Figure 12.6).

Tanks, canals and other water structures should be re-excavated to retain more water that might help in recharging groundwater and also reducing our dependency on submersible irrigation. More water harvesting structures should be constructed under the NREGA (National Rural Employment Guarantee programme). (Men FGD, Polsanda village)

Adaptation Strategies Beyond Agriculture

The strategies discussed above are concerned with changing the way agriculture is being done. However, to be insured against the uncertainties of agricultural income due to climate change, an increased number of farmers have withdrawn from agriculture and started nonagricultural livelihoods. The overall survey establishes, for example, that 11.7 percent of the small households decided to keep their land fallow in the five last years (Figure 12.7).

These strategies, in some cases, might induce fundamental changes in lifestyle. We discuss some of those strategies in the following paragraphs. Due to the vulnerability of agricultural production to climatic shocks, people

Figure 12.7:

Percentage of households that drew back from agriculture in the last five years, by land size

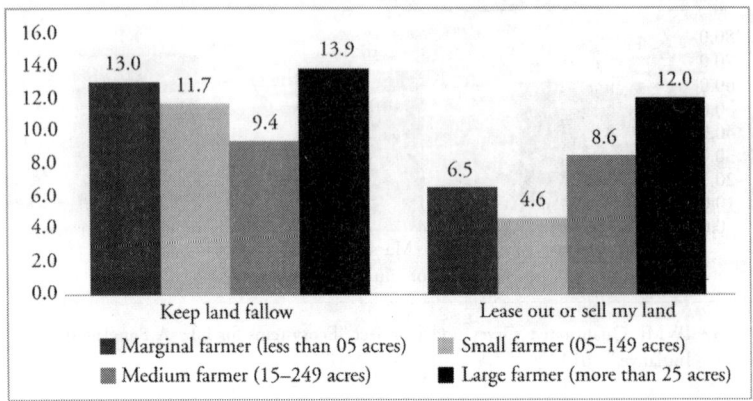

Source: IWMI. Quantitative survey under project "Ecosystems for Life: A Bangladesh-India Initiative," 2013.

have understood the importance of alternative sources of income to support their livelihoods. Indeed, 28 percent of the households have started a new activity in the last five years.

However, creating a new source of income often requires some investments, or training and a strategy is consequently more often implemented by richer households as shown in the graph (Figure 12.8) and the quotes below. While

Figure 12.8:

Percentage of households that started a new activity in the last five years, by land size

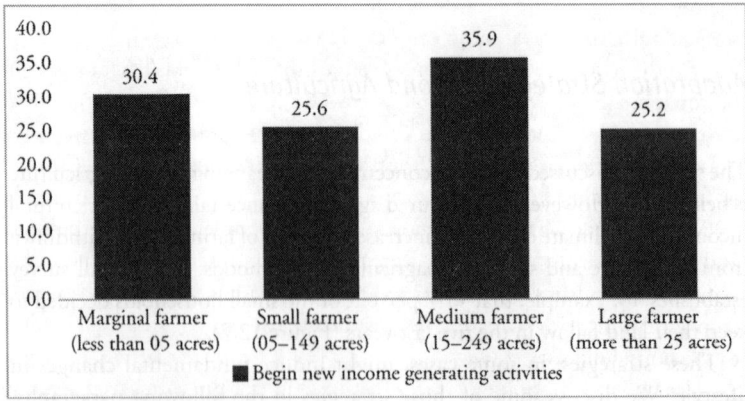

Source: IWMI. Quantitative survey under project "Ecosystems for Life: A Bangladesh-India Initiative," 2013.

35.9 percent of the families of the medium farmers have started a new activity in the last five years, for the small farmers it is only 25.6 percent.

> Definitely poor people and marginal farmers of the lower castes are worst affected from climate change. We do not have the capacity to avail other job or business facilities even when agricultural works are being less profitable and more risky. (KII, Polsanda village)

This is also a way to smooth the stream of income throughout the year and be prepared for lean seasons. Again, this is also a way to diversify the members involved in earning activities and to provide some income sources to female members.

MIGRATION

Studies around the world indicate that when livelihoods are subjected to continuous stress, migration, either seasonally, temporarily, or permanently, is considered the most immediate coping strategy by the farmers (Afifi and Warner 2008). In our study site too, considering the small number of job opportunities in the rural villages out of the main agricultural tasks, migration is often required to diversify the sources of income of the households. Some of these migrations are seasonal, whereas some are more permanent like household members, especially the younger migrating permanently out of the village. In both cases, this is a way to diversify the sources of income and be protected against variability in agricultural income.

Many farmers, especially the marginal farmers, are used to seasonal migration for job. In addition to the normal agricultural activities, some members of the family are involved in some jobs outside the village during some months of the year, most often during the dry season. When the survey was conducted, in dry season, 14 percent of the households had at least one person temporarily out of the house. However, the reasons for migration are different for different households: poorer farmers migrate first for working opportunities, whereas in richest households migrations are also for studies or for personal visits.

Migrating for work is then mostly a strategy implemented by the small and marginal farmers (Figure 12.9) to earn a complementary income and to avoid the agricultural uncertainties. These households mainly practice food crop agriculture; migrating for work is therefore a way to bring cash income to the household.

Permanent migration is then more important for the younger generation who is less interested in farming and can more easily find job out of the villages. The kind of jobs is generally cash for work; however, some are employed in

Figure 12.9:
Percentage of households that migrated for work in the last five years, by land size

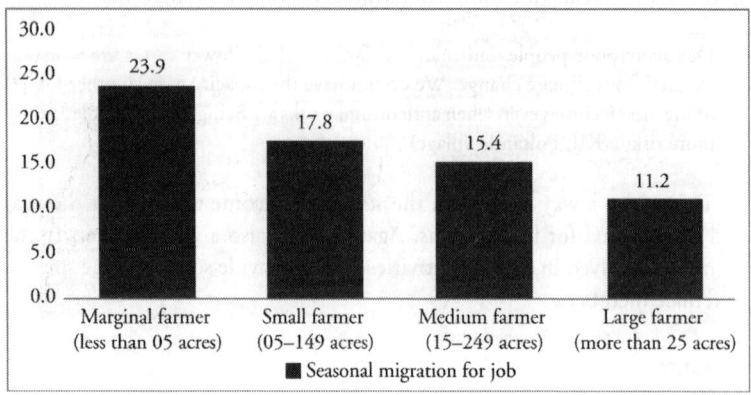

Source: IWMI. Quantitative survey under project "Ecosystems for Life: A Bangladesh-India Initiative," 2013.

the government sector also. So, rather than all members staying in village and doing agricultural work, some members earn through work outside village. In both the cases, remittances are sent to the members who live in the village.

> Migration of village youths has increased in the recent years as they are not finding a bright future in agriculture. They are becoming more and more reluctant to take up the cultivation works and going outside even for doing manual works that are more remunerative. (KII, Polsanda village)

Considering the trends of agriculture and the increasing number of young leaving the villages for better jobs, some households consider the education of their children an important strategy. Indeed, these households consider that in the long run, agriculture alone would not be able to provide adequate livelihood and alternative stable sources of income are required.

> The youths of our village, after studying up to class VIII–IX–X, are migrating to far away cities leaving family behind in search of livelihood and this trend of migration is increasing. They are somehow managing the families by sending money from outside. (Women FGD, Kuli village)

In our sample, 74 percent households promoted better education for their children. Although in richer families, the emphasis is greater still. Also, in some households, they encourage their children not to be involved in

Figure 12.10:
Percentage of households that promoted better education or encouraged their children to leave the village in the last five years, by land size

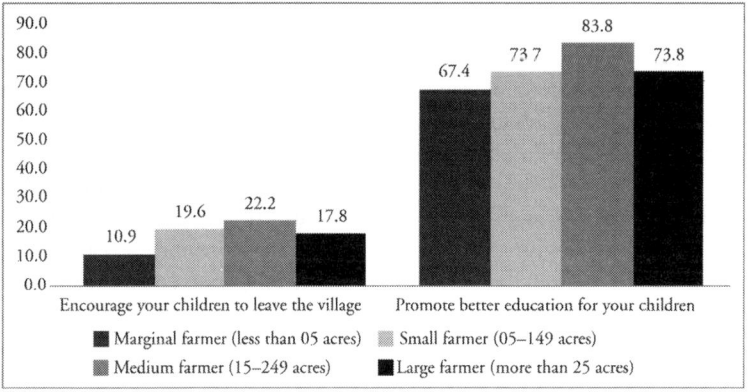

Source: IWMI. Quantitative survey under project "Ecosystems for Life: A Bangladesh-India Initiative," 2013.

agriculture and to leave the village to find better jobs. Only 18 percent of the households encourage their children to leave the village (Figure 12.10).

Conclusion

The effect of climate change is being felt by the rural communities, which are vulnerable in terms of coping capacity against climate change, which has a strong consequence on their agricultural practices. The households that are most vulnerable to the types of crises caused by climate change are inescapably those with low incomes and few assets. Impact of climatic changes includes the decrease in rainfall, increase in temperature, and more uncertainty on the timing of rains. Thence, impacts of rainfall variability lead to losses of livelihood. This leads to change in the natural resources on which the local communities depend for livelihood. The farmers have their own perception of the climate changes in terms of rain, storm, temperature, water, land fertility, river course, and pest. Interestingly, farmers with different status in terms of land and water access have different perception about climate change and implement different adapting strategies. Accordingly, they have adopted strategies to cope with climate change. Nonetheless, perception of climate change leads to long- and short-term mitigation, which at times are individual vis-à-vis collective strategies, which could be in the form of migration, crop diversification, change in groundwater use, promotion of education, water harvesting, etc. The study reveals that landholding size determines the farmers coping strategies for climate change. Indeed, larger farmers, taking advantage of their social and economic capital, are able to change the agricultural behavior and to reduce their agricultural vulnerability to climate. On the contrary, small and marginal to be insured against the uncertainties of agriculture have to withdraw from agriculture and start nonagricultural livelihoods. This is a way to diversify the sources of income and be protected against variability in agricultural income. It should be kept in mind that the most vulnerable to climate change are the poor in India and unfortunately their assets and livelihoods are tied to climate-sensitive factors of production. Adaptation to climate change requires farmers to notice that the climate has changed and then identify adaptations, and in this, various strategies/measures prompted by the local governments or international organizations should emphasize on examining how, when, why, and under what conditions adaptations actually occur and work best in a given economic and social system.

As the adaptive capacity to adjust to climate change is determined through a varied range of factors, including technological options, economics resources, human and social capital, and governance, we can conclude in our research that the way different farmers adapt strategies to climate change proves that adaptation is a kind of development toward making them more resilient against the climate change; however, this is again time and space context. As the impacts of climate change are felt more immediately by individuals in a society, adaptation is typically viewed as obeying the everyday self-interest of individuals. Keeping this as an instrument point, the objective of sustainable development could be achieved if the new and local sociocultural specific development strategies are promoted by the policy-makers at large.

However, we have adopted the case study method that is specific to some villages in West Bengal, India; nevertheless, our findings are relevant to any rural social class, gender, ethnicity, and caste group that faces similar kind of climate change and has similar agrarian relations and set-up.

Hence, greater political and bureaucratic attention is needed to give access to long-term adaptation strategies to the most vulnerable farmers. Therefore, this chapter would help the policy-makers to plan responses to climate change in India by understanding the local people's perception of climate change, its effects, and their adaptation to it. This would lead to a better guideline for more equity in the implementation of agricultural policies and investments in sectors other than agriculture.

References

Afifi, T. and J. Jager, eds. 2010. *Environment, Forced Migration and Social Vulnerability.* Heidelberg: Springer.

Afifi, T. and K. Warner. 2008. "The Impact of Environmental Degradation on Migration Flows across Countries." Working Paper No. 5/2008. Bonn: United Nations University-Institute for Environment and Human security (UNU-EHS). Available at http://www.ehs.unu.edu/file/get/3884 (Accessed on May 13, 2014).

Aggarwal, P.K., S.K. Bandyopadhyay, H. Pathak, N. Kalra, S. Chander, and S. Kumar. 2000. "Analysis of the Yield Trends of Rice–Wheat System in North-Western India." *Outlook on Agriculture* 29 (4): 259–268.

Balasubramanian, M. and D.P.J. Kumar. 2014. "Climate Change, Uttarakhand and the World Bank's Message." *Economic and Political Weekly* 49 (1): 65–68.

Banuri, T. and S. Gupta. 2000. *The Clean Development Mechanism and Sustainable Development: An Economic Analysis.* Manila: Asian Development Bank.

Bryant, R.C., B. Smit, M. Brklacich, R.T. Johnston, J. Smithers, Q. Chiotti, and B. Singh. 2000. "Adaptation in Canadian Agriculture to Climatic Variability and Change." *Climatic Change* 45 (1): 181–201.

Cohen, S., D. Demeritt, J. Robinson, and D. Rothman. 1998. "Climate Change and Sustainable Development: Towards Dialogue." *Global Environmental Change* 8 (4): 341–371.

Fujisawa, M., K. Kobayashi, P. Johnston, and M. New. 2015. "What Drives Farmers to Make Top-down or Bottom-up Adaptation to Climate Change and Fluctuations? A Comparative Study on 3 Cases of Apple Farming in Japan and South Africa." *PLoS ONE* 10 (3): 1–16.

Gbetibouo, G.A. 2009. "Understanding Farmers' Perceptions and Adaptations to Climate Change and Variability." IFPRI Discussion Paper No. 849. Environment and Production Technology Division, 1–40, IFPRI, Washington DC.

IPCC (Intergovernmental Panel on Climate Change). 2001. *Climate Change 2001: Synthesis Report*, edited by R.T.Watson and the Core Writing Team. Contribution of Working Groups I, II, and III to the Third Assessment Report of the Intergovernmental Panel on Climate Change. Cambridge: Cambridge University Press.

Livelihoods and Forestry Programme. 2010. *Participatory Tools and Techniques for Assessing Climate Change Impacts and Exploring Adaptation Options.* LFP Publisher: Kathmandu.

Long, N. 2001. *Development Sociology: Actor Perspectives.* London: Routledge.

Murali, J. and T. Afifi. 2014. "Rainfall Variability, Food Security and Human Mobility in the Janjgir-Champa District of Chhattisgarh State, India." *Climate and Development* 6 (1): 28–37.

Ray, S. and I.A. Ray. 2012. "Impact of Climate Change on Food Security in India." *Advances in Asian Social Sciences* 2 (2): 461–468.

Robinson, J. and D. Herbert. 2001. "Integrating Climate Change and Sustainable Development." *International Journal of Global Environmental Issues* 1 (2): 130–148.

Sharma, A. 2008. "Climate Change to Impact Indian Agriculture: IARI." *The Financial Express*, January 28.

Stern, N. 2007. *The Economics of Climate Change: The Stern Review.* Cambridge: Cambridge University Press.

United Nations. 2011. "UN Data: Country Profile—India." Available at: http://data.un.org/
CountryProfile.aspx?crName=INDIA (Accessed on January 14, 2014).

United Nations Development Programme. 2011. "India Factsheet: Economic and Human
Development Indicators." Available at: http://www.undp.org/content/dam/india/docs/
india_factsheet_economic_n_hdi.pdf (Accessed on January 6, 2014).

Wheeler, T. and J. von Braun. 2013. "Climate Change Impacts on Global Food Security."
Science 341 (6145): 508–513.

World Bank. 2010. "Poverty Headcount Ration." Available at: http://data.worldbank.org/
indicator/SI.POV.DDAY/countries (Accessed on January 14, 2014).

———. 2013. *Turn Down the Heat: Climate Extremes, Regional Impacts, and the Case for
Resilience.* Washington, DC: World Bank. Available at: http://documents.worldbank.org/
curated/en/2013/06/17862361/turn-down-heat-climate-extremes-regional-impacts-case-
resilience-full-report (Accessed on January 6, 2014).

13

Adaptive Capacity of Marginalized Urban Women to Climate Change: National Capital Territory of Delhi

Sakshi Saini and Savita Aggarwal

Introduction

Climate change brings with it droughts, floods, heat episodes, deforestation, and scarcity of natural resources, making the lives of poor women in developing countries much harder since they have to struggle much more to fulfill their roles and responsibilities. Women spend large amounts of time in accession and management of prime resources such as food, fodder, fuelwood, and freshwater for their families. Climate change coupled with urbanization, modernization, and industrialization is expected to increase the existing shortfalls in water and fuelwood, thus increasing the time taken by women in the accession of these resources. Besides this, climatic stresses and extremes magnify the caregiving burdens of women due to increased disease load, thereby diminishing their role in nontraditional activities of income generation further confining women to the home (Nellemann et al. 2011; UNDP 2008). Thus, there are both direct and indirect impacts of climate change on women.

The direct impacts of climate change constitute increased droughts and accompanied water shortages and increased weather events such as cyclones, floods, heat waves, and hurricanes. The indirect impacts of climate change are increased epidemics, decreased food security due to decreased crop production, and loss of species due to decreased biodiversity. A combination of all these factors leads to poorer nutritional and overall status of women in the society and has been an impediment in the nation's ability to achieve its own national goals as well as the Millennium Development Goals. These factors will continue to be major stumbling blocks to the attainment of a large majority of the Sustainable Development Goals by 2030. Figure 13.1 summarizes the likely impacts of climate change on women.

Figure 13.1:
Likely impacts of climate change on women

CLIMATE CHANGE IMPACTS	IMPACTS EXACERBATE GENDER INEQUITIES
CROP FAILURE	Women experience increased agricultural work and overall household food production burden
FUEL SHORTAGE	Many women in developing countries can spend between 2 and 9 hours a day collecting fuel and fodder, and performing cooking chores
WATER SCARCITY	Increased burden on women walking further distances to access safe water, impacts the education and economic stability
NATURAL DISASTER	Women have a higher incidence of mortality in natural disasters; women can suffer from an increased threat of sexual violence
DISEASE	As caregivers women often experience an increased burden for caring for young, sick, and elderly as well as lack of access to health care facilities
DISPLACEMENT	Forced migration could exacerbate women's vulnerability
CONFLICT	While men are more likely to be killed or injured in fighting, women suffer greatly from other consequences of conflict, such as rape, violence, anxiety, and depression

Source: WEDO (2012).

Women have shared an intimate relationship with water since it is at the core of the traditional household responsibilities of women. Collection of water and other resources for the families already takes a very large chunk of women's time. The condition may deteriorate further due to climate change–induced shortfalls of resources leading to magnified traditional roles of women, thereby diluting their economic role.

According to Alexander (2010), water is an important source contributing to the socioeconomic advantage of any society, both as a means and an end in itself. Water scarcity caused by negative changes in the environment is detrimental for women's lives. This is due to the fact that in most developing countries, water accession and management are perceived to be the woman's responsibility. An estimated 200 million hours are spent each day by women globally collecting water (WHO/UNICEF 2010). A survey from 45 developing countries has shown that women and children bear the primary responsibility of water collection in a large majority (76 percent) of the households. This is also the time that is not spent attending school or on income generation (WHO/UNICEF 2010). The decreased availability and quality of water in future due to climate change and other factors will therefore negatively impact the quality of life of people and more so of women and children, who are hitherto responsible for procurement and management of water. It is the women who are most vulnerable because of their dependence on resources at maximum risk due to climate change; therefore, it is important to engage them in areas of climate change preparedness, risk reduction, adaptation, and mitigation, and to strengthen their capacity to adapt to climate change. Isolated studies conducted in different parts of the world including India have shown limited knowledge of people toward climate change and related aspects. In a poll conducted in 2007 across 128 countries, it was found that people from developing countries were much less aware of climate change and did not perceive it as a threat (Brett 2009). In parts of rural India, though the women were not aware of the term 'climate change', an amazing similarity was found between the anecdotal evidence of changes in weather over several years provided by them and data provided by climate scientists (JU, UPCAR, CRIDA/ANGRAU, NBPGR, RS/GBIHED quoted in Kapoor 2011). There are hardly any studies reflecting on the climate-related literacy of the urban poor women as well as their vulnerabilities to different kinds of environmental stresses including climate change. By and large, poor women both in rural and urban areas are very vulnerable to the impacts of climate change but are completely ignorant about the issue.

At the same time, women are far more concerned about environmental issues because of their close association with the environment and have tremendous innovative and social skills that position them uniquely to minimize the harmful impacts of climate change on their families. They have the potential in terms of experience and a strong body of indigenous knowledge to combat the increased disaster risks and enable their families to cope with climate change. Given the knowledge and skills, women can find sustainable solutions to reduce the vulnerability of their families to climate change. The capacity of women in participating in the process of adapting to climate change can be enhanced by increasing their knowledge of technical know-how in the areas of adaptation and mitigation. This is supported by a lot of evidence such as in the case of Honduras in 1998, when Hurricane Mitch struck, a community by the name of La Masica reported no casualties. This was because six months earlier, a disaster agency had provided gender-sensitive community education on early warning systems and hazard management (GGCA 2009). Also, an innovative ActionAid project in Nepal had seen women's empowerment making rapid progress through the use of video discussions about climate change (Khamis et al. 2009). These clearly indicate the positive role that awareness and knowledge play in enhancing the adaptive capacity of women in dealing with climate stresses and extremes. As impacts of climate change will be affecting the citizens, even the solutions of the problem will emerge from the grassroots level. The present study envisions capacity building of poor urban families especially women to make informed choices to participate in bottom-up planning and implementation of adaptive and mitigative measures to deal with climatic stresses and extremes.

Both top-down and bottom-up approaches are used for climate change adaptation planning. Both have their advantages and limitations. On the one hand, the top-down approach is based on climate projections more suitable for developed countries where infrastructure development is already in place to adapt to climatic stresses and extremes. Its limitation is the high level of uncertainty of climate projections as well as the long duration over which the projections are made since most governments are interested in short-term impacts and adaptation solutions. On the other hand, the bottom-up approach considers vulnerability as a characteristic of social and ecological systems generated by multiple factors and processes such as education, social equity, housing, and availability of other resources to populations. The bottom-up approach captures the geopolitical, social, and economic contexts, which may be the cause of vulnerability (Brooks et al. 2005; O'Brien et al. 2007). Despite some limitations such as limited applicability outside the specific context and inability to consider the long-term implications of climate change, the bottom-up approach to climate change adaptation planning is very suitable especially for developing countries and can promote well-being of people through stakeholder involvement.

Adaptive capacity can be enhanced by a combination of factors such as access to information, knowledge and skills, and infrastructure and technology (IPCC 2007). Studies conducted in different parts of the world have shown that lack of education- and gender-related development are constraints that contribute to vulnerability (Aggarwal et al. 2014; Paavola 2008). At the same time, education and knowledge are important factors that determine how people select adaptation options that ensure climate resilience of people to changed socio-ecological systems (Chinowsky et al. 2011; Sovacool et al.

2012). Enhancing knowledge leads to building human capital, which along with economic, social, physical, and natural capitals determines the adaptive capacity of individuals at the household and community levels (Elasha et al. 2005; Vincent 2007). This can be done by using the Devcom approach that implies engaging key stakeholders for supporting sustainable change in development operations. The Devcom approach can establish a conducive environment for establishing risks and opportunities, disseminate information, and induce behavior and social change (Mefalopulos 2008). It can play a critical role in reaching out to different target audiences and enabling them to adapt to climate change by providing them information, knowledge, and skills and promoting positive behavior. The emphasis of Devcom is on people's empowerment using participatory approaches through use of a variety of media and tools ranging from rural radio to information and communication technologies (ICTs) (FAO 2010).

The present study was, therefore, designed to first assess the current awareness level of urban poor women residing in slums and related settlements across the National Capital Territory (NCT) of Delhi toward various aspects of climate change, and to study their water accession and management practices in normal times as well as the coping strategies adopted by them during periods of environmental stresses. The second objective of the study was to design a communication module to raise the adaptive capacity of women to deal with climatic- and water-related stresses and to undertake suitable mitigation measures through strategically planned and targeted communication. The third objective of the study was to assess the impact of the communication intervention on the climate literacy of urban poor women in terms of change with respect to climate change mitigation and water-linked adaptation strategies. The final objective of the study was to make an attempt to understand the key factors that impact the climate change-linked change in behavior of poor urban women with respect to their immediate near and far environment comprising of their family, community, and the society.

Methodology

The study was conducted using the Devcom methodology (Mefalopulos 2008) on a statistically defined sample of 300 women drawn from slums of all the five regions of NCT of Delhi. Both quantitative and qualitative data were collected through a primary survey using interview schedules, focus group discussions (FGDs), and in-depth interviews. A two-stage stratified sampling technique was used to select families residing in the slums. The perceptions of different stakeholders including women, other family members, and local leaders of the area with respect to the issues under study were considered. Information was gathered from the poor women residing in slums related to their knowledge of climate change, water accession and sanitation practices, and the common coping strategies followed by them during environmentally stressful periods. The survey was also a useful tool to gauge the information needs of women as well as the level of complexity of information desired for planning appropriate communication strategies for them in order to build/enhance their adaptive capacity to climate change.

Various traditional and modern media were developed in consultation with experts on climate change, gender, grassroots-level trainers as well as people from the community to impart knowledge and skills related to climate change and various adaptation strategies to women residing in slums. ICTs such as short films, documentaries, public service announcements, and camera-mediated exercises were scripted and produced. These were combined with print and traditional media such as posters, flip charts, flash cards, and puppetry and were packaged into a communication module. The standardized communication module on 'enhancing adaptive capacity of women to climate change' was then administered in 10 slums in the five regions of Delhi. The group size of each campaign was 15–20 women to enable face-to-face interaction. The impact of the communication intervention on the awareness, knowledge, attitude, and potential behavior (AKAB) of the women was assessed by a specially designed tool.

Results

The Profile of the Urban Poor Women

Most women reported that their families had been living in Delhi for more than three years. Almost 95 percent of the families had been identified as below poverty line (BPL) as per the norms of poverty line set by the local government. They were therefore recipients of some benefits such as receiving cereals, sugar, and medical treatment at subsidized rates. Large majorities (75 percent) of the women had never been to school and were illiterate portraying a dismal state of women's education. Among the rest, 18 percent women had education up to primary level, 6 percent up to secondary level, and less than 1 percent women had any form of higher education. Most women who had been to primary school reported they were either semi-literate or illiterate as they had dropped out of school in first, second, or third grades. There were hardly any differences in education level of slum women across the five regions of Delhi.

About 25 percent of the women were gainfully employed as domestic helpers, petty sellers, or daily wagers in construction or in factories. The rest (75 percent) were housewives and had not taken up paid employment since they could not get suitable employment coinciding with their free time.

Source of Water

A large majority of the respondents (80 percent) used piped water supply from the taps installed at community points, while a much smaller number (7 percent) drew water from community hand pumps. Some families (7 percent) had installed illegal water taps at home using plastic pipes as extensions and therefore did not have to walk to the community point. However, these families lived in fear of municipal authorities since the water connections were illegal. The rest of the families relied on other sources such as water tanker or purchased packaged water from private vendors (Figure 13.2).

Responsibility of Water Collection

The overall responsibility of water collection was that of the women in 95 percent households. This was apparent during the visits to the different slums as women moved around with buckets and jerrycans and queued up for water. The women reported that their lives revolved around water accession, storage, and management. A small percentage of women (10 percent) said that their children did help them in water collection. Young daughters-in-law were responsible for water collection in a large majority (70 percent) of the households. In the other households, the mothers-in-law and children were responsible. The participation of men in water collection was very low (5 percent households) across different regions in Delhi. In such families, young women were not present and the older women were not physically fit to carry water. Buckets and jerrycans were the common mode of carrying

Figure 13.2:
Available sources of water for household activities

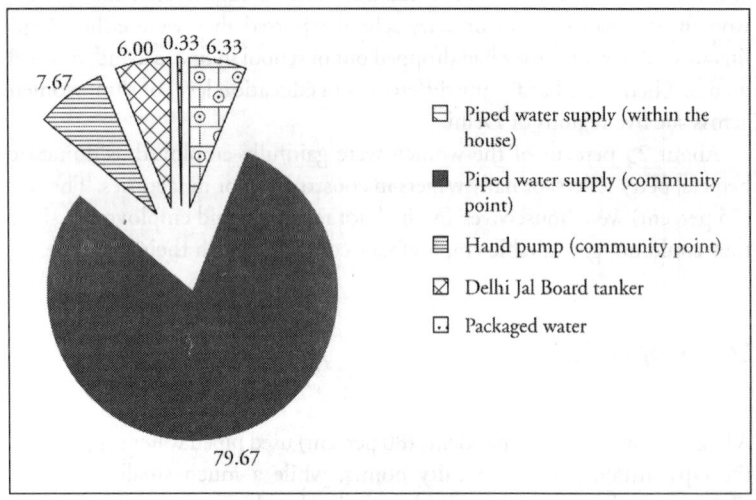

6.00 0.33 6.33
7.67
79.67

☐ Piped water supply (within the house)

■ Piped water supply (community point)

▤ Hand pump (community point)

☒ Delhi Jal Board tanker

⊡ Packaged water

Source: Authors.
Note: n = 300.

water for a large majority of the households. Only few of the respondents said that they used cycles/carts for carrying the water home (Figure 13.3).

More than 56 percent of the women reported that if they were not well, they received help from the men in water collection in the form of filling the buckets with pipes and/or carrying the cans/buckets to the houses. However, the in-depth interviews and FGDs revealed that this help was indeed very occasional. In practice, when the women were sick, children (both male and female) helped their mothers in water collection. Thus, the prime responsibility of water collection on a daily basis in the slum areas rested with the women of the households reinforcing the fact that water collection is a deeply gendered task.

Time Spent in Water Collection

A large majority of the women (73.33 percent) were spending up to two hours every day on water collection for fulfilling different domestic needs (Figure 13.4). They said that the timings of water supply were very erratic

Figure 13.3:
Persons normally responsible for water collection

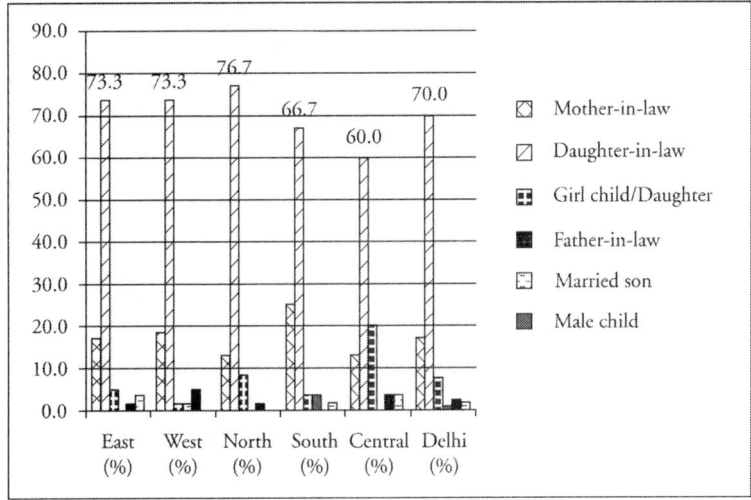

Source: Authors.

Figure 13.4:
Time spent by women in water collection

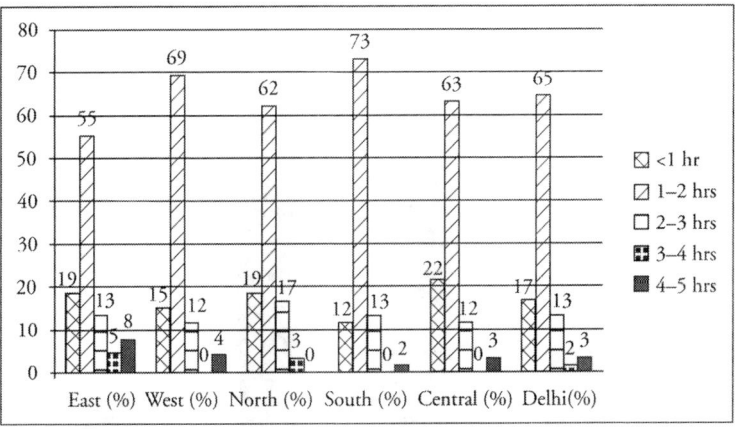

Source: Authors.

forcing them to spend even greater amounts of time in the entire water management process. About 13–17 percent women were spending 2–3 hours in water collection across different regions of the city, while a much smaller number of women (2–13 percent) were spending as many as 3–5 hours in water collection. The women reported that the timings of water collection clashed with the timings of working outside the home for gainful employment and they thus lost out on viable opportunities of income generation.

The women reported that they had to make several trips every day to the water source since they were primarily responsible for meeting the domestic water requirements for the entire family. They also did not have much storage space in the house. The buckets in which they carried water had 15–20 liters capacity beyond which the women found it difficult to carry the weight. Thus, a large majority of the respondents (81 percent) made four or more trips every day to collect water for the family. Out of these, 42 percent women had to make more than six trips to fetch water (Figure 13.5). During FGDs, the women said that there was an increase in the number of trips when water was in short supply such as during climatically stressful (extreme heat, late/no/erratic monsoon) periods. Several trips had to be made just to find out whether water supply was coming in the community taps or to place the cans/buckets in the queue.

Figure 13.5:
Number of trips made by women for water collection per day

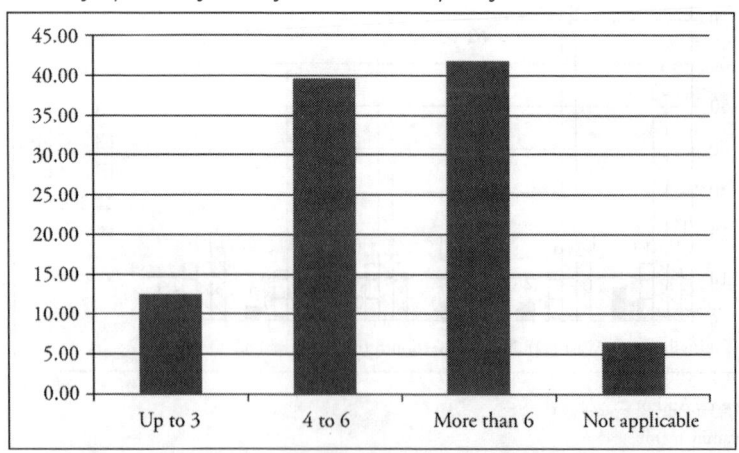

Source: Authors.

Coping Strategies During Water Shortage

The slum women very commonly experienced periods of moderate as well as extreme water scarcity. The women adopted several strategies to tide over such times. More than half of the respondents reported that they spent less time on household work, cut down on leisure time (19 percent), sought help of children (16–17 percent), or took help of other family members (10 percent). Of the 25 percent women who were engaged in paid employment, one-fifth women reported cutting down time on income-generating activities (Table 13.1). There were some differences across various regions in Delhi in terms of the coping strategies practiced by the women. Whereas, 35 percent women from West Delhi said that they had to cut down on time spent on household work and child care in order to fill up water for the family, 73 percent women from Central Delhi reported doing so. They had to travel to nearby colonies, local markets, or railway stations where water taps or hand pumps were installed and were accessible to them. The help of both male and female children was taken across all the slums. The women reported that they often asked the older children especially daughters to skip school

Table 13.1:
Coping measures at the time of water shortage (n = 300)

| Coping Strategies to Water Shortage | Regions of Delhi (percentages) | | | | | Average (%) |
	East	West	North	South	Central	Delhi
Spending less time on household work/child care	54.33	35.00	51.67	38.33	73.33	50.53
Cutting down on income-generating activities	5.00	8.33	3.33	7.67	1.67	5.20
Cutting down on leisure time	17.67	10.00	22.67	21.67	25.00	19.40
Taking help of female children	12.67	25.00	12.67	12.67	15.67	15.74
Taking help of male children	13.33	26.67	13.33	13.33	16.67	16.67
Taking help of other family members	12.67	10.00	8.33	10.33	10.10	10.29

Source: Authors.

and look after the siblings while they themselves collected water. A relatively higher number of families in West Delhi took the help of children in water accession since the water sources were few and far off, making it difficult for the women alone to fill up water for the family. The FGDs revealed that the slum residents were requesting the local leaders/politicians to call for water tankers, while many families purchased water from the private vendors operating in the area.

Quality of Water Supply

Almost one-third of the respondents reported inferior quality of water. Despite this, a large majority of the families (95 percent) did not purify the water for drinking or cooking and were content as long as water was available to them.

Sanitation Facilities

The sanitation in the slums was abysmal with almost half of the dwellers practicing open defecation (Table 13.2), makeshift bathrooms for bathing, and highly inadequate system of disposing off solid wastes. Heaps of garbage was a common sight and the drains were choked with solid wastes in most of the slums.

Since the availability of water is closely linked to the state of sanitation, the slum dwellers were extremely vulnerable to poor sanitation facilities. Since women and children spend greater amounts of time at home as compared to men, their vulnerability is enhanced. Most families did not want to invest in improving the water and sanitation facilities since they lived

Table 13.2:
Defecation pattern of the community (n = 300)

Toilet Facilities Used by Females	Regions of Delhi (%)					Average (%)
	East	West	North	South	Central	Delhi
Flush latrine	0.00	0.00	0.00	0.00	6.67	1.33
Pit latrines	8.33	0.00	6.67	3.33	1.67	4.00
Community toilets	86.67	63.33	50.00	41.67	53.33	59.00
Open defecation	5.00	36.67	43.33	55.00	38.33	35.67

Source: Authors.

either in temporary houses or in rented accommodation that could be demolished any time by the municipal authorities.

Awareness of Women About Climate Change

The present study revealed dismal state of awareness and knowledge of urban poor women about various aspects of climate change, including causes and impacts, adaptation as well as mitigation strategies. As many as 98 percent women had never heard of the term 'climate change' and said that if they had been educated, perhaps then they would have known about it. On further probing about changes in weather over the past years, a very small number (2 percent) said that they experienced longer summers and hot days. They also reported that back home in their villages, the rainfall had become erratic and droughts and floods were more frequent. The women however had no idea about the causes or impacts of climate change.

Considering the high vulnerability of women due to a combination of climatic and nonclimatic factors and large amounts of time and effort spent in collection of water and other resources, a problem likely to intensify in future, it was considered necessary to enhance climate literacy of the urban poor women to enable them to enhance their overall as well water-linked adaptive capacities.

A two-day communication intervention based on Devcom methodology (Mefalopulos 2008) was developed focusing on causes of climate change, linkages with water and women, and adaptation and mitigation strategies for dealing with climate change. The participatory intervention comprised of use of traditional, print as well as modern media. Total 10 communication interventions were administered in small groups to slum women conducted across the five regions of Delhi.

Impact Assessment of the Communication Intervention

CHANGES IN AWARENESS TO CLIMATE CHANGE

Before the intervention, the women had not heard of the term climate change through either the mass media (radio/television) or interpersonal networks and were also not aware of its causes and impacts. After administering the communication module, there was a dramatic change in overall awareness of women about the phenomenon of climate change and its related aspects. All the respondents (100 percent) said that climate change was indeed a reality to which they themselves could now relate. The results of the paired samples t-test showed that there was a significant difference in the awareness scores of the women before the intervention (M = 0.32, SD = 0.648) and after the communication intervention (M = 2, SD = 0.0, t (149) = 31.734, $p < 0.001$), suggesting that the communication intervention had a significant impact on the awareness of women about climate change.

CHANGE IN KNOWLEDGE OF WOMEN ABOUT CLIMATE CHANGE AND RELATED ISSUES

There was a considerable increase in the knowledge of women regarding the causes of climate change, its impacts as well as adaptation and mitigation strategies. Excessive use of petrol, electricity, kerosene, diesel, and burning of coal (46–53 percent) was cited as the major reason for climate change. This was followed by deforestation and setting up of cement and steel industries (25–37 percent). Figure 13.6 shows knowledge of respondents about the causes of climate change.

More than 46 percent had knowledge of at least three impacts of climate change and another 49 percent knew about 4–6 impacts and the rest (4.67 percent) knew of seven or more impacts of climate change. Changes in temperature and rainfall were cited as two major impacts of climate change as cited by women (67–73 percent). These were followed by changes in water availability, quality, and the occurrence of floods and droughts (23–41 percent

Figure 13.6:
Knowledge of women about the causes of climate change

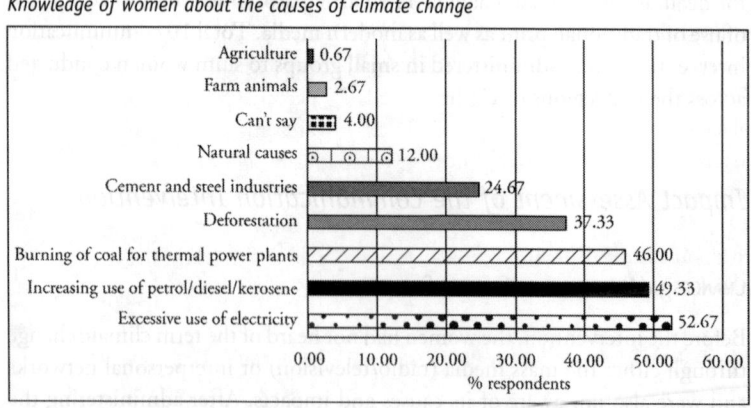

Figure 13.7:
Knowledge of respondents about the impacts of climate change

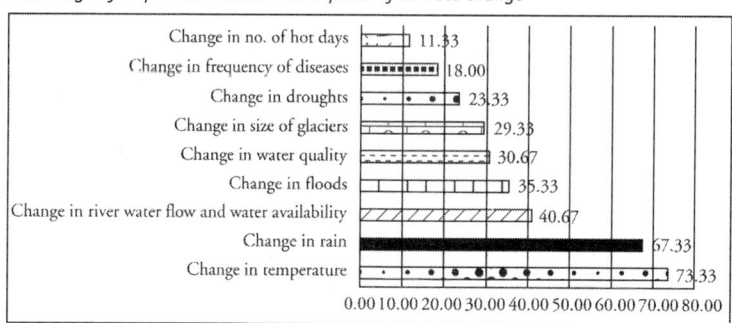

Source: Authors.

women). Figure 13.7 shows knowledge of respondents about the impacts of climate change.

Almost three-fourths of the women said that saving electricity, stopping deforestation, and planting more trees would be the best strategies to slow down the pace of climate change. This was followed by other measures such as the use of public transport and CNG vehicles and discontinuing the use of plastic bags (21–36 percent women). With regard to water-mediated adaptation strategies to climate change, a large majority of women (81–100 percent) reported that conserving water, purifying water, use of simple technologies for water accession, and storage were very important. The involvement of other family members in water accession and joining Self-help groups for accessing finance required to purchase water management-related equipment were the other adaptation strategies cited by the women (61–75 percent). A large majority of women (71–75 percent) felt that filtration and solar disinfection were the best methods of water purification and were also easy to practice. The results showed a statistically significant change in the mean knowledge scores before (M = 27.42, SD = 3.61) and after (M = 44.72, SD = 5.408, $t(149) = 31.467$, $p < 0.001$) the communication intervention. The communication intervention therefore had a significant impact on enhancing the knowledge of the urban poor women

CHANGE IN ATTITUDE OF WOMEN ABOUT CLIMATE CHANGE AND RELATED ISSUES

There was a significant change in the attitude of women to climate change after the intervention. Total 94 percent women felt that climate change would pose a threat in future, especially if the present rate of burning of coal and oil continues. Total 86 percent women (as compared to 10 percent earlier) disagreed after the communication intervention that climate change was due to God's will and felt that dealing with climate change was not the government's responsibility alone but a collective responsibility of people. A higher number of women (90 percent as compared to 9 percent earlier) felt after the intervention that the availability of water would be adversely affected by climate change in future, thus affecting the health of women and children much more negatively. A greater number of women (38 percent as compared to 14 percent earlier) felt after the intervention that water management should not be only women's responsibility but should be shared by all the family members. With regard to adaptation strategies, almost all women felt before as well as after the intervention that water conservation, purification, and appropriate storage were very important. About three-fourths of women as compared to 26 percent earlier agreed that in order to improve their lives, the families needed to invest more money in the purchase of water management equipment.

An analysis of the attitudinal scores also showed a significant difference before (M = 76.68, SD = 5.222) and after the communication intervention (M = 85.36, SD = 3.816, $t(149)$ = 17.041, $p < 0.001$), suggesting that the communication module had a statistically significant impact on changing the attitude of the urban poor women to various aspects of climate change, including its causes, impacts, adaptation as well as mitigation strategies.

Intent to Behavior Change

Even before the communication intervention, a large majority of the women and their families were already practicing several mitigation strategies suggested to them in the present study such as minimizing the use of electricity by switching off electrical equipment when not in use, using CFLs, coloring walls with light colors, using public transport, and using CNG vehicles. This was because of very low levels of income of families and very poor availability of natural as well as other economic resources to families, which strictly called for minimizing their use. These families used cycles or public transport to travel long distances and walked to the nearby destinations. They could not afford to buy or hire petrol/diesel-driven private vehicles.

After the communication intervention, the other respondents (who were small in number and were practicing the above mitigation strategies to some extent) said that they would use these mitigation strategies in future. In terms of change in behavioral intent, 28 percent women said that they would like to try using solar equipment (as compared to none before), 12 percent women who were nonusers of CFLs wanted to switch over to them, and another 14 percent women thought that they would like to use CFLs. The constraint felt by most women in the use of CFLs was their high cost. As many as 26 percent women said that they would like to adopt better waste disposal techniques in the future, while 24 percent women said that they would use less plastic bags in future (Table 13.3).

Some mitigation strategies were considered as difficult to practice by the women, both before and after the intervention either due to lack of finance, lack of space, temporary and rented housing or lack of willingness of other family members. These were the use of solar equipment, composting and planting more greenery. Composting and plantations were considered very difficult by the women due to lack of space and therefore, not practiced by almost all the families. Similarly, the use of solar equipment was also considered difficult to practice because it involved a lot of initial investment and space for installation. The houses in most slums could be demolished any time by the municipal authorities and the families could not risk their money by making such investments. The women said they were not likely to adopt these mitigation strategies (composting, planting more plants/trees and using solar equipment) in future also.

Similarly, a large majority of the families were already using several adaptation strategies even before the communication intervention such as judicious use of water, use of covered and clean utensils for water storage (as per their perception of cleanliness), use of pipes for water accession (to some extent), use of cycles and/or carts for transporting water (occasionally), and taking help of other family members in water collection (to some extent). Water purification by filtration and the use of water storage tanks were found to be easy but not practiced by the women since these involved high investment of money. Rainwater harvesting was also found to be difficult by women due to lack of finance and space and was therefore not practiced by households. Even before the intervention, almost 60 percent women were members of some women's group, which collected money from every member on a monthly basis and provided lump sum money to a member by rotation. The women said that they used this money for meeting personal expenses (clothes, jewelry, and gifts). Some women said that they could think of utilizing the money for purchase of water management-related equipment, which could improve the quality of their day-to-day life. After the intervention, more women (24 percent) as compared to (4 percent) earlier felt that they would

Table 13.3:
Behavior of women toward climate change mitigation strategies (n = 150)

Mitigation Strategies	How Easy Is It to Practice It?			Do You Practice It?			Will You Practice It in Near Future?		
	Easy	Fair	Difficult	Yes	To Some Extent	No	Yes	To Some Extent	No
Save electricity									
Switch off electrical equipment when not in use	81	13	6	78	22	0	22	0	0
Use solar equipment	0	7	93	0	0	100	0	28	72
Use CFLs	57	17	26	62	12	26	12	14	0
Use cold water for cleaning as far as possible	66	24	10	62	28	10	0	10	0
Color walls with light colors	84	16	0	94	6	0	0	0	0
Use resources judiciously									
Save petrol, diesel, or kerosene	64	36	0	24	68	8	8	0	0
Use fuel judiciously at home	66	34	0	62	38	0	0	0	0
Use public transport	36	58	6	100	0	0	0	0	0
Use CNG transport	22	68	10	100	0	0	0	0	0
Greening									
Plantation	18	66	16	0	8	92	0	10	82
Stop deforestation	76	24	0	100	0	0	0	0	0
Dispose waste properly	64	24	2	64	24	2	2	0	0
Reuse	36	62	2	34	66	0	0	0	0
Use less plastic	10	52	38	4	42	54	24	30	0
Use cloth/jute bag	64	24	12	10	78	12	12	0	0
Composting	2	36	62	0	0	100	0	0	100

Source: Authors.

like to install a water tank in their houses to provide relief to them as well as use cycles/carts to bring water at least to some extent. For adopting these strategies, the women said that they could not decide on their own and would have to request/motivate the male members of the family. With regard to purification/disinfection of water, even after the communication intervention, more than two-thirds women still said that they would not like to invest in the purchase of water filter and would also not use boiling as a method of water purification. They preferred the use of solar disinfection technique and chlorine tablets for water purification/disinfection since these did not involve much cost (Table 13.4).

In case of change in the behavioral intent of women to climate change adaptation and mitigation, the scores before the intervention were (M = 59.49, SD = 6.303) that changed marginally to (M = 60.88, SD = 11.420), *t* (149) = 1.323, *p* < 0.188 after the communication intervention. The change in behavioral intent of the urban poor women to climate change adaptation and mitigation was not statistically significant.

Overall, the AKAB scores before the intervention were (M = 163.91, SD = 10.56) that, after the communication intervention, changed to (M = 193.02, SD = 12.80), *t* (149) = 20.21, *p* < 0.001. A breakup of the scores indicates the maximum change occurred in the awareness component (74.66 percent) followed by the knowledge level of women (59.66 percent). The percent change in the attitude of women toward various issues pertaining to climate change was much lower, (13.4 percent) followed by change in their behavioral intent toward climate change adaptation and mitigation practices (4.1 percent). It is important to note that as a result of this communication intervention, despite a significant gain in the awareness and knowledge components, there was a limited change in the attitude and intent to behavior change of the women (Figure 13.8).

This can be attributed to several reasons. One of the reasons for the differential change in AKAB was that the women had almost no awareness and knowledge about various aspects of climate change prior to the communication intervention and had negligible scores in both these dimensions, thus depicting high gain in scores after the intervention. By contrast, the women had scored substantially in the attitude and behavior components even before the communication intervention, thus limiting the scope for gain after the communication intervention. Several families despite not having knowledge of climate change and related aspects were already practicing several of the suggested mitigation (cutting down fuel/electricity consumption, use of public transport) and adaptation strategies (saving water) owing to their very low levels of income and a general lack of availability of resources.

Table 13.4:

Change in behavioral intent of women toward water-linked climate change adaptation strategies (n = 150)

Adaptation Strategies	How Easy Is It to Practice It?			Do You Practice It?			Will You Practice It in Near Future?		
	Easy	Fair	Difficult	Yes	To Some Extent	No	Yes	To Some Extent	No
Water Conservation									
Use water judiciously; do not waste	56	17	27	62	11	27	13	14	0
Join local SHG for loan facility	66	24	10	62	28	10	0	10	0
Rainwater harvesting	4	34	62	0	0	100	0	0	100
Purchase water storage/purification equipment									
Use pipe for filling water	84	16	0	54	36	10	0	10	0
Use covered utensil to store water	100	0	0	100	0	0	0	0	0
Use water tank for storage	10	52	38	4	42	54	24	30	0
Use water filter	74	24	2	10	3	87	12	8	67
Use cycle or cart to carry water	57	17	26	11	22	67	12	55	0
Seek help of family members	66	34	0	62	38	0	0	0	0
Purify water									
Filtering by cotton cloth	26	72	2	0	0	100	0	22	88
Use filters available in market	74	24	2	10	3	87	12	8	67
Boil water	42	26	32	0	14	86	0	14	72
SODIS (Solar Disinfection Technology)	0	0	100	0	0	100	62	24	14
Use chlorine tablets	20	80	0	0	0	100	82	18	0
Use moringa seeds	0	0	100	0	0	100	0	10	90
Use alum	0	0	100	0	0	100	6	36	58
Store water in clean utensil	100	0	0	100	0	0	0	0	0
Keep the water covered	100	0	0	100	0	0	0	0	0
Install tap in storage container to draw water	36	34	30	12	0	88	88	0	0
Practice hand washing	100	0	0	100	0	0	0	0	0

Source: Authors.

Figure 13.8:

Changes in AKAB scores after the communication intervention

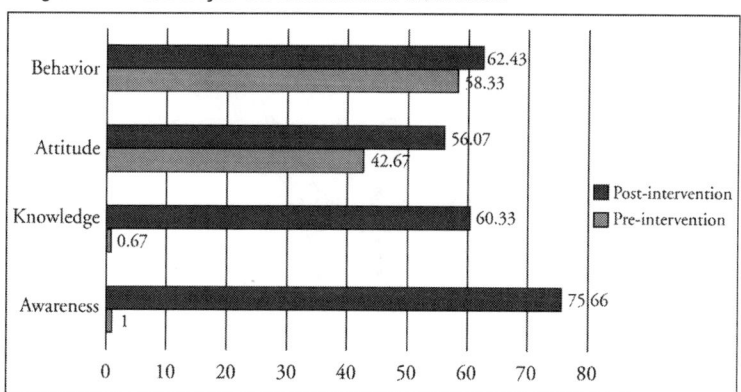

Source: Authors.

With respect to the other adaptation strategies suggested, the women had several apprehensions related to finance, lack of support of other family members, including older women and men, lack of space, and poor level of motivation due to temporary and rented housing, which were under constant threat of demolition. The women were therefore not ready to adopt these adaptation strategies on their own without the consent of or agreement with the male and other senior members of the households.

Figure 13.9 depicts the likely factors responsible for limited behavioral change amongst urban poor women toward appropriate mitigation and water-related adaptation strategies to climate change. These factors have emerged from both qualitative and quantitative approaches used for inquiry in the present study. The interactions with women after the communication intervention revealed that the women found it relatively easy to change certain factors in their immediate environment, which were within their control and possible within the resources that were available to them. These were adoption of simple practices, which did not entail possibility of conflict with the male or senior family members. However, there were several hindrances in the women's intermediate and larger environment, which acted as barriers in the adoption of suitable adaptation as well as mitigation strategies to climate change, many of which were also highlighted during the review of literature. These were larger requirement of resources such as money, time, and space, all of which were limited in availability to the urban poor. In addition, societal norms, strict gendered distribution of roles, low/no decision-making power

Figure 13.9:
Model of behavior change of slum women with respect to mitigation and water-linked adaptation strategies to climate change

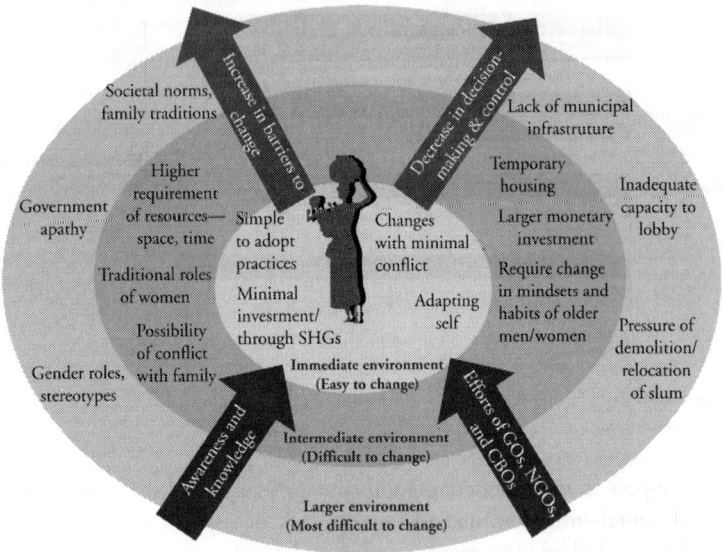

Source: Authors.

of women, traditional mindsets of people toward the plight of women, temporary nature of housing, fear of demolition of the slum as well as the apathy of the government toward their condition further add to their vulnerability. To facilitate effective behavioral change amongst urban poor women, all these issues need to be addressed holistically.

This is also supported by several studies in different parts of the world that though some adaptations may be technologically possible. they may not be economically feasible or culturally desirable (IPCC 2007). The inequalities in society and differential access to resources may also limit the use of technology for adaptation. At the same time, financial poverty of people especially the low-income groups, communities, and nations may pose an additional constraint in climate change adaptation practices. These constraints have been widely reported in the fields of agriculture as well as health (Appendini 2001; Taylor et al. 2006; Yanda et al. 2006).

Psychological research has also shown that the individual's awareness, knowledge, personal experience, and sense of urgency are important conditions for behavior change, yet these are not sufficient. The perception of people to risk their own vulnerability, level of motivation, and capacity to adapt will also affect their behavior. Since these attributes vary among individuals and groups, these act as barriers to behavior change. In addition, there may be social and cultural barriers due to diverse understandings of prioritization of climate change issues, which differ across social and cultural groups and can limit their adaptive responses (Ford and Smith 2004). It is therefore important to integrate cognitive (knowledge), experiential (perception of impact), and normative (moral and societal norms) factors in order to maximize the potential of people to change behavior (Van der Linden 2012).

Conclusion

The present study has shown that the Devcom-enabled approach to climate change adaptation planning was instrumental in enhancing the awareness, knowledge, and positive attitude of the urban poor marginalized women toward various aspects of climate change adaptation and mitigation. There was, however, a limited impact on changing the behavioral intent of the women due to a number of factors discussed above. Overall, the communication intervention is likely to enhance the adaptive capacity of women since studies have shown that the more the members of a community understand the problem of climate change, the more will be the likelihood of supporting the changes brought about by various sectors, including the government and nongovernmental organizations. Knowledge-based initiatives may also work as a precursor to behavior-based initiatives (Curnow and Spehr 2011). Communities are more likely to cope well with climate change if they have knowledge of potential threats and resources to adapt to them. At the same time, it is important to overcome sociocultural barriers, which often constraint the adoption of sound adaptation strategies. It is necessary not only to integrate cognitive, experiential, and normative factors but also to invest in furthering the enabling factors such as access to finance and technology in order to catalyze change in behavior.

The study has also shown that bottom-up approach to adaptation planning is extremely important since it takes into account the perception of various stakeholders and considers existing social, economic, cultural, and other factors affecting vulnerability. The best practices of the community members also provide insights into suitable adaptation options that can be further improved and outscaled.

The vulnerability of poor women may worsen in future due to rapid urbanization adding to the number of marginalized population in cities. It is estimated that in developing countries, almost 40 percent of the population will come to reside in cities by 2030. The urban population in India is expected to grow to 590 million by 2030, a large chunk of which will reside in the slum areas (Bouselly et al. 2006; Sankhe et al. 2010). The poor women residing in slums are already very vulnerable to climatic stresses because of high dependence on natural resources for household management, low levels of education, socioeconomic barriers that limit their mobility, and participation in productive activities that exacerbate the gender-based inequalities. It is therefore extremely important to pay attention to the most vulnerable section of society and enhance its adaptive capacity to deal with climatic changes. Such efforts have the potential to enable the most vulnerable groups of society to take appropriate action toward climate change mitigation and adaptation and lead more climate resilient and empowered lives.

References

Aggarwal, S., G. Punhani, and J. Kher. 2014. "Hotspots of Household Water Insecurity in India's Current and Future Climates: Association with Gender Inequalities." *Journal of Gender and Water, University of Pennsylvania* 3 (1): 34–41.

Alexander, N. 2010. *The Country Policy and Institutional Assessment (CPIA) and Allocation of IDA Resources: Suggestions for Improvements to Benefit African Countries,* 43. Washington, DC: Heinrich Boll Foundation.

Appendini, K. 2001. *De la Milpa a los Tortibonos: La Restructuración de la Política Alimentaria en México* (2da edición). México, DF: Colegio de México.

Bouselly, L., S. Gupta, and D. Ghosh. 2006. "Water and Urban Poor." Working Paper No. 06-11. National Institute of Urban Affairs, New Delhi.

Brett, W. 2009. "Awareness, Opinions About Global Warming Vary Worldwide." Retrieved from: http://www.gallup.com/poll/117772/Awareness-Opinions-Global-Warming-Vary-Worldwide.aspx (Accessed on December 21, 2015).

Brooks, N., W. Adger, and P.M. Kelly. 2005. "The Determinants of Vulnerability and Adaptive Capacity at the National Level and the Implications for Adaptation." *Global Environmental Change* 15: 151–163.

Chinowsky, P., C. Hayles, A. Schweikert, N. Strzepek, K. Strzepek, and C.A. Schlosser. 2011. "Climate Change: Comparative Impact on Developing and Developed Countries." *Engineering Project Organization Journal* 1 (1): 67–80.

Curnow, R.C. and K.L. Spehr. 2011. "Influencing Social Change by Engaging the Community in a Holistic Evaluation Methodology." Paper presented at the Australasian Evaluation Society International Conference, Sydney, August 29, 2011–September 2, 2011.

Elasha, B. Osman, Nagmeldin Goutbi Elhassan, Hanafi Ahmed, and Sumaya Zakieldin. 2005. "Sustainable Livelihood Approach for Assessing Community Resilience to Climate Change:

Case Studies from Sudan." AIACC Working Paper No. 17, London: Overseas Development Institute.

FAO (Food and Agriculture Organization). 2010. *Advancing Adaptation through Communication for Development*. Proceedings of the Technical Session on Communication, Third International Workshop on Community-based Adaptation to Climate Change, Dhaka, Bangladesh, 2009.

Ford, J. and B. Smit. 2004. "A Framework for Assessing the Vulnerability of Communities in the Canadian Arctic to Risks Associated with Climate Change. *Arctic* 57: 389–400.

Sankhe, Shirish, Ireena Vittal, Richard Dobbs, Ajit Mohan, Ankur Gulati, Jonathan Ablett, Shishir Gupta, Alex Kim, Sudipto Paul, Aditya Sanghvi, and Gurpreet Sethy. 2010. *India's Urban Awakening: Building Inclusive Cities, Sustaining Economic Growth*. Global McKinsey Report. McKinsey and Company.

Gender and Climate Alliance, (GGCA). 2009. *Training Manual on Gender and Climate Change*. San José, Costa Rica: Marzo.

IPCC (Intergovernmental Panel on Climate Change). 2007: *Climate Change 2007: Impacts, Adaptation and Vulnerability*, edited by M.L. Parry, O.F. Canziani, J.P. Palutikof, P.J. van der Linden, and C.E. Hanson, 976. Working Group II Contribution to the Fourth Assessment Report of the Intergovernmental Panel on Climate Change. Cambridge: Cambridge University Press.

Kapoor, A. 2011. *Engendering the Climate for Change: Policies and Practices for Gender—Just Adaptation*. New Delhi: Alternative Futures and Heinrich Boll Foundation (HBF).

Khamis, M., T. Plush, and C.S. Zelaya. 2009. "Women's Rights in Climate Change: Using Video as a Tool for Empowerment in Nepal." *Gender and Development* 17 (1): 125–135.

Mefalopulos, P. 2008. *Development Communication Sourcebook: Broadening the Boundaries of Communication*. Washington, DC: The International Bank for Reconstruction and Development/The World Bank.

Nellemann, C., R. Verma, and L. Hislop, eds. 2011. *Women at the Frontline of Climate Change: Gender Risks and Hopes*, 119–125. A Rapid Response Assessment. United Nations Environment Programme and GRID-Arendal, Norway.

O'Brien, K., S. Eriksen, L.P. Nygaard, and A. Schjolden. 2007. "Why Different Interpretations of Vulnerability Matter in Climate Change Discourses." *Climate Policy* 7: 73–88.

Paavola, A. 2008. "Livelihoods, Vulnerability and Adaptation to Climate Change in Morogoro, Tanzania." *Environmental Science and Policy* 11 (7): 642–654.

Sovacool, B.K., A.L. D'Agostino, H. Meenawat, and A. Rawlani. 2012. "Expert Views of Climate Change Adaptation in Least-developed Asia." *Journal of Environmental Management* 97: 78–88.

Taylor, M., A. Chen, S. Rawlins, C. Heslop-Thomas, A. Amarakoon, W. Bailey, D. Chadee, S. Huntley, C. Rhoden, and R. Stennett. 2006. *Adapting to Dengue Risk—What to Do?* AIACC Working Paper No. 33, International START Secretariat, Washington, DC, 31 pp.

UNDP. 2008. *Human Development Report 2007: Fighting Climate Change—Human Solidarity in a Divided World*. United Nations Development Programme, New York: Macmillan.

Van der Linden, S. 2014. "Towards a New Model for Communicating Climate Change. In *Understanding and Governing Sustainable Tourism Mobility: Psychological and Behavioural Approaches*, edited by S. Cohen, J. Higham, P. Peeters, and S. Gössling, 243–275. Routledge: Taylor and Francis Group.

Vincent, K. 2007. "Uncertainty in Adaptive Capacity and the Importance of Scale." *Global Environmental Change* 17 (1): 12–24.

WEDO. 2012. *Gender, Climate Change and Human Security: Lessons from Bangladesh, Ghana and Senegal.* New York: WEDO.

————. 2010. *Joint Monitoring Programme Report. Progress on Sanitation and Drinking Water for Water Supply and Sanitation.* New York and Geneva: UNICEF and WHO.

Yanda, P., S. Wandiga, R. Kangalawe, M. Opondo, D. Olago, A. Githeko, T. Downs, R. Kabumbuli, A. Opere, F. Githui, J. Kathuri, L. Olaka, E. Apindi, M. Marshall, L. Ogallo, P. Mugambi, E. Kirumira, R. Nanyunja, T. Baguma, R. Sigalla, and P. Achola. 2006. *Adaptation to Climate Change/Variability-Induced Highland Malaria and Cholera in the Lake Victoria Region.* AIACC Working Paper 43, International START Secretariat, Washington, DC, 37 pp.

14

Community-based Climate Change Adaptation in Coastal India: A 'Bottom-up' Approach Using 3P Model

Rachna Arora, Ashish Chaturvedi, Manjeet Saluja, Nikita Mundra, and Arushi Sen

Introduction

It is being increasingly understood and accepted, both theoretically and in practice, that actions that may fulfill aspirations for sustainable development and those that combat climate change are critically linked to each other (see Chapter 1 of this book). The reason is straightforward but complex—sustainable development and climate change interact in myriad cross-cutting ways. Climate risks and vulnerabilities are often aggravated by threats to natural resources and have the potential for causing social, economic, and environmental damage impinging on sustainable development. In such a context, any policy initiative or action taken without reference to climate change has the potential to underestimate the vulnerabilities of the affected population. Consequently, enhancement of adaptive capacity to climate change can be regarded as one component of broader sustainable development initiatives (Ahmad and Ahmed 2000; Munasinghe 2000; Robinson and Herbert 2000; Tompkins and Adger 2005). It is therefore critical for sustainable development initiatives to explicitly consider hazards and risks associated with climate change (Apuuli et al. 2000; IPCC 2001). The need to combat these negative climatic impacts and make communities more resilient to the erratic patterns posed by the climate has hence become imperative. An entire body of literature has emerged in the recent years based on the implementation of projects on community-based climate change adaptation (CCA) at the grassroots level (Chishakwe et al. 2012; Reid et al. 2009; Sekine et al. 2009). The process is bottom-up and community-driven, placing a strong emphasis on incorporating indigenous knowledge, social capital, and local context in adaptation planning.

According to Thaler and Sunstein (2008), people can be nudged toward taking a particular decision depending on how the choices are presented to them. They term this concept 'choice architecture', that is, a person's decision can be influenced if the way of presenting the issues/alternatives/solutions is improved. The actors presenting the choices are called libertarian paternalists; these "Libertarian paternalists want to make it easy for people to go their own way ... it is legitimate for choice architects to try to influence people's behavior in order to make their lives longer, healthier, and better" (ibid.: 16). The 3P model (the Ps denote Perception, Pilot, and Planning)—developed by us in this study—can be considered a type of choice architecture: the facilitators of the model act as choice architects at the Perception stage for the target community by 'nudging' the community not only in their attitude toward climate change but also in the way they understand and approach their vulnerabilities. At the Planning stage, the results of the pilot measures showcase to the decision-makers the benefits and co-benefits of the technical solutions in climate proofing and nudge them to incorporate similar projects into development planning.

In what follows, we describe the process of how we developed the elements of the 3P model and their significance through case studies from the AdaptCap project in coastal communities of Andhra Pradesh and Tamil Nadu (India), which has been able to influence the perceptions of communities toward climate change and establish the necessity of mainstreaming CCA into the local planning processes. In conclusion, we discuss the possibilities, challenges, and processes required to achieve the three Ps in practice. We also recommend that despite several challenges, the dynamics of the adaptation systems and the developmental planning processes need to be initiated at the local level to reflect the felt needs of the community so as to reduce vulnerabilities and achieve sustainability in the social, economic, and environmental indices.

The 3P Model

As stated in the recent report of IPCC (2014), adaptation and mitigation have the potential to both contribute to and impede sustainable development, and sustainable development strategies and choices have the potential to both contribute to and impede climate change responses. The specific model that is discussed in this chapter is an attempt to theoretically construct a frame of reference for scoping sustainable methods aimed at integration of CCA, climate change mitigation (CCM), and disaster risk reduction (DRR).

Reid et al. (2009) explain that "[i]t is now increasingly recognized that, for poor communities, adaptation approaches that are rooted in local knowledge and coping strategies, and in which communities are empowered to make their own decisions, are likely to be far more successful than top-down initiatives." However, working only with local communities and implementing need-based projects is a temporary and an incomplete solution to the problem. Although the crisis-coping culture has naturally evolved in communities, their perceptions of disasters caused due to climate change and its adaptation mechanisms need to be incorporated in local, regional, and national governance mechanism.

Adaptation to climate change cannot be treated as an isolated process—it requires mainstreaming into the local development planning in order to strengthen the capacities of communities and authorities to make it more effective. Top-down adaptation approaches have been found rigid and have missed important elements of the solution to the problem (Shukla et al. 2003). This calls for an integrated approach that encourages policy-makers to account for climate change risks in their local plans to mitigate the vulnerabilities of communities to environmental and climatic risks. We refer to this integrated approach as the 3P model that has been developed during the course of our work with the coastal communities of Andhra Pradesh and Tamil Nadu. The three Ps of the model—Perceptions (how the local communities understand and address climate change and its risks), Pilots (implementing technical measures to reduce their vulnerabilities and address the risks), and Planning (incorporating the results of the projects into the local development planning)— lay the foundation for mainstreaming adaptation to climate change in the planning process. In the 3P model, the pilot implementation is aimed at addressing the needs of the target group and the beneficiaries, following which the success of the pilots is measured.[1] This depends on whether or not the beneficiaries and target group were affected and if the interventions assisted to mitigate the risks posed by climate change in the form of strengthening the capacities of the people and building more resilience in the community. The pilots also consider the integration of CCA, CCM, and DRR to demonstrate mutual benefits as well as co-benefits with developmental planning. This creates a platform for communicating the success of the pilots to the key stakeholders and policy-makers. Hence, the model uses pilots as evidence for climate

---

[1] Pilots are hard measures using indigenous technology that serve a dual purpose in our model. First, they serve as an evidence of climate change–vulnerability linkages and how to overcome such linkages in a practical manner making communities more resilient. Second, they serve as an evidence for the policy-makers to take action.

pro-planning to policy-makers enabling them to use similar projects to address climate change issues in their region (OECD 2005). Also, inclusion of climate risks in designing developmental initiatives leads to upscaling and replication in other regions as well. The 3P model, therefore, is the translation of needs and requirements of target groups into initiatives, the success of which is communicated to the stakeholders in order to create a policy-level impact with enhanced capacities for management of climatic risks. Building the capacity of individuals, communities, and governance systems to adapt to climate impacts is a function of both dealing with developmental deficits (e.g., poverty alleviation, reducing risks related to famine, and food insecurity) and improving risk management (e.g., alert systems, disaster relief, crop insurance, seasonal climate forecasts, and risk insurance) (IPCC SREX 2012; Mirza 2003; Schipper and Pelling 2006; Warner et al. 2013).

The 3P model is based on the understanding of the perceptions of communities toward vulnerabilities arising due to dependence on natural resources. As the next step, the link between the vulnerabilities and climate change is established using theoretical tools as well as practical experiences. This framing of perceptions of the community on risk-related impacts and suggestive adaptation measures is achieved through a variety of community-based participatory tools that lead to a planning document entitled the 'Local Adaptation and Mitigation Guide' (LAMG) (OECD 2009). As communities engage in this visioning and planning process regarding climate variability and extremes, they also work toward potential approaches that need to be implemented in order to make them more resilient. This unfolds the need for building capacities of the affected communities and paves the path for the launch of pilots. The success of pilots, measured in terms of the benefits to the target group and beneficiaries, can then be showcased to the policy-makers as a potential case for replication and uptake in development planning processes. The pilots, in the 3P model, play a dual role of demonstrating solutions for addressing perceptions and creating evidence for planning that accounts for climatic risks and associated vulnerabilities. The pilot projects as prioritized by community members cover a variety of options for enhancing adaptive capacities, namely livelihoods resilience; DRR to minimize the impact of hazards; capacity development of local civil society; government institutions and community members; advocacy; and social mobilization to address the underlying causes of vulnerability, including poor governance, inequitable control of resources and limited access to basic services, discrimination, and other social injustices.[2]

---

[2]   Also concurs with Huq and Reid (2007) and Dodman and Mitlin (2011).

The model is an attempt to shift adaptation work from a reactionary approach to a more proactive approach by suggesting inclusion of local needs in designing responses and strategies to enhance adaptive capacities and foster linkages with developmental planning for mainstreaming CCA. This is done by situating climate change-related stress on communities, namely rising sea level, extreme weather conditions, coastal erosion, etc. in the wider frame of development stress, namely livelihoods, infrastructure, etc. across various sectors. Thus, it lays great emphasis on assessing and addressing vulnerabilities of communities at risk and develops short-term and long-term strategies to significantly move toward climate resilient pathways. Schipper (2007) argues that the "vulnerability reduction approach" to development is more desirable than the "adaptation approach." In this view, the development processes help reduce vulnerability to climate change, which reduces the impacts of climate hazards. The 3P model positions itself in the ambit of CCA and CCM, sustainable development as well as DRR. The framework provides methodological tools to vulnerable communities as well as institutions in charge of policy formulation for 'climate-proofing' regions and people susceptible to both slow onset effects of climate change and drastic climatic events.

In this section, we elaborate the theoretical frame of each of the stages of the model individually and then provide the linkages between them that make it an effective model for CCA. Although the model has been conceptualized based on experiences from the ground, it however includes certain issues that have been topics of discussion in scientific and policy-making circles for a long time. Thus, in an attempt to theoretically legitimize the model, scientific references have been made that in turn lend credibility to the rationale behind the links that bind the three Ps.

Perceptions

As the word Perception suggests, this first stage of the model is all about 'understanding and addressing' the needs and problems of the targeted community. "Societies have inherent capacities to adapt to climate change," which is why communities are the best decision-makers on what they need to do to improve their realities (Adger 2010). The facilitators of the model implement the theoretical concept of acting as "choice architects" (Thaler and Sunstein 2008).

Through various participatory methods that include all the relevant stakeholders in the project, the facilitator or choice architect (who does not necessarily have to be an outsider) introduces the concept of climate change through real-life experiences of the target community and gets people to acknowledge the changes from their own understanding of observable changes. These local experiences and traditional knowledge on climatic hazards are then superimposed on the scientific projections of the regions to enunciate scientific evidences for the decision-makers. This in turn paves the way for acknowledgment and identification of their vulnerabilities and the linkages shared between various climate change-induced problems. This two-way dialogue enables a situation assessment by all involved and in the process catches the attention of the benefactors, in terms of both awareness and ownership of responsibility. It also provides a platform for building capacities and communicating evidence—local communities are informed about their rights and policy-makers enable the process of exerting these rights in an appropriate way. These information-sharing sessions also strengthen the accountability of the facilitators as these introduce in them accountability and ownership toward helping communities become more resilient. It is important therefore that members of policy-making institutions, namely local institutions and government bodies at various levels, are also included in the project from this stage.

Pilots

The technology-oriented pilots will be essential building blocks to showcase vulnerability reduction measures for climate-proofing communities. Based on the climate change-related impacts identified while framing perceptions, potential solutions are developed with the participation of the community for bottom-up measures. The documentation of the climate-related risks, key impacts across sectors, and adaptation options is defined as LAMG/strategy in the project intervention. The logical steps adopted in the formulation of the local strategies are similar to the OECD (2009) guidelines for mainstreaming CCA in developmental planning. In order to demonstrate that vulnerability reduction change is possible in practical terms, one specific climate change impact is selected with community engagement process for reducing the related threats. All project partners, with their expertise in technology development, engage in community discussions for prioritization of the adaptation options by reviewing the benefits and co-benefits of the suggested measures with careful assessment of affordability, technology transfer options, and operation and maintenance (O&M) aspects. This selection is also done based on a set of criteria that help to identify opportunities for climate proofing that would ensure long-term sustainability with the least amount of external inputs required. The immediate needs of the community are given significance in the decision-making process and they are empowered on prioritizing the solutions that could enhance their resilience. Community ownership also is made clearly visible in the pilot implementation stage by ensuring community contribution (in kind/cash), which leads to sustainability of the implemented measures.

Planning

The third stage essentially showcases the results of the implemented pilot measures for the purposes of mainstreaming, replication, and upscaling. As mentioned previously, the planning authorities are involved in the 3P model from the very beginning. However, the development of local solutions encourages them to climate-proof and localize sustainable development policy formulation and implementation. The results of the pilots provide all involved stakeholders a lived concrete case for engaging in climate-proofing initiatives that are driven from within, substantiating the argument by Adger (2010): "As individuals and groups interact synergistically with the state, so the institutions of the state also evolve in a process of policy learning." The real tangible benefits of the pilots are pertinent to engage policy-makers to develop strategies that allow for developing climate-resilient strategies to reduce the impacts. As the pilots allow for local-level innovation and responses, the replication and upscaling of the technical measures are undertaken by the community and local authorities. Any action by them is a reflection of the acceptance of the solution and a recognition of its suitability to address climate concerns.

Taking the Three Ps Together

For a successful implementation of this model for vulnerability reduction, all three stages are essential and important. The model meets the goals of reducing the vulnerabilities of communities to climate change and strengthening capacities of local authorities and the population on CCA, CCM, and DRR. The three separate stages of the model function in a manner whereby one stage naturally follows the next and results in addressing climate

Box 14.1:
Product of joint knowledge building: LAMGs

The LAMG, a living document that was compiled jointly with each community where a project was initiated, not only served as a resource and vulnerability guide but also contained details about pilots implemented in a region and actions needed to make them sustainable. It was the product of inputs from the communities (perceptions), proposed and implemented solutions (pilots), and development of adaptation measures taken. This document can be used by policy-makers and local administration to help them effectively integrate the project into their regular development (planning). In fact, the solutions and recommendations mentioned in the LAMG can be used as a resource by policy-makers across the board. Evidence of this as seen from the AdaptCap experience include:

■ The Revenue Divisional Officer (RDO) of Ponneri in the Thiruvallur district requested the LAMG exercise to be implemented in all coastal villages of the district.

■ National Institute of Disaster Management (NIDM) worked with the LAMG to review and develop a robust base for its disaster plans at the village and district levels (Gupta et al. 2014).

In addition, such a document becomes a ready-made handbook of solutions for other communities that face similar climate change-related problems and are situated in a similar context.

change-related issues from the wider perspective of sustainable development. From the very outset, the model works to build a sense of ownership and equity among the community members and local institutions to sustain the interventions (Thomas and Twyman 2005).

The pilots, in the 3P model, play a dual role. On the one hand, the pilots demonstrate solutions for addressing perceptions/solving problems by getting the beneficiaries to buy into the climate-proofing project. This creates enfranchised message bearers of the proposed solutions, showcasing that it can be done on the ground in specific circumstances, and ensures that the community is engaged with pilot implementation (documented in Local Adaption and Mitigation Guides [LAMGs]). The LAMG exists as a strategy document with the community and local authorities for climate-sensitive planning decisions (see Box 14.1). On the other hand, the pilots also become responsible for creating evidence for the Planning stage, and this is essential for taking climate-proofing measures further for upscaling and replication. An important aspect of the 3P model is communication and information sharing on the successful implementation of the pilot through the community members and local institutions. This is done by establishing a community-based communication network that not only acts as a vehicle of dissemination but also creates demands for change from the bottom-up. As per Adger et al. (2005),

> Environmental issues are defined by society to be appropriately tackled at a particular scale: ultimately the choice of how an environmental governance problem is handled within a jurisdiction is a reflection of the strength of the interests and power of the actors who define the problem.

This argument provides the rationale for choosing the district level as the nodal level of implementation of action on climate change in the intervention. In a federal and decentralized polity like India, the district level is the one that ultimately implements the locally situated policies and mobilizes funds for policy implementation. Thus, for successful implementation, it is the district that needs to take ownership for incorporating climate-proofing policies in its overall development plans. The focus for mainstreaming should also begin at the district level, and the local-level capacities for implementation should also be strengthened for effective mainstreaming.

AdaptCap[3]—A Case Study

As mentioned above, the 3P model is inspired by the lessons learnt while implementing the AdaptCap project that aimed to reduce vulnerabilities of coastal communities and strengthened capacities of local authorities to prepare and plan for coping with climate change in the coastal communities of Andhra Pradesh and Tamil Nadu.

Coastal communities are especially vulnerable to risks of extreme weather events and are also exposed to the immediate impacts of rising sea levels and their complex effects. Rising sea levels and storm surges result in saltwater incursion, which challenges crop productivity. Unpredictable rainfall and higher than average temperatures increase the difficulties in ensuring water availability. Flooding affects water quality and human health. Infrastructure

---

[3] For more information about AdaptCap and further references, please visit http://www.adaptcap.in

Figure 14.1:
3P model

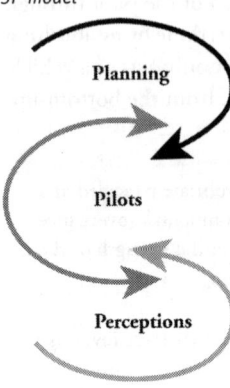

Source: Author.

damage, due to increasingly intense weather events, robs communities of what little physical assets they have, affecting productivity and community employment. Climate change events have multidimensional impacts that threaten the already fragile realities of coastal communities. AdaptCap interlinked the activities of CCA, DRR, and CCM, introduced the central role of partnership, and integrated holistic planning. The 3P model serves as a framework around which the AdaptCap project has been woven to create policy incidence and mainstream these policies into the local planning process.

As discussed in the previous section, understanding the perceptions of the target group forms an important part of the community-based approach and builds the first pillar of the model. AdaptCap took steps that involved communities to identify issues faced by them and their source. These steps, termed as Vulnerability and Needs Assessments (V&NA) and Participatory Rural Appraisals (PRA), constituted transect walks, interviews, and social mapping with the selected communities allowing them to examine their own experiences of climate change and discuss the triggers to the problems faced. As a result, the communities bridged their small stories of the changing landscape to the larger issue of climate change and understood how it affected their livelihoods, health, and village economic situation. For example, they were able to link their experience of reduced fish catch to the changing weather patterns and reduced crop yield to the advancing sea level and coastline. This helped them deduce the actual issue at hand and the importance of adapting to climate change.

In order to have a long-term effect, the AdaptCap process equipped communities with the skills and language required for further engagement with all levels of government, including the village panchayats, in order to give them a better understanding of the grassroots-level scenario. Since the request for village project initiatives goes to the district level through the panchayats and subdistrict levels, the need for them to be involved in this dialogue, to better understand the communities' requirements, was amplified. The outcomes of this were manifold.

- The project outline planned the V&NA for one ward of Ponneri Town Panchayat, Tamil Nadu, with meetings and shared dialogues with the city officials and local authorities. This process led to continuous engagement with the officials for assessing the climatic risks in the region and options for strategizing risk resilience. This engagement and capacity development led to the upscaling of the V&NA approach and its replication in the 18 wards as the officials realized the need of developing a strategy to enhance adaptive capacity of the city wards. Also, the plan is to share the vulnerability assessment studies with the chief minister of Tamil Nadu for effective allocation of resources for further work.

- The pilot initiatives that were selected for implementation were outcomes of a community-based approach as well. Based on their perceptions, communities were actively involved in suggesting solutions in order to adapt to climate change, that is, the pilots they wanted to be implemented in their respective villages. This list of solutions, which also formed a part of LAMGs, was then taken up for discussion at the next level. Apart from this, communities jointly participated in the cost–benefit analysis in order to assess the feasibility of pilot implementation with the AdaptCap partners and the beneficiaries. Selection of a pilot was based on its climatic benefits and feasibility in terms of cost, acceptance by the communities and their interest to contribute in and sustain the project, hence showcasing value addition. The sensitization of the local people brought about the following results to take the project forward.

- Workshops and trainings on pilot O&M raised community contribution to sustain them for a longer period. The fact that they were involved in the pilot selection process from the beginning brought about a sense of ownership for the project and motivation for it to succeed. It also eliminated seeking funds from other stakeholders for pilot sustenance.

- Members came forward to form task force committees that ensured community participation in pilot interventions and collection of beneficiary fees, hence filling the lacuna between the larger team of project actors and the local communities. Similarly, at the city level, city task forces were formed to carry out the same activities in the urban landscape.

At the final stage, AdaptCap ensured that the CCA pilots were implemented and showcased to planning officials. AdaptCap pilots were solutions that demonstrated benefits within a short span of time ensuring that communities kept up the momentum of action as well as financial contributions (Chaturvedi et al. 2013). The results and continuous involvement boosted the communities to continue their contributions to the project. The project further ensured that the beneficiaries benefited from the pilots in order for them and the state partners to appreciate and spread its success stories on a common platform. For instance, the following success stories emerged from the project:

- The pilot to replace firewood for cooking by smokeless cooking stoves and introducing new latrines in the village of Govupeta, Visakhapatnam district of Andhra Pradesh, reduced smoke by 90 percent and reclaimed 25 acres of land that was previously used for open defecation. This positively impacted 80 families and is thus taken up for replication in two more villages. Also, improved cook stoves led to better fuel efficiency and reduced indoor health hazards.

Twelve of the implemented 18 pilots have been chosen for replication. Most importantly, the costs of further replication were not borne by the project or other organizations but by the communities, and the local and state governments supported replication through local funds at the panchayat, district, and state levels by tapping the developmental schemes and plans. For instance, consider the following initiative: resource allocation for source improvement work for drinking water provision by the Rural Water Supply and Sanitation (RWSS) department in the Visakhapatnam district of Andhra Pradesh was supported by local government to enhance resilience of community. As a pilot measure, the project had facilitated discussions with the RWSS department on the community need for the provision of drinking water that is affected due to erratic rainfall and extreme weather events. The project had prepared the concept of the pilot measure with technical specifications, which was shared with the RWSS department. Since source improvement work involved other nearby communities as well, the RWSS department upscaled the approach for the entire region, and the project was funded by the Government of Andhra Pradesh. Similar success stories on resource allocation and mobilization are evident in other AdaptCap communities as well.

Conclusions

Human social adaptation entails a process of sustainable and permanent adjustment to new or changing environmental circumstances ... [i]t is an on-going process that has taken place since humanity's first appearance on Earth. An additional call for adaptation has been made recently to address increasing impacts of anthropogenic climate change and growing vulnerability to and risk of such impacts in developing countries. (Schipper 2007)

Although developed from ground experiences, the 3P model is based on concepts that have been discussed and debated in academic and scientific literature (Chishakwe et al. 2012; Dodman and Mitlin 2011; Reid et al. 2009). However, a large part of the research on dealing with climate change often does not provide practical guidance for policy-makers, which therefore hinders the capacity of planning institutions to practically address the stresses that vulnerable communities face. This model acknowledges the fact that "adaptation is a continuous stream of activities, actions, decisions and attitudes that informs decisions about all aspects of life, and that reflects existing social norms and processes" (Adger et al. 2005). Hence, a bottom-up approach that from the very outset gives the target community and policy-makers equity in the project and planning is essential to build response capacity.

The model also reflects that climate stresses are just one part of the overall development stresses that vulnerable communities face. Keeping this as the starting point, it makes beneficiaries talk, for real impact. These lived experiences are then converted to action through pilots, which are then upscaled and replicated. The climate change-related problems identified through detailed discussions with the local communities over the course of the implementation of the model and their solutions are documented in the LAMG, which becomes a guiding document for future climate-proofing projects. The LAMG is also a tangible product for policy-makers to use in planning sustainable development projects, which includes climate resilience in its ambit for other regions of the country or even the world for that matter. This model integrates the bottom-up knowledge of the community, which can thus be used for addressing development issues across the board with necessary changes that contextually situates the model for its application and implementation.

References

Adger, W.N. 2010. "Social Capital, Collective Action, and Adaptation to Climate Change." In *Der Klimawandel—sozialwissenschaftliche Perspektiven*, edited by M. Voss, 327–345. Wiesbaden: Verlag für Sozialwissenschaften.

Adger, W.N., W.A. Nigel, and L.E. Tompkins. 2005. "Successful Adaptation to Climate Change Across Scales." *Global Environmental Change* 15 (2): 77–86.

Ahmad, Q.K. and A.U. Ahmed. 2000. "Social Sustainability, Indicators and Climate Change." In *Climate Change and Its Linkages with Development, Equity and Sustainability: Proceedings of the IPCC Expert Meeting held in Colombo, Sri Lanka, 27–29 April, 1999*, edited by M. Munasinghe and R. Swart, 95–108. Colombo, Bilthoven, and Washington, DC: LIFE, RIVM, and World Bank.

Apuuli, B., J. Wright, C., Elias, and I. Burton. 2000. "Reconciling National and Global Priorities in Adaptation to Climate Change: With an Illustration from Uganda." *Environmental Monitoring and Assessment* 61 (1): 145–159.

Chaturvedi, A., R. Arora, M.S. Saluja, H. Mewes, R. Chakrabarti, K. Kumar, and G.G.K. Murthy. 2013. *Piloting People's Climate Adaptation.* New Delhi: Deutsche Gesellschaft für Internationale Zusammenarbeit (GIZ) GmbH. Available at: http://www.preventionweb. net/files/35836_35836pilotingpeoplesclimateadaptati.pdf (Accessed on February 27, 2017).

Chishakwe, N., L. Murray, and M. Chambwera. 2012. *Building Climate Change Adaptation on Community Experiences: Lessons from Community-based Natural Resource Management in Southern Africa.* London: International Institute for Environment and Development (IIED).

Dodman, D. and D. Mitlin. 2011. "Challenges for Community-based Adaptation: Discovering the Potential for Transformation." *Journal of International Development.* doi:10.1002/jid.1772

Gupta, A.K., S.S. Nair, S. Singh, A. Chaturvedi, R. Arora, M.S. Saluja, N. Mundra, and H. Mewes. 2014. "Strengthening Climate Resilience through Disaster Risk Reduction: Approach in Andhra Pradesh and Tamil Nadu in India—Experience and Lessons." Special Technical Paper, P 36. New Delhi: GIZ-IGEP and NIDM P 36. Available at: http:// nidm.gov.in/PDF/pubs/Strengthening%20Climate.pdf (Accessed on February 27, 2017).

Huq, S. and H. Reid. 2007. "Community-based Adaptation: A Vital Approach to the Threat Climate Change Poses to the Poor." IIED Briefing Paper (17005IIED). London: IIED.

IPCC (Intergovernmental Panel on Climate Change). 2001. "Adaptation to Climate Change in the Context of Sustainable Development and Equity." In *Climate Change 2001: Impacts, Adaptation and Vulnerability,* edited by J.J. McCarthy, O.F. Canziani, N.A. Leary, D.J. Dokken, and K.S. White, 877–912. Working Group II Contribution to the Third Assessment Report (AR 3) of the Intergovernmental Panel on Climate Change. Cambridge: Cambridge University Press.

———. 2014. "Climate-Resilient Pathways: Adaptation, Mitigation, and Sustainable Development." In *Climate Change 2014: Impacts, Adaptation, and Vulnerability* edited by C.B. Field, V.R. Barros, D.J. Dokken, K.J. Mach, M.D. Mastrandrea, T.E. Bilir, M. Chatterjee, K.L. Ebi, Y.O. Estrada, R.C. Genova, B. Girma, E.S. Kissel, A.N. Levy, S. MacCracken, P.R. Mastrandrea, and L.L. White. Working Group II Contribution to the Fifth Assessment Report (AR 5) of the Intergovernmental Panel on Climate Change. Cambridge: Cambridge University Press.

IPCC SREX. 2012. *Managing the Risks of Extreme Events and Disasters to Advance Climate Change Adaptation,* edited by C.B. Field, V. Barros, T.F. Stocker, D. Qin, D.J. Dokken, K.L. Ebi, M.D. Mastrandrea, K.J. Mach, G.K. Plattner, S.K. Allen, M. Tignor, and P.M. Midgley, 582. A Special Report of Working Groups I and II of the Intergovernmental Panel on Climate Change. Cambridge and New York, NY: Cambridge University Press.

Mirza, M.M.Q. 2003. "Climate Change and Extreme Weather Events: Can Developing Countries Adapt?" *Climate Policy* 3: 233–248.

Munasinghe, M. 2000. "Development, Equity and Sustainability (DES) in the Context of Climate Change." In *Climate Change and Its Linkages with Development, Equity and Sustainability: Proceedings of the IPCC Expert Meeting held in Colombo, Sri Lanka, 27–29 April, 1999,* edited by M. Munasinghe and R. Swart, 13–66. Colombo, Bilthoven, and Washington, DC: LIFE, RIVM, and World Bank.

OECD. 2005. "Conclusions of the Chair." Global Forum on Sustainable Development on Development and Climate Change, November 11–12, 2004, Paris. Available at: www.oecd. org/dataoecd/60/7/34393852.pdf (Accessed on February 27, 2017).

OECD. 2009. *Integrating Climate Change Adaptation into Development Co-operation: Policy Guidance*. Paris: OECD.

Reid, H., M. Alam, R. Berger, T. Cannon, S. Huq, and A. Milligan. 2009. "Community-based Adaptation to Climate Change: An Overview." In *Participatory Learning and Action: Community-based Adaptation to Climate Change*. London: IIED. Available at: http://pubs. iied.org/pdfs/14573IIED.pdf (Accessed on February 27, 2017).

Robinson, J.B. and D. Herbert. 2000. "Integrating Climate Change and Sustainable Development." In *Climate Change and Its Linkages with Development, Equity and Sustainability: Proceedings of the IPCC Expert Meeting held in Colombo, Sri Lanka, 27–29 April, 1999*, edited by M. Munasinghe and R. Swart, 143–162. Colombo, Bilthoven, and Washington, DC: LIFE, RIVM, and World Bank.

Schipper, L. and M. Pelling. 2006. "Disaster Risk, Climate Change and International Development: Scope for, and Challenges to Integration." *Disasters* (Special Issue: Climate Change and Disasters) 30 (1): 19–38.

Schipper, F.L.E. 2007. "Climate Change Adaptation and Development: Exploring the Linkages." Tyndall Centre for Climate Change Research Working Paper 107, 13. Available at: http://www.preventionweb.net/files/7782_twp107.pdf (Accessed on February 27, 2017).

Sekine, H., K. Fukuhara, A. Uraguchi, C.K. Tan, M. Nagai, and Y. Okada. 2009. "The Effectiveness of Community-based Adaptation (CBA) to Climate Change—From the Viewpoint of Social Capital and Indigenous Knowledge." GEIC Working Paper Series 2009–00. Mitsubishi Research Institute & United Nations University–Institute of Sustainability and Peace (UNU–ISP), Tokyo.

Shukla, P., R. Nair, M. Kapshe, A. Garg, S. Balasubramaniam, D. Menon, and K. Sharma. 2003. *Development and Climate: An Assessment for India*. Ahmedabad: Indian Institute of Management, Ahmedabad.

Thaler, R.H. and C.R. Sunstein. 2008. *Nudge: Improving Decisions About Health, Wealth, and Happiness*. New Haven, CT: Yale University Press.

Thomas, D.S.G. and C. Twyman. 2005. "Equity and Justice in Climate Change Adaptation amongst Natural-resource-dependent Societies." *Global Environmental Change* 15 (2): 115–124.

Tompkins, E.L. and W.N. Adger. 2005. "Defining Response Capacity to Enhance Climate Change Policy." *Environmental Science & Policy* 8 (6): 562–571. Available at: http://www. ecologyandsociety.org/vol9/iss2/art10/ (Accessed on February 27, 2017).

Warner, K., S. Kreft, M. Zissener, P. Höppe, C. Bals, T. Loster, C. Linnerooth-Bayer, S. Tschudi, E. Gurenko, A. Haas, S. Young, P. Kovacs, A. Dlugolecki, and A. Oxley. 2013. "Insurance Solutions in the Context of Climate-change-related Loss and Damage: Needs, Gaps and Roles of the UNFCCC in Addressing Loss and Damage." In *Policy, Diplomacy and Governance in a Changing Environment*, edited by O.C. Ruppel, C. Roschmann, and K. Ruppel-Schlichting. Baden-Baden: Nomos.

15

Local Knowledge, Social Capital, and Governance of Climate Change Adaptation in Bangladesh

Md. Masud-All-Kamal

Introduction

Climate change governance has gained bourgeoning attention in recent years because of frequent extreme climatic events already appeared as threat in many regions of the world. Societies respond to climate change through mitigation and adaptation. Mitigation focuses on reducing greenhouse emission, which causes climate change. It is evident that climate change mitigation policies have failed to curb emissions over the past two decades. The fifth assessment report of Working Group 1 of the Intergovernmental Panel on Climate Change (IPCC) has reiterated its earlier findings with an additional emphasis for taking stronger mitigation actions if the world wishes to keep global temperature below 2°C over the next century. Recognizing failure in the mitigation efforts, adapting to these changes has become an increasing policy concern in the present world. Adaptation involves involuntary spontaneous changes and deliberative adaptation strategies. It is important to consider both spontaneous responses evolved when a social system intrinsically reacts and responses and planned policy interventions by the national and local governments. The unprecedented pace of present climate change and the increasing complexity of societies suggest that autonomous, self-regulated societal adaptation alone is not sufficient; governments have to play an active role (Berkhout 2005). National governments are expected to provide general framework and guidance on how to adapt to climate change, and expected to raise awareness and co-fund adaptation projects (Bauer and Steurer 2014). Other entities such as civil society, business, and individuals are also involved in facilitating climate change adaptation (Adger et al. 2005). The key responsibility of climate change adaptation, however, lies to country's government as it is accountable of coordinating across different agents and institutions.

It is evident that adaptation is a dynamic process in which multiple actors and agents in society are involved. The affected communities respond to climate variability autonomously that are conceptualized as community-based adaptation. As such, adaptation to climate change is inherently local, and its efficacy depends on local and extra-local institutions through which incentives for individual and collective actions are structured (Agrawal 2010). This argument suggests that a bridge between top-down and local knowledge is indispensable to address climate challenges. Societies have intrinsic capacities to adapt to climate change, but these capacities are tied with their ability to act collectively (Adger 2003). In other words, the adaptive capacity to climate change in part depends on how different actors respond together. Evans (1996) has proposed a state–society synergy view of social capital that can be a catalyst for development. The Government of Bangladesh (GoB) and civil society organizations especially nongovernmental organizations (NGOs) have taken a number of strategies and actions to adapt the challenges of climate change and achieve the development objectives. The success of these efforts largely depends on their collective actions, which recognizes that social capital is central to adaptation and community resilience. As Adger (2003) argues, social capital is a necessary glue for adaptive capacity, particularly in dealing with climate change risks. Bangladesh is relatively advance to enhancing adaptive capacity as well as building resilience, though it does not have any comprehensive climate change policy (Haq and Ayers 2008). Bangladesh is one of the first least developed countries (LDCs) that have developed a National Adaptation Programme of Action (NAPA) and then carried out ambitious Bangladesh Climate Change Strategy and Action Plan (BCCSAP). However, the plans encounter a set of new challenges including when these adaptation plans start to move from planning to practice. They further face challenges with regard to integrating or mainstreaming them into development plans. In addition, the integration of local knowledge into the policies and strategies has been inadequate.

This chapter critically reviews climate change adaptation policies of Bangladesh with a focus on the prevailing concepts, perspectives, and institutional structure. Using social capital perspective, this chapter aims to answer the following questions: To what extent do existing adaptation policies, strategies, and actions that are taken by the government respond to reduce vulnerability of the real victims? What is the institutional and legal landscape for climate change adaptation in Bangladesh? Do the adaptation policies and actions build on local knowledge? Are the government initiatives effective and sustainable in curbing the climate risks of susceptible population? The chapter begins by describing a relationship between climate change, adaptive capacity, and social capital. This is followed by a discussion of the country characteristics of Bangladesh that make it vulnerable to climate change. It then examines the government's policy and program responses to the climate change. The final section concludes the chapter by summarizing the main findings and calls to redesign the country's adaptation efforts that may lead to successful outcomes.

Adaptation, Adaptive Capacity, and Social Capital

Adaptive capacity is an important aspect of resilience indicating to what extent people are capable to withstand the natural extremes. Chapin et al. (2009: 23) contend that adaptive capacity is "the capacity of actors, both individuals and groups, to respond to, create, and shape variability and change in the state of a system." It is evident that every society has some inherent capacity to adapt. But this capacity is far outstripped compared to unprecedented climate change effects. It is therefore dependent on the wider governance regimes and different kinds of assets/capitals (White et al. 2004). Among different types of capital, social capital is central, as other capitals such as human, physical, and financial are largely produced and activated in a certain social milieu.

Notwithstanding social capital is a recent development in theory and research; it has different dimensions and perspectives. The concept of social capital is therefore defined in diverse ways. Putnam (1993: 167) has defined social capital as "features of social organization, such as trust, norms, and networks, that can improve the efficiency of society by facilitating coordinated action." Fukuyama (2001) sees "social capital [a]s an instantiated informal norm that promotes co-operation between two or more individuals." From a more developmental aspect, Woolcock and Narayan (2000) maintain that social capital is composed of the norms and networks that enable people to act collectively. Despite these divergences in conceptualization, academicians and practitioners alike agree that features such as trust, networks, norms of reciprocity, and exchange are common in social capital. Some consensus also exists for the analytical purpose in the dimensions of social capital: bonding, bridging, and linking. *Bonding* social capital refers to the strong social relationship among homogeneous ties of the family, neighbors, and close friends. It reinforces specific reciprocity and mobilizes solidarity (Putnam 2000: 22). *Bridging* social capital is composed of socially heterogeneous ties that connect people across social groups by which they can access wider resources. Economic sociologist Granovetter (1973) calls it 'weak ties' and suggests that 'weak ties' are more important than 'strong' interpersonal ties in enhancing community cohesion and collective action. Putnam (2000) sees that bridging social capital is essential for generating broader identities and reciprocity. *Linking* social capital refers to link between masses and people in power. Szreter and Woolcock (2004: 655) refer to linking social capital as "norms of respects and networks of trusting relationships between people who are interacting across explicit, formal and institutionalized power or authority gradients in society." This extended network therefore provides people access to the scarce resources. These three types of social capital are essential for facilitating collective action, which, in turn, is crucial to climate change adaptation and community resilience.

Among the different perspectives of social capital, synergy view receives the most attention in public policy debates. Evans (1996: 1119) argues that "Norms of cooperation and networks of civic engagement among ordinary citizens can be promote[d] by public agencies and used for development ends." He proposes that synergy between government and citizen action is depended on complementarity and embeddedness. *Complementarity* refers to mutually supportive relations between public and private actors, while *embeddedness* refers to the nature and extent of the ties connecting citizens and public authorities. Woolcock and Narayan (2000) have summarized the synergistic perspective as follows: (a) Neither the state nor societies are inherently good or bad; governments, corporations, and civic groups are variable in the impact they can have on the attainment of collective goals; (b) Identifying the condition under which these synergies emerge is a central task of development research and practice; and (c) The role of the state, among different actors, in facilitating positive developmental outcomes is the most important and problematic. This indicates that government policies and actions have a crucial role in shaping good state–society relations. As Evans (1996) observes, "Creative action by government organizations can foster social capital; linking mobilized citizens to public agencies can enhance the efficacy of government." Therefore, the major role of public policy and institutions is to engage ordinary citizens and include their knowledge and experiences into actions. Such ties across the state–society transform traditional ties into "developmentally effective social capital" (Evans 1996: 1122). This synergistic social capital, as Adger (2003) argues, promotes the adaptive capacity of societies to cope with climate change and promotes sustainability and legitimacy of any adaptation strategy. The construction of synergy is therefore essential to achieve better outcomes in any given adaptation program.

The governance of climate change adaptation comprises a range of issues, among others: the state of adaptation preparedness; institutional arrangements and capacities; the scale of funding required for adaptation; the best ways to administer development cooperation support; effective mechanisms for delivery; and mechanisms to ensure that adaptation efforts target and benefit the most vulnerable sectors of society (Madzwamuse 2010). Nevertheless, appropriate adaptation policies and planning play a crucial role in addressing the unavoidable and inevitable adverse impacts of human-induced climate change. Madzwamuse (2010) has argued that adaptive capacity is dependent on policies and strategies that are formulated in response to the needs as well as enhance the resilience of the most vulnerable systems and groups in society. In doing so, adaptation policy-making process needs to be deliberate and participatory (Ayers 2011). Besides, the capacity of social groups to act in their collective interest depends on the quality of the formal institutions under which they reside (Woolcock and Narayan 2000: 234). Effective adaptation to climate change therefore requires new governance approach that is able to bridge or even transcend governmental levels and social domain (Adger et al. 2005; Bauer et al. 2012). In sum, adaptation policy and strategy based on local knowledge and social capital can enhance community resilience to climate change.

The Context

The Vulnerability Context of Bangladesh

Bangladesh is globally recognized as one of the most vulnerable countries to climate change and it is predicted that the impacts of climate change can endanger the entire progress toward achieving development. The country is already facing a number of extreme climatic events such as sever flood, drought, salinity intrusion, and cyclone. These impacts have been threatening people's lives and livelihoods over the past decades. Between 1980 and 2010, for instance, 234 major disasters were recorded in Bangladesh, resulting in 191,836 deaths and causing about US$17 billion in economic damage with high losses in agriculture and infrastructure.[1] Climate scientists affirm that

---

[1] www.preventionweb.com

recent weather-related extreme events are clear evidence of climate change impacts, and in the near future, Bangladesh will face more drastic consequences. A recent study shows that Bangladesh is one of the potential climate change impact hotspots in South Asia threatened by extreme floods, more intense tropical cyclone, rising sea levels, and very high temperatures due to global warming (World Bank 2013). This vulnerable condition of Bangladesh is caused by its geographical position and socioeconomic conditions.

The geographical location and hydro-geological characteristics, such as dominance of flood plains and low elevation from the sea, have made Bangladesh one of the most vulnerable ones to climate change. In other words, the country's prospect is trapped between the melting Himalayas in the north and the encroaching Bay of Bengal in the south. Most of the country is low-lying land comprising mainly the delta of rivers Ganges, Brahmaputra, and Meghna and more than 230 major rivers and their tributaries and distributaries. The floodplain comprises 80 percent of the land, and only in the extreme do land elevations exceed 30 m above mean sea level, making the majority of Bangladesh prone to flooding for a part or whole of the year (MoEF 2005). The floodplains of northwestern, central, south-central, and northeastern regions are subject to regular flooding, while coastal plains are subject to cyclone, storm surges, salinity intrusion and coastal inundation (MoEF 2005).

With nearly 152 million inhabitants on a landmass of 147,570 sq. km, Bangladesh is among the most densely populated countries in the world. According to the UNDP Human Development Report 2013, Bangladesh ranked 146 out of 187 countries on the Human Development Index (HDI), with a score of 0.515, and remains a less developed country. Yet, despite lower levels of economic growth, Bangladesh has achieved greater gains in human development than its neighbors such as India, Pakistan, and Nepal. Bangladesh has achieved massive development, particularly in food production, life expectancy, population control, maternal mortality, and primary education. Despite these successes, an overwhelming majority (47 million) of its population lives below the poverty line and climate change–induced risks pose tremendous challenges to the efforts of poverty reduction.

Climate Change and Development: Bangladesh Perspective

Climate change is considered one of the most severe threats to sustainable development, with adverse impacts on food security, human health, physical infrastructure, natural resources, and environment. This will impact severely to achieve Bangladesh's development aspirations and its roadmap for sustainable development. Bangladesh is still an agriculture-based society; almost 60 percent people's livelihood directly and indirectly depends on this sector, which makes the country's economy relatively sensitive to climate variability and change. The IPCC in its fourth assessment report (2007) stated that the production of rice and wheat might drop in Bangladesh by 8 percent and 32 percent respectively by the year 2050. It is argued that due to sea level rise, costal Bangladesh has already experienced the worst impacts, especially in terms of costal inundation, saline intrusion, deforestation, and loss of biodiversity. About 2.38 ha of arable land is affected by varying degrees of soil salinity (IPCC 2007). With a 1 meter rise in sea level, the Sundarbans mangrove forest is likely to be lost, and 1,000 sq. km of cultivated land will become salt marsh. Moreover, the intensity and frequency of flood have clearly increased over the past decades, causing damage of people's lives and livelihoods. According to a primary estimation by Ministry of Environment and Forests (MoEF) in Bangladesh, in 2007 flood 32,000 sq. km of land was inundated, over 85,000 house destroyed or partially damaged, and damage estimated over US$1 billion (GoB 2008a).

The GoB is committed to the Millennium Development Goals (MDGs), including halving poverty and hunger by the year 2015. Climate change is challenging the country's ability to achieve the high rates of economic growth and efforts to MDGs targets. Over the last 10 years, the annual cost to the national economy of disaster is estimated to be between 0.5 and 1 percent of GDP (Climate Change Cell 2007). In the coming years, it is predicted that there will be increasingly frequent and severe floods, tropical cyclones, and droughts, which will disrupt the life of the nation and the economy. In the worst-case scenario, sea level rise could result in the displacement of millions of people from costal region and have huge adverse impacts on the livelihoods and long-term health of a large proportion of the population. It is now essential to enhance the country's adaptive capacity and to safeguard its future economic well-being.

Development depends on the improvement of its economy, environment, and society over the time. Any effort to achieve development goals is and will be challenged in Bangladesh due to climate change. Adaptation is therefore considered best option to encounter such problems, which meets also the development challenges. Several researches suggest that adaptation is inseparable from development (Ayers 2011). Development policies and actions have been taken in considering adaptation to climate change. Poverty Reduction Strategy Paper (PRSP), for instance, recognizes direct link between poverty and vulnerability to natural hazards and has proposed an anticipatory approach to reduce vulnerability: to natural, environmental, and human-induced hazards through community empowerment and integration of sustainable risk management initiatives in all development programs and projects (GoB 2008b). In addition, specific policies have been formulated to deal with climate change challenges. The following section reviews the Bangladesh's adaptation policies.

Government's Strategies and Actions to Climate Change

Bangladesh Climate Policies

Bangladesh government's first major climate change-related policy document is National Adaption Programmes of Action (NAPA) prepared by MoEF in 2005, which was a response to the decision of the Seventh Session of the Conference of the Parties (COP7) of the United Nations Framework Convention on Climate Change (UNFCCC). The aim of NAPAs is to identify priority activities that respond to their urgent and immediate needs to adapt to climate change, those for which further delay would increase vulnerability or lead to increase in costs at a later stage (UNFCCC 2013). Like other LDCs, NAPA is the principal mechanism through which Bangladesh has identified its adaptation needs and programs. This adaptation plan drew upon the understanding gathered through the discussion with relevant stakeholders in four regional workshops, one national workshop, and background papers prepared by six sectoral working groups. NAPA identifies priority activities to address urgent and immediate needs to adapt existing and anticipated adverse effects of climate change. Although NAPA was an action plan, it ended up proposing 15 immediate adaptation projects, categorized into two major types, that is, intervention and facilitating, in order that those can be financially supported by development partners. NAPA was formulated on the basis of project-based approaches and it was drawn up with little participation by affected communities (Kamaluddin et al. 2006). NAPA was endorsed in November 2005, but over the following four years, it had been resulted in seminars, meetings, and workshops in Dhaka and outside Bangladesh. Given the experiences, MoEF had updated NAPA in 2009 focusing on four basic national security issues: food security, energy security, water security, and livelihood security. This also set 18 priority projects without any ranking for design and implementation at short and medium terms (Islam et al. 2013). Unfortunately, only a handful adaptation projects have started at community level since NAPA was endorsed. For that matter, the results attained so far from first climate change adaptation policy are not much convincing. Reviewing 18 NAPAs including Bangladesh's one, Agrawal and Perrin (2009) conclude that not only little attention was paid to incorporate local communities and institutions in adaptation plans, but little evidence of consultation and coordination between the local and national levels can also be seen in the descriptions of selected high-priority projects. The assessment of UN Development Programme (UNDP) finds two serious shortcomings in identifying adaptation needs of LDCs:

> First, they provide a limited response to the adaptation challenge, focusing primarily on "climate proofing" through small-scale projects…. Second, the NAPAs have, in most countries, been developed outside the institutional framework for national planning on poverty reduction. The upshot is a project based response that fails to integrate adaptation planning into the development of wider policies for overcoming vulnerability and marginalization. (cited in Martin 2010)

Considering the drawbacks of NAPA, the GoB prepared BCCSAP in 2008 and revised it in 2009. BCCSAP was developed due to the specific interest of UK government and presented in London so that UK government commits to financially support to implement this plan. Consultants were given the authorization by the government to draft this plan and UK government funded for the consultant (Hossain 2009). Most of the time, governments like Bangladesh with little financial and technical capabilities cannot frame their policies internally and these kinds of policies do not give the result as expectation. Thus, donor organizations and governments expose their internal affairs in developing countries.

BCCSAP is designed as a 10-year program from 2009 to 2018 to build the capacity and resilience of the country to meet the challenges of climate change (GoB 2008a). It has six broad themes, and under these pillars, 120 actions and 37 programs have been identified for the first five years (2009–2013). BCCSAP's six themes are (a) food security, social protection, and health (9 programs and 26 actions); (b) comprehensive disaster management (4 programs and 10 actions); (c) infrastructure (7 programs and 28 actions); (d) research and knowledge management (5 programs and 19 actions); (e) mitigation and low carbon development (7 programs and 22 actions); and (f) capacity building and institutional strengthening (5 programs and

15 actions). Indeed, this policy illustrates almost all extent to cope with the climate change challenges. However, BCCSAP did not specify year-wise programs and actions, which could be helpful to evaluate the achievements.

BCCSAP has given highest emphasis on infrastructure development as a means of climate change adaption. The action plan even did not propose any Environmental Impact Assessment (EAI) before implementing those new interventions. Hossain (2009) contends that many infrastructure projects in the past created environmental problems in many areas of Bangladesh and reduced the capacity to adapt with the climate change impacts. This policy therefore may jeopardize the whole adaptation process and create problems to people's livelihood. BCCSAP was drafted for financial support from the development partners, and donor demands were prioritized. Hossain (2009: 11) observes that the actions are "perceived as a costly activity with less chance of getting support from the development partners." This donor convincing policy and action do not represent the needs of the real sufferers.

BCCSAP was formulated through a fully consultative process involving government, civil society, and development partners. As a consequence, the views of affected communities were largely ignored. Moreover, political parties, private sector, and some of the renowned climate experts of Bangladesh were also not consulted during the formulation process. During the formulation of BCCSAP, the government did not carry out any need assessment. This is revealed in the financial plan, where it is stated that

> It is estimated that a $500 million program will need to be initiated in years 1 and 2 (e.g., for immediate actions such as strengthening disaster management, research and knowledge management, capacity building and public awareness programs, and urgent investments such as cyclone shelters and selected drainage programs) and that the total cost of programs commencing in the first five years could be of the order of $5 billion. (GoB 2008a: 29)

Without any need assessment and necessity, this estimation is not unambiguous. So, this kind of strategy may not be as effective as expected to adapt with the risk of climate change.

Bangladesh has minimal contribution to global climate change, but it is highly vulnerable to risks surfaced for climate change. Addressing such a problem, actions to reduce greenhouse emissions in Bangladesh have never been considered an important measure. Rather, the country can strongly argue in global climate change negotiation for immediate and drastic mitigation activities in highly emitting developed countries. On the contrary, BCCSAP has proposed actions under the theme 'migration and low carbon development' without indicating the utmost necessity of mitigation in developed countries, which is far more important to reduce climate change impacts in Bangladesh. Thus, the policy has failed to provide the clear vision for Bangladesh regarding its stand in global climate change negotiation. The most important point Bangladesh should consider is limitless adaption in absence of mitigation activities. Furthermore, putting mitigation in the policy is unnecessary, as there would be eventual question of allocating resources for that. Nicholas Stern (2007) observes that the developing countries' low income makes the adaptation difficult. In spite of the need for billions of dollars for adaptation activities, which is very scarce for a poorest country like Bangladesh, it is unwarranted at the moment to allocate resources for mitigation activities.

Government Actions

The GoB has taken a number of initiatives to adapt to climate change. The government has established a Climate Change Cell in the Department of Environment (DoE) under the MoEF in 2004 to especially address current impacts of climate change and manage future risks and variability at all levels. Its objective is to enable the management of long-term climate risks and uncertainties as an integral part of national development planning. The Climate Change Cell promotes partnership with both government and non-government agencies to serve long-term and immediate needs of the climate-vulnerable people. In addition, the main activities have been the implementation of the BCCSAP through the two climate change funds, namely the Bangladesh Climate Change Resilience Fund (BCCRF) and Bangladesh Climate Change Trust Fund (BCCTF). The former has several hundred million US dollars from donor countries such as the United Kingdom, Sweden, Australia, Denmark, Switzerland, and others, while the latter has several hundred million from Bangladesh's own resources. Both funds have been supporting activities by the government agencies and civil society organizations as part of the BCCSAP. The BCCRF has invested over US$170 million grant funds to build their resilience to the effects of climate change since its establishment in 2010. These investments in adaptation include coastal embankment, foreshore afforestation, multipurpose cyclone shelters, and further strengthening of the early warning systems to improve climate resilience. The BCCTF-funded projects include providing benefits to the climate-vulnerable communities through providing cyclone-tolerant houses in Cyclone Aila-affected areas, ensuring safe drinking water, improving sanitation, constructing rubber dams for irrigation, repairing and constructing embankments in coastal and river erosion-prone areas, protecting river banks, improving drainage system, excavating and re-excavating canals and rivers, improving cropping system in drought and coastal saline areas, reclaiming land through construction of cross-dam, creating coastal greenbelt to protect coastal areas, conserving biodiversity, and reducing greenhouse gas emission. The main challenge remains to ensure transparency, accountability, and governance of funds. Studies show that lack of good governance and corruption are hindering utilization of these funds. Transparency International Bangladesh (TIB) has assessed several projects funded by BCCTF and published as a report titled 'Climate Finance in Bangladesh: Challenge of Good Governance and Way Forward' in the last quarter of 2013. It finds that political consideration got priority as 13 out of 63 projects were given to nongovernment agencies closely linked to political parties. In addition, their study does not find the existence of 10 NGOs that received funding from BCCTF. The TIB also finds that only 17 NGOs had some prior experience with environment, climate change, and disasters. According to the report, out of the total BCCRF fund of US$170 million, the government has already allocated US$146.19 million for 11 projects. But, the assessment reveals that despite the most climate-vulnerable areas of the country as per the risk map of BCCSAP (GoB 2009) are Khulna, Satkhira, and Bagerhat districts, they received only 6.5 percent, 1.2 percent, and 0 percent allocation respectively. This indicates that there are inadequacy and mismanagement regarding implementation of climate change adaptation actions.

The government adaptation actions are mainly infrastructure based and implemented through top-down approach, which sometimes damage people's livelihoods and environment. Following the devastating flood of 1988, the Bangladesh Flood Action Plan (FAP), which proposed the construction of several thousand kilometers of flood embankment along the country's three rivers, was implemented in 1990. This plan displaced millions of people, damaged inland fisheries, failed to protect from cyclone-induced tidal surge, and failed to address the potential impacts of climate change (Thomalla et al. n.d.). Bangladesh government has been built around 2,500 cyclone shelters in costal districts so far in order to protect people from storms and surges, which is one of the six key mitigating actions together with embankments, afforestation, early warning systems, awareness building, and communications. Nevertheless, without implementing the latter measures the former will not be successful. For example, in Cyclone 'Sidr' many people died due to lack of awareness and communications, though there were places to take shelters. Thus, emphasizing more on physical capital cannot secure lives and livelihoods of vulnerable people; actions should address the other capitals equally. Following the Cyclone Aila, households of affected area were predominantly relied on natural, human, and social capital to recover the loss and damage (Masud-All-Kamal 2013). There is a need to focus on 'soft technologies' as an important component of climate change adaptation. Although the GoB is also expanding community-based climate change adaptation, the large share of climate change funds has been spending on top-down structural measures.

The government recognizes that tackling climate change requires an integrated approach involving many different ministries and agencies, civil society, and the business sector (GoB 2008a). Apart from the government, NGOs have appeared as an indispensable actor in climate change adaptation. From policy-making to implementation of adaptation programs, they are considered an important actor. NGOs are responding to climate shocks and implementing programs in anticipating future climate risks to enhance resilience of communities. Over time, NGOs have changed their approaches on climate change and disaster management in relation with international dominant approaches. It is argued that NGOs contribute to transfer vulnerable communities into resilient ones with a more participatory approach. They have pioneered community-based approaches to reduce vulnerability to climate change. NGOs claim their success in working at grassroots–national levels mainly to build awareness and capacity to address the climate change challenges. They are also playing an important role in generating knowledge through research and blending scientific knowledge into practice. Along with the donors, the government has been channeling climate change adaptation funds through NGOs. Although NGOs and the government have similar objectives of climate change adaptation, there is a lack of coordination between them, particularly at local level. It is also criticized that NGOs' adaptation policies are driven by the donors and ineffective to promote sustainable livelihoods to the vulnerable. Although they work at the community level and participate with community people, their programs and projects are designed by the funding agencies and project experts. In short term, these projects could be helpful to increase adaptive capacity, but evidence is scarce on the longer term impacts. Both the government and NGOs are dependent on project approach to build climate-resilient communities. These 'projectized' climate change adaptation strategies and actions have been decried for its inability to make results self-sustaining as these are discrete activities, aimed at specific objectives with earmarked budgets and limited time frames (Honadle and Rosengard 1983). Honadle and Rosengard also argue that in contrast with 'programs' that occupy a more permanent status in an institutional setting, projects have a tendency to either bypass or fragment local institutions and therefore to neglect the need for local capacity building.

Conclusion

Achieving adaptive capacity to tackle climate change hazards is one of the fundamental development challenges for Bangladesh. The government and NGOs are working to enhance the capacity to climate change adaptation at all levels. International communities have also been extending their support by identifying it as one of the most vulnerable countries. However, the government's policies and programs are not representing the needs of the most vulnerable groups. The government actions also have lack of understanding of people's risks and livelihoods, which jeopardized the country's efforts to achieve climate resilience development. Unless the policies of the government are reshaped based on local experiences and participation of the most vulnerable people, the expected results may not be achieved as planned. There is a need to start planning for adaptation at local, regional, and national levels in an integrated way so as to decrease the vulnerability of the people from climate change. In doing so, state–society linkages are important. The synergy approach of social capital may be supportive to increase state–society interaction through embeddedness and complementarity. This social capital lens is more likely to result in a more comprehensive and productive adaptation actions.

One of the most fundamental aspects of adaptation is the capacity of local people and institutions. The government has a key role in planned adaptation to climate change through increasing the capacity of local communities. In doing so, as some observers note, adaptation must consider adaption risks and development as strongly complementary. This approach leads to focus on developmental needs such as access to livelihood and productive assets in order to enhance the adaptive capacity of more vulnerable people. Although the government has pronounced the importance of local knowledge in climate change adaptation and disaster management, there is little reflection of the recognition in policy and actions. This constraint is resulted in limited success of addressing vulnerability of local people. Without understanding the attributes of social organization, any planned climate change adaptation initiatives may not be successful. Social capital perspective allows realizing the constraints as well as dynamics of local social institutions. In other words, this leads toward a more comprehensive understanding of social and political capacities of communities that climate change strategies and actions intend to enhance. The inclusion of local knowledge is a precondition for successful policy-making as climate variability is locally and contextually specified, as well as communities' responses mediated by local social structure.

Adaptation is a dynamic process, and consequently combined efforts determine in part the success of tackling climate change. In Bangladesh context, the effectiveness of climate change adaptation depends largely on well coordination and partnership between government institutions and NGOs. The state can act proactively to achieve such synergistic interaction. In doing so, the forms of social capital in a particular society need to be understood, and its promotion and facilitation are the critical resource for sustainable adaptation planning. Policy formulation for climate change adaptation can thereby be viewed as policy learning.

References

Adger, N.W. 2003. "Social Capital, Collective Action, and Adaptation to Climate Change." *Economic Geography* 79 (4): 387–404.

Adger, N.W., N.W. Arnell, and E.L. Tompkins. 2005. "Successful Adaptation to Climate Change Across Scale. *Global Environmental Change* 15 (2): 77–86.

Agrawal, A. 2010. "Local Institutions and Adaptation to Climate Change." In *Social Dimension of Climate Change: Equity and Vulnerability in a Warming World*, edited by R. Mearns and A. Norton, 173–197. Washington, DC: The World Bank.

Agrawal, A. and N. Perrin. 2009. "Climate Adaptation, Local Institutions and Rural Livelihoods." In *Adaptation to Climate Change: Thresholds, Values, Governance*, edited by W.N. Adger, I. Lorenzoni, and K.L. O'Brien, 350–367. Cambridge: Cambridge University Press.

Ayers, J. 2011. "Resolving the Adaptation Paradox: Exploring the Potential for Deliberative Adaptation Policy-making in Bangladesh." *Global Environmental Politics* 11 (1): 62–88.

Berkhout, F. 2005. "Rationales for Adaptation in EU Climate Change Policies." *Climate Policy* 5 (3): 377–391.

Bauer, A., J. Feichtinger, and F. Steurer. 2012. "The Governance of Climate Change in 10 OECD Countries: Challenges and Approaches." *Journal of Environmental Policy and Planning* 14 (3): 289–304.

Bauer, A. and R. Steurer. 2014. "Multi-level Governance of Climate Change Adaptation Through Regional Partnership in Canada and England." *Geoforum* 51: 121–129.

Chapin, III, F.S., G.P. Kofinas, and C. Folke, (eds). 2009. *Principles of Ecosystem Stewardship: Resilience-based Natural Resource Management in a Changing World.* New York: Springer.

Climate Change Cell. 2007. *Climate Change and Bangladesh.* Dhaka: Department of Environment, Government of Bangladesh. Available at: http://www.climatechangecell-bd. org/publications/13ccbd.pdf (Accessed on February 28, 2017).

Evans, P. 1996. "Government Action, Social Capital and Development: Reviewing the Evidence on Synergy." *World Development* 24 (6): 1119–1132.

Fukuyama, F. 2001. "Social Capital, Civil Society and Development." *Third World Quarterly* 22 (1): 7–20.

GoB (Government of Bangladesh). 2008a. *Bangladesh Climate Change Strategy and Action Plan (2008).* Dhaka: Ministry of Environment and Forest. Available at: http://www.moef.gov. bd/moef.pdf (Accessed on June 25, 2013).

———. 2008b. *Moving Ahead: National Strategy for Accelerated Poverty Reduction II (FY 2007–11).* Dhaka: General Economic Division, Planning Commission.

Granovetter, M. 1973. "The Strength of Weak Ties." *American Journal of Sociology* 78 (6): 1360–1380.

Haq, S. and J. Ayers. 2008. *Climate Change Impacts and Responses in Bangladesh.* Brussels: European Parliament.

Honadle, G.H. and J.K. Rosengard. 1983. "Putting 'Projectized' Development in Perspective." *Public Administration and Development* 3 (4): 299–305.

Hossain, M.K. 2009. "Birth of Climate Change Policy and Related Debates: Analyzing the Case of Bangladesh." Paper presented at "Environmental Policy: A Multinational Conference on Policy Analysis and Teaching Methods," Seoul, June 11–13. Available at: http://umdcipe.org/conferences/epckdi/6.pdf (Accessed on February 28, 2017).

IPCC (Intergovernmental Penal on Climate Change). 2007. *Climate Change 2007: Synthesis Report.* Geneva: IPCC. Available at: http://www.ipcc.ch/pdf/assessment-report/ar4/syr/ ar4_syr.pdf (Accessed on February 28, 2017).

Islam, A., R. Shaw, and F. Mallik. 2013. "National Adaptation Programme of Action." In *Climate Change Adaptation Actions in Bangladesh,* edited by R. Shaw, F. Mallik, and A. Islam, 93–106. Tokyo: Springer.

Kamaluddin, A.M., A.U. Ahmed, N. Haque, A. Islam, M. Reazuddin, I.S. Rector, M.S. Ali, Z. Haque, and R. Ernst. 2006. *Climate Resilient Development: Country Framework to Mainstream Climate Risk Management and Adaptation.* Dhaka: Climate Change Cell.

Madzwamuse, M. 2010. *Climate Governance in Africa: Adaptation Strategies and Institutions.* Cape Town: Heinrich Böll Stiftung.

Martin, S. 2010. Climate Change, Migration, and Governance." *Global Governance* 16 (3): 397–414.

Masud-All-Kamal, M. 2013. "Livelihood Coping and Recovery from Disaster: The Case of Coastal Bangladesh." *Current Journal of Social Sciences* 5 (1): 35–44.

MoEF (Ministry of Environment and Forests). 2005. *National Adaption Programme for Action (NAPA).* Dhaka: GoB.

Putnam, R. 1993. *Making Democracy Work: Civic Traditions in Modern Italy.* Princeton, NJ: Princeton University Press.

———. 2000. *Bowling Alone.* New York, NY: Touchstone.

Stern, N. 2007. *The Economics of Climate Change: The Stern Review.* Cambridge: Cambridge University Press.

Szreter, S. and M. Woolcock. 2004. "Health by Association? Social Capital, Social Theory, and Political Economy of Public Health." *International Journal of Epidemiology* 33 (4): 650–667.

Thomalla, F., T. Cannon, S. Haq, R.J.T. Klein, and C. Schaerer. n.d. "Mainstreaming Adaptation to Climate Changes in Costal Bangladesh by Building Civil Society Alliances." Available at: http://iiedtest.merfa.co.uk/pubs/pdfs/G00016.pdf (Accessed on May 28, 2013).

UNFCCC. 2013. "Chronological Evolution of LDC Work Programme and Introduction to the Concept of NAPA." Available at: https://unfccc.int/adaptation/knowledge_resources/ ldc_portal/items/4722.php (Accessed on February 28, 2017).

White, P., M. Pelling, K. Sen, D. Seddon, S. Russell, and R. Few. 2004. *Disaster Risk Reduction: A Development Concern.* Norwich, UK: University of East Anglia.

Woolcock, M. and D. Narayan. 2000. "Social Capital: Implications for Development Theory, Research and Policy." *The World Bank Observer* 15 (2): 225–249.

World Bank. 2013. *Turn Down the Heat: Climate Extremes, Regional Impacts, and the Case for Regional Bangladesh.* Washington, DC: The World Bank.

SECTION 4

Synergizing 'Top-down' and 'Bottom-up'

16

Neither 'Top-down' nor 'Bottom-up': A 'Middle-out' Alternative to Sustainable Development

Stephen Zavestoski and Pradip Swarnakar

We know that a data revolution is unfolding, allowing us to see more clearly than ever where we are and where we need to go, and to ensure that everyone is counted in. We know that creative initiatives across the world are pioneering new models of sustainable production and consumption that can be replicated. We know that governance at both the national and international levels can be reformed to more efficiently serve twenty-first century realities. And we know that today our world is host to the first truly globalized, interconnected and highly mobilized civil society, ready and able to serve as a participant, joint steward and powerful engine of change and transformation. (United Nations General Assembly 2014: 7)

This statement by the UN General Assembly reflects an understanding of the scale of the sustainable development challenge we face and contends that given this scale, volumes of data as well as the best possible technologies for big data management and analysis must be brought to bear. We conclude this volume by suggesting that the scale of the challenge, and perhaps more importantly the depth of the solutions required, calls not just for the quantitative big data implied in the UN General Assembly's statement but also for 'qualitative big data'. By 'qualitative big data', we emphasize the local and grassroots-level knowledge of the stakeholders—knowledge without a priori structure that tends to be documented in ways through case studies that produce 'unstructured' data. Case studies, like the research presented here in this volume, allow us to capture the important depth and richness of cultural contexts that must be understood for sustainable development 'solutions' to work at the local level. If harnessed at a large enough scale, this depth and richness derived from specific case studies can meaningfully inform policy, economic, technological, social, and other innovations across the infinitely diverse communities and cultures endeavoring to achieve sustainable development.

Beginning with Chapter 1, we have emphasized the need to bridge bottom-up and top-down approaches to sustainable development. Section 1 consisted of case studies calling to question strictly top-down approaches. On the one hand, from Chatterjee's illustration of the shortcomings of the Forest Rights Act to Pushkaran's account of the limitations of science to enumerate India's tigers or Yadav and Sahu's description of the failure of conservation measures in the Chambal Wildlife Sanctuary to meet the expectations of local people, these chapters point out how top-down approaches can be undermined by unanticipated or misunderstood dimensions of local contexts. On the other hand, as Chandrashekar and Seshaiah demonstrate in Chapter 4, top-down approaches do have the potential to improve weak or ineffectual institutions, for example, land governance, especially when they introduce innovative hybrid forms of governance such as those described in the context of 'comanagement' in Bangladesh's wetlands by Chowdhury in Chapter 6.

The chapters in Section 2 explore experiments with bottom-up approaches to sustainable development. Whether flood management in Assam, community-based water resource management in the Indian Sundarbans, or community-based ecotourism in Goa, these chapters point to the potential for bottom-up approaches to produce relevant solutions in which affected communities are invested. But they also point to the limitations of bottom-up approaches. Lack of access to financing, a problem of rural local bodies discussed by Rajanayagam in Chapter 9, is one example. Likewise, the deployment of solar home lighting systems in rural Indian villages, as demonstrated by Singh in Chapter 10, can suffer when local communities lack access to technology, infrastructure, or knowledge required to maintain solar home lighting systems.

Finally, in Section 3 we complicate matters by introducing the imperative of climate change adaptation in a series of chapters that explore the opportunities and challenges of achieving sustainable development goals while adapting to climate change. These chapters tell us that local knowledge and perceptions of climate change matter in how people choose to adapt and that adaptive capacities can be enhanced through education. They also highlight the need for bridging top-down and bottom-up approaches. Collectively, these chapters reveal that local and grassroots sustainable development strategies are widespread throughout India and Bangladesh and increasingly well documented by case study methodology. They also show us that innovative efforts to integrate local knowledge with institutions linked to the sustainable development apparatus are also occurring. These lessons point to the importance of developing strategies for aggregating and disseminating models of innovation and adaptation practiced at the local level so that high-level institutions can expand their knowledge of sustainable development strategy and diffuse this knowledge across the sustainable development apparatus.

To date, academics and policy-makers have insufficiently utilized the extensive knowledge being produced in case studies of sustainable development and climate change adaptation. This is partially due to the nature of case studies that are seen as having limited usefulness because idiosyncratic differences across cases make generalizations problematic. This is precisely the tension identified in Chapter 1 between the homogeneous and universal goal of 'sustainable development' and the unique and site-specific sustainable development goals of communities. In this concluding chapter, we aim to make a case for the power of case studies through accumulating knowledge that is only available when a large number of case studies are aggregated and then analyzed. We call for scaling up the aggregation of case study research to a regional and then global level so that knowledge accumulation and dissemination can happen more rapidly and efficiently than these currently do, all the while respecting local knowledge and epistemology.

Adapting Case Studies for a 'Middle-out' Bridge Between Top-down and Bottom-up

Given the scale and speed of global climate change and its attendant socio-ecological disruptions, there is an urgent need to begin identifying successful sustainable development strategies across multiple social, geographical, and temporal scales. This volume makes the case that neither top-down approach nor bottom-up approach alone has the ability to produce the needed insight. And yet, given the demands of climate adaptation, community-based sustainability experiments—often little more than tinkering or muddling in livelihood and survival strategies—are occurring on a daily basis around the world. The case study approach represents one possible method for beginning to tap into the knowledge being produced through these experiments. Furthermore, the information technology now exists to facilitate compilation of qualitative reports of community-based sustainability experiments into a vast corpus with the potential to be mined by both communities and individual actors and high-level institutions and decision-makers within the sustainable development apparatus.

Figure 16.1:
'Middle-out' coordination of sustainable development efforts

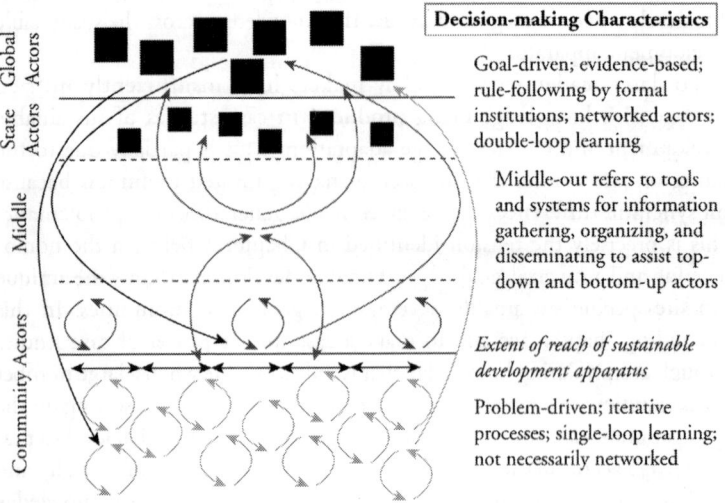

Source: Authors.

Figure 16.1 depicts the relationship between the state and global actors currently controlling the sustainable development apparatus and the community actors engaged in iterative, and typically closed, processes of experimentation to achieve local sustainable development goals. The 'middle-out' approach does not call for new institutions bridging these extremes. Middle-out refers instead to a system of information flow wherein lessons learned in the previously isolated and closed-loop community-based experiments are aggregated and organized. With access to this information, top-down decision-makers can more effectively support strategies relevant to local contexts and community-based actors, who previously relied on ad hoc networks to learning anecdotally from one another, can more systematically learn directly from one another.

In short, the middle-out approach relies on aggregation of case studies and other reports of community-based sustainability experiments to develop a knowledge base for community-based sustainability experimentation processes that can be leveraged by a wide range of decision-makers and actors at multiple levels. The 'middle-out' platform will create a space for both top- and bottom-level stakeholders to combine the power of the descriptive richness of qualitative case studies with the analytical power of natural language processing and other information science tools. This strategy holds the potential to bring to bear substantial yet untapped knowledge in sustainable development strategies. The virtual 'middle-out' space is possible because of present rapid diffusion and access to information and communication technologies (ICTs) in developing countries such as India and Bangladesh. Brabham (2009: 243) has identified that "the medium of the Web enables us to harness collective intellect among a population in ways face-to-face planning meetings cannot." The task of creating a structured and user-friendly 'middle-out' platform is no doubt challenging as it involves vast amounts of unstructured data. However, we believe that if successful, this platform can open up a plethora of rich local-level knowledge, which can be utilized from local- to national- and global-level stakeholders.

The middle-out approach begins with the assumption, explained above, that small experiments in climate adaptation and sustainable development are producing extensive yet underutilized knowledge vital to the social learning required for societies to transition to a sustainable future. It draws on the 'small experiment framework', an approach to behavior change developed by psychologists (Irvine and Kaplan 2001; Kaplan 1996), that is premised on the notion that the human tendency to 'muddle through' problem situations can be modified into a form of 'adaptive muddling' that can result in more rapid social learning and change (De Young and Kaplan 2012). In 'muddling through', humans constantly take small steps but without straying far from the results of past changes. The problem is that "[m]uddling is a process characterized by … a tendency to compromise, and an avoidance of significant bold or visionary steps" (De Young and Kaplan 2012: 290). Key to adaptive muddling is that it emphasizes not small steps but small experiments.

According to De Young (1999: 602),

> It offers a way of simultaneously exploring several possible solutions, thus avoiding the sluggishness that plagues one-solution-at-a-time approaches. People are empowered to apply local or personal knowledge to a situation. Different people applying different knowledge to the same situation creates a variety of potential solutions.

Small experiments are a framework to support problem-solving, as De Young (2014: 9) explains, "that is based on people's innate inclinations to explore and understand (Kaplan and Kaplan 2003, 2009) and on their brain having evolved to prospect the future not just track the past (Seligman et al. 2013)." As important, small experiments support behavioral innovation and have the potential for rapid dissemination of findings. Finally, the ubiquity of small experiments makes them an ideal unit of analysis. "Small experiments are going on all the time," contends De Young (2014). "Consider the many pilot programs, demonstration sites, field tests, and trial runs regularly reported in both popular and scientific publications, as well as [in] neighborhood, community, and village examples." Many of the chapters in this volume exemplify small experiments—whether in rural electrification or local governance of ecotourism, wildlife, or water.

Novelty of Middle-out Approach

What is unique about the middle-out approach is its focus on what is essentially an information management system capable of capturing knowledge emerging out of bottom-up small experiments and integrate it with top-down knowledge, strategies, and resources. In this manner, the middle-out approach aims to overcome the gap between the historically dominant top-down approach to sustainable development and the more recent emphasis on bottom-up approaches in which solutions emerge and are implemented by communities.

Our middle-out approach proposes utilizing a whole new set of tools only recently available to network formal and informal knowledge, thus moving beyond a process orientation (e.g., inserting NGOs as mediators between top-down and bottom-up actors) toward an actual platform to facilitate knowledge integration that informs more effective decision-making. We opened this chapter by pointing out that the UN General Assembly is quite aware of both the scale of the challenge ahead of us and the emerging technological potential to scale data and knowledge to meet the challenge. Many of the same ICT advances that have ushered in the era of highly quantitative 'big data' to which the UN General Assembly refers can be deployed to harness the power of 'qualitative big data' central to a middle-out approach. The sources of these data, we propose, are the individuals working independently within their communities to develop innovative approaches in the struggle to meet everyday needs. Even at the lowest socioeconomic strata, increased access to mobile and Internet technologies makes these individuals potential data producers. Their knowledge and experiences can be documented, aggregated, and networked so that grassroots knowledge at the bottom of the pyramid can become more accessible to policy-makers, development professionals, and others working from the top-down.

This newly networked knowledge, which exists in a conceptual 'middle', is valuable not only to the extent that it is pushed upward into the top-down decision-making process but also to the extent that it can be accessed by grassroots actors to drive learning and innovation. In our proposed middle-out approach, individuals and communities would be able, using the same mobile and Internet technologies used to upload local sustainable development knowledge, to access the networked knowledge resulting from the synthesis of wide-ranging case studies, reports, and informal documentation of community-level sustainable development efforts. The primary advantage of utilizing ICTs to begin networking otherwise isolated knowledge is that the same technologies can be utilized to overcome both vertical communication barriers between bottom-up and top-down actors and horizontal communication barriers preventing geospatially dispersed communities from learning from one another. For example, a community on the Tamil Nadu coast might recover from flood damage through a combination of top-down government intervention and bottom-up initiatives. An academic researcher or NGO involved in the recovery might next produce a case study documenting the successes and failures of the efforts and their associated learning. Normally, such a report winds up isolated in academic journals or might gain modest circulation among development professionals. Our middle-out approach proposes utilizing ICT tools to aggregate this case study with thousands of others, apply machine learning to organize and categorize the knowledge, and package it for presentation through an online platform. Through this, middle-out approach, in the future communities in West Bengal, Bangladesh, or elsewhere, can learn from the specific experiences of the community in Tamil Nadu as well as from the knowledge that emerges by networking these experiences in a database of thousands of other experiences.

In this manner, the middle-out platform has the following advantages:

- Geographically isolated communities can more easily exchange information and knowledge.
- National- or state-level policy-makers can be more strategic in applying top-down strategies by learning from previously inaccessible and disconnected knowledge that emerges from communities.
- The World Bank, Asian Development Bank (ADB), and other financial institutions that direct funding through state and national governments toward poverty eradication and sustainable development can see more transparently the justifications of decisions about why, when, where, and how to fund specific initiatives, especially at the grassroots level where large institutions seldom apply their resources directly.
- The middle-out approach can reduce the costs of obtaining community-level feedback because the middle-out framework itself generates reports

on the experiences of communities engaging in sustainable development initiatives.

- The middle-out approach complements rather than replaces the existing top-down and bottom-up approaches to sustainable development governance. Case studies are valuable in providing micro-level contextual knowledge, but at the same, generalizing across cases is problematic because each case study represents a sample size of one. Our middle-out approach proposes to aggregate large numbers of case studies and then apply qualitative big data analysis techniques that can produce topics, categories, and tags to represent them, that can make generalization across the cases possible.
- The most important feature of middle-out approach is that it accumulates indigenous and traditional knowledge that would otherwise be underutilized or even lost.

While the middle-out approach has great potential, it also faces the following challenges:

- Even though the penetration of ICTs toward the bottom of the pyramid is occurring even faster than earlier (Chib et al. 2015; Chuang and Schechter 2015), the penetration is not even and has not benefited all sections of the society (Dutta and Das 2016).
- Even if the citizens of developing countries are getting more and more access to ICTs, personal skills and abilities may limit one's potential to benefit from the presence of the technology. One example is the problem of illiterate mobile users, for whom technological innovations around audio, graphical, or other non-text interfaces are needed.
- The middle-out approach is based on the premises of structuring text-based qualitative data that is by its nature unstructured. For example, micro case studies provide nuanced facts that are highly heterogeneous in nature. Advanced text analytics procedures can be deployed for such a challenge but have not been tested in this context.

Middle-out Approach: Beyond Participatory Development

Participation, empowerment, and partnership are the key concepts related to development discourse. Participation 'by the people' is defined not just as a basic need but also as a fundamental human right, which is a tool for empowerment (Cornwall 2002, 2003). According to the World Bank (1996), participatory development is "a process through which stakeholders influence and share control over development initiatives, and the decisions and resources which affect them." The overarching objective of participatory development is to involve people and communities actively in identifying problems, formulating plans, and implementing decisions over their lives (DFID 2003). In this process, all the stakeholders have a critical role in the success of the project (Chopra et al. 1990; Mansuri and Rao 2004). A stakeholder is "any individual, community, group or organization with an interest in the outcome of a programme, either as a result of being affected by it positively or negatively or by being able to influence the activity in a positive or negative way" (DFID 2003: 2.1). Yet nowhere in the participatory development literature is discussion of how to aggregate dispersed and isolated knowledge that exists in communities and individuals throughout the world so that participation is not just a process but also a crucial mechanism for accessing and applying networked knowledge. The middle-out approach we describe here fills this gap.

The middle-out approach presumes participatory processes as a starting point and then proposes tools for aggregating the cumulative knowledge of communities to be accessed through participatory processes, thus making them more effective. If scaled up appropriately, the middle-out platform can create a significant amount of data, which can be utilized across communities. As such, the middle-out approach enhances participatory development. We introduce the middle-out approach in this concluding chapter with the hope of provoking readers into considering the possibilities, within this new conceptual model, for ICTs to provide tools that address the gaps between top-down and bottom-up approaches captured in the preceding chapters. As a conceptual model, we do not have details of actual implementation to share with readers. A small pilot project aggregating case study and other bottom-up knowledge in single country would be a logical next step for testing proof of concept.

In a modest way, this volume points to the potential value of building a middle-out approach by aggregating case studies across bottom-up and top-down approaches to sustainable development. Scaled up 100 or even 1,000 times from the 14 case studies compiled here, new insights will emerge and new strategies for locally relevant and contextualized sustainable development strategies will follow. Local innovators and leaders could have access to extensive and systematic case study-derived knowledge to adapt and apply to local experiments, while top-down institutions can more sensitively and strategically place resources to facilitate local strategies. The decentralized structure and horizontal rather than vertical linkages will provide peer-to-peer communication and diffusion of knowledge with the potential to overcome imbalances of power, economic inequality, cultural barriers, and other social factors that prevent the dissemination, flow, and implementation of sustainability solutions. Governance for sustainable development in the anthropocene will require nothing less.

References

Brabham, D.C. 2009. "Crowdsourcing the Public Participation Process for Planning Projects." *Planning Theory* 8 (3): 242–262.

Chib, A., M.H. van Velthoven, and J. Car. 2015. "mHealth Adoption in Low-Resource Environments: A Review of the Use of Mobile Healthcare in developing Countries." *Journal of Health Communication* 20 (1): 4–34.

Chopra, K., G.K. Kadekodi, and M.N. Murty. 1990. *Participatory Development: People and Common Property Resources*. New Delhi: SAGE Publications.

Chuang, Y. and L. Schechter. 2015. "Social Networks in Developing Countries." *Annual Review of Resource Economics* 7 (1): 451–472.

Cornwall, A. 2002. *Beneficiary, Consumer, Citizen: Perspectives on Participation for Poverty Reduction*. SIDA. Available format: http://www.sida.se/contentassets/232a6ee3f3504314 b13dc2d793933b03/beneficiary-consumer-citizen–perspectives-on-participation-for-poverty-reduction_1627.pdf (Accessed on August 20, 2016).

———. 2003. "Whose Voices? Whose Choices? Reflections on Gender and Participatory Development." *World Development* 31 (8): 1325–1342.

De Young, R. 1999. "Tragedy of the Commons." In *Encyclopedia of Environmental Science*, edited by D.E. Alexander and R.W. Fairbridge, 601–602. Hingham, MA: Kluwer Academic Publishers.

———. 2014. "Some Behavioral Aspects of Energy Descent: How a Biophysical Psychology Might Help People Transition Through the Lean Times ahead." *Frontiers in Psychology* 5 (1255): 1–16.

De Young, R. and S. Kaplan. 2012. "Adaptive Muddling." In *The Localization Reader: Adapting to the Coming Downshift*, edited by R. De Young and T. Princen, 287–298. Cambridge, MA: MIT Press.

DFID (Department for International Development). 2003. *Tools for Development: A Handbook for Those Engaged in Development Activity*. Performance and Effectiveness Department, DFID. Available at: http://webarchive.nationalarchives.gov.uk/+/http:/www.dfid.gov.uk/ Documents/publications/toolsfordevelopment.pdf (Accessed on August 20, 2016).

Dutta, U. and S. Das. 2016. "The Digital Divide at the Margins: Co-designing Information Solutions to Address the Needs of Indigenous Populations of Rural India." *Communication Design Quarterly Review* 4 (1): 36–48.

Irvine, K.N. and S. Kaplan. 2001. "Coping with Change: The Small Experiment as a Strategic Approach to Environmental Sustainability." *Environmental Management* 28 (6): 713–725.

Kaplan, R. 1996. "The Small Experiment: Achieving More with Less." In *Public and Private Places*, edited by J.L. Nasar, and B.B. Brown, 170–174. Edmond, OK: Environmental Design Research Association.

Kaplan, S. and R. Kaplan. 2003. "Health, Supportive Environments, and the Reasonable Person Model." *American Journal of Public Health* 93: 1484–1489. doi:10.2105/AJPH. 93.9.1484.

———. 2009. "Creating a Larger Role for Environmental Psychology: The Reasonable Person Model as an Integrative Framework." *Journal of Environmental Psychology* 29: 329–339. doi:10.1016/j.jenvp.2008.10.005.

Mansuri, G., and V. Rao. 2004. "Community-based and -driven Development: A Critical Review." *The World Bank Research Observer* 19 (1): 1–39.

Seligman, M.E.P., P. Railton, R.F. Baumeister, and C. Sripada. 2013. Navigating into the Future or Driven by the Past." *Perspectives in Psychological Science* 8 (2): 119–141.

United Nations General Assembly. 2014, December 4. "The Road to Dignity by 2030: Ending Poverty, Transforming All Lives and Protecting the Planet." Report No. A/69/700. Available at: http://www.un.org/ga/search/view_doc.asp?symbol=A/69/700&Lang=E (Accessed on Jan 29, 2016).

World Bank. 1996. *The World Bank Participation Sourcebook* (online). Available at: http://documents.worldbank.org/curated/en/289471468741587739/The-World-Bank-participation-sourcebook (Accessed on August 20, 2016).

About the Editors and Contributors

Editors

Pradip Swarnakar is Associate Professor of Sociology at the ABV-Indian Institute of Information Technology and Management Gwalior, India. Dr Swarnakar's'research areas include environmental sociology, climate change, social networks, and sustainability transition. He has completed projects funded by the Indian Council of Social Science Research and the National Science Foundation. He was a visiting scholar at the Department of Sociology, the University of San Francisco, USA; the Department of Social Research, the University of Helsinki, Finland; and the Department of Urban and Environmental Sociology, the Helmholtz Centre for Environmental Research GmbH-UFZ, Germany. He was a recipient of the Fulbright-Nehru Academic & Professional Excellence Award, the Kone Foundation Senior Researcher grant, and the Deutsche Forschungsgemeinschaft (German Research Foundation) grant. His work on climate change has been published in the *British Journal of Sociology*.

Stephen Zavestoski is Professor of Sociology and Environmental Studies at the University of San Francisco, USA. Dr Zavestoski's research areas include environmental sociology, social movements, sociology of health and illness, and urban sustainability. He is co-editor of *Social Movements in Health* (2005) and *Contested Illnesses: Citizens, Science, and Health Social Movements* (2012). Dr Zavestoski's current work explores strategies to address both sustainability and public health through urban and transportation planning. This work has culminated in *Incomplete Streets: Processes, Practices and Possibilities* (2014), co-edited with Julian Agyeman. Dr Zavestoski is co-editor of the book series *Equity, Justice, and the Sustainable City*.

Binay Kumar Pattnaik is Professor of Sociology at the Indian Institute of Technology Kanpur, India, and former Director of the Institute for Social and Economic Change, Bangalore, India. He has contributed handsomely to the areas of sociology of science and technology, social movements, and development studies. More particularly, he has contributed to the studies on technology transfer and R&D in the Indian industry, scientific productivity, effects of globalization and liberalization on science and technology regimes in developing countries, science technology and social stratification, etc. His recent contributions include papers in People's Science Movement in India, Appropriate Technology Movement in India (journal titled *Sociology of Science and Technology*), and ICT Revolution in India (*Polish Sociological Review*). His latest contribution pertaining to sustainable technologies from the developing societies based on grassroots-level innovations featured in the *Technology in Society* (2014). In addition to a good number of papers/articles, he has contributed nine books (both written and edited, including guest-edited volumes of international journals) of which the latest one is entitled *Sociology of Science and Technology in India* (2013, SAGE).

Contributors

Savita Aggarwal is Associate Professor at the Institute of Home Economics, University of Delhi with more than 25 years of experience in teaching, research and extension. She specializes in the field of Community Resource Management and Extension from the University of Delhi and is currently Head, Department of Development Communication and Extension at the Institute. Her major research interests are in the areas of sustainable development, development communication, and gender issues including gender-specific impacts of climate change. She has been the principal investigator for several research projects and has authored two books, *Quality of Life of Farm Women and Media for Effective Communication*. She is actively involved in planning and production of both traditional media and information and communication technologies (ICTs) for development communication. She is currently guiding PhD students undertaking research in the area of gender, ICTs, health, climate change adaptation, and water poverty.

Rachna Arora is Senior Technical Advisor of Resource Efficiency & Management of Secondary Raw Materials at Deutsche Gesellschaft für Internationale Zusammenarbeit (GIZ) GmbH, Germany. She is working as a deputy team leader in the resource efficiency project of the GIZ on issues related to fostering resource efficiency and secondary resource utilization in the construction and mobility sector. She has been working with

GIZ since 2007 under bilateral and combi-finance projects of the European Commission, and partnership projects between the private sector and relevant ministries and government departments. Her areas of expertise are electronic waste management, climate change, sustainable consumption, environmental fiscal instruments, and circular economy. She has a doctorate degree in environmental chemistry from the Indian Institute of Technology Roorkee, India. She has several articles and papers published in national and international journals, has book publications, and is a part of the research and development (R&D) committee set up by the Department of Information Technology and Communication, GoI, on electronic waste management.

Marie-Charlotte Buisson holds a PhD in economics from the CERDI (Centre for Studies and Research on International Development), France. Her research interests cover impact evaluation, agricultural microeconomics, natural resources management, vulnerability, and households' behavior. Along with her research, Marie-Charlotte has conducted several impact evaluations and socioeconomic surveys around the world (Africa, Central Asia, and Southeast Asia) with the purpose of enlightening the decision-making process at the micro level in order to understand how vulnerability may be alleviated. Marie-Charlotte joined IWMI, New Delhi office, in June 2012, and is now International Researcher in Economics and Impact Evaluation. She leads several research projects on community-led organizations for improved water access and better water governance.

Jyotiprasad Chatterjee is Associate Professor at the Department of Sociology, Barrackpore Rastraguru Surendranath College, West Bengal, India, and pursues research in the field of social movements, ethnicity, and ethnic movements under the broader scope of democracy–development interface. He has co-authored a book and contributed articles to different journals, edited volumes, and newspapers.

Ashish Chaturvedi is Director-Climate Change at GIZ. Ashish specializes in climate change mitigation and adaptation, waste management, sustainable consumption and production, and environmental policy. His current focus is on policy and knowledge management aspects of climate change. During 2014–16, Ashish worked as a fellow in the Green Transformations Cluster of the Institute of Development Studies (IDS) at the University of Sussex, UK, and as a senior fellow with adelphi, a German think tank. From 2006 to 2014, he led the Policy for Environment and Climate component of the bilateral Indo-German Environment Programme. As a senior advisor, he advised policy-makers at various levels of government and implemented projects in the area of waste/resource management, climate change, and economic instruments. Ashish has published extensively in books, journals, and general interest magazines. A PhD in economics from the University of California, USA, Ashish's PhD thesis explored the role of influence activities on democratic policy-making processes.

Mohammad Abu Taiyeb Chowdhury is Professor and former Chairman of the Department of Geography and Environmental Studies, University of Chittagong, Bangladesh. He has held academic positions at the University of Western Ontario (UWO), London, UK, and Ryerson University, Toronto, Canada. He received BSc (Honors) and MSc degrees in geography from the University of Dhaka, Bangladesh, and later obtained advanced degrees—MA and PhD—in the same from UWO. He is a recipient of the International Development Research Centre (IDRC, Ottawa, Canada) Award for Doctoral Studies and an elected fellow of the Royal Geographical Society (FRGS), London, UK, with the Institute of British Geographers. Prior to joining the University of Chittagong, he worked in Indonesia for four years (1997–2001) under a prescribed development assistance program of the Government of Canada, that is, a program funded by the Canadian International Development Agency (CIDA). As part of the program, he served the UN-ESCAP Centre for Poverty Alleviation through Sustainable Agriculture (CPASA, Bogor, Indonesia) as an agro-economic training specialist. He has published widely in the areas of environment and sustainable development with a focus on rural resources.

Arpita Das has a doctorate in social work from the Tata Institute of Social Sciences, Mumbai, India. She is currently the Education Coordinator at Shakti Community Council, Auckland, New Zealand, where she is working toward developing a program in sustainable social work practices. Prior to this, Arpita was a corporate social responsibility professional at Bharat Oman Refineries Limited in India. In this capacity, she implemented programs in education, skill development, and sustainable development. She has taught students of social work in India and Finland. She has been involved in research projects funded by the UKIERI, ICIMOD, University Grants Commission, and Caritas India, to name a few. Her research interests include indigenous knowledge, disaster management, environment and ecology, social work theory, and ethnographic studies.

Partha Jyoti Das is an environmental scientist with more than 15 years of experience of research, advocacy, and education in the fields of hydro-meteorology, water resources management, climate change, disaster risk mitigation, adaptation and resilience, water conflict, and transboundary water governance mostly in the Brahmaputra river basin in India. He has carried out more than 20 research and action research projects funded by various national and international donor agencies. He has about 20 publications comprising research papers in national and international journals, technical reports, monographs, and articles. He has edited two books, one on sustainable natural resources management and the other on water conflicts in Northeast India. He did his MPhil and PhD from Gauhati University, Assam, India. Dr Das is presently working as Head, 'Water, Climate & Hazard Division' in Aaranyak, a front-ranking environmental NGO of India. Dr Das works closely with communities and civil society organizations, as well as government and nongovernment agencies with a mission to promote environmental sustainability, disaster resilience, and environmental equity and justice for vulnerable and marginalized people and endangered ecosystems. Dr Das is a member of the Steering Committee of the Forum for Policy Dialogue on Water Conflicts in India, a reputed water think tank of India, based in Pune. He is a life member of the Indian Society for Ecological Economics and the India Water Partnership. He is also working as a researcher in several national and international research consortiums.

Satabdi Datta is currently pursuing her doctoral research at the Department of Economics, Jadavpur University, Kolkata, India. Her present research interests revolve around areas of natural resource management, disaster vulnerability, risk, and resilience in coupled human–environmental systems. She is also working with the Global Change Programme, Jadavpur University (GCP-JU), as a project fellow. She completed her undergraduate and post-graduate degrees in economics from the University of Calcutta, India. She has done MPhil in development studies from the Institute of Development Studies, Kolkata.

Rohini Fadte is Associate Professor at the Department of Sociology, K.J. Somaiya College of Arts and Commerce, Mumbai, India. She has been associated with the field of teaching for the last 20 years. She is pursuing doctoral research at the University of Mumbai in the field of tourism. Her research interests include gender issues, environmental concerns, sociology of education, and communities studies. She was awarded the Asiatic Society of India's scholarship for her project on 'Experiments in Ecotourism'.

Md. Masud-All-Kamal is Assistant Professor at the Department of Sociology, University of Chittagong, Bangladesh, and a doctoral researcher at The University of Adelaide, Australia. He has obtained a Master of Science degree from Lund University, Sweden. Masud has several articles published in reputed academic journals in the areas of climate risk management and public policy. His current research interests are social capital, governance, climate change adaptation, and coastal management.

Archisman Mitra is Research Associate at the International Maize and Wheat Improvement Center (CIMMYT), New Delhi, India. His current research relates to sustainable intensification of cereal-based farming systems in South Asia. His previous position was at the International Water Management Institute where he worked on impact evaluation studies of the water policy change in West Bengal and community-based water management practices in Bangladesh. Archisman is an economist by training and has a master's degree in economics from New York University (NYU), USA.

Nikita Mundra is Certification Specialist, Fair Trade USA. Nikita pursues interests in sustainable agriculture, labor rights, waste management, and environmental policy. Her work currently focuses on trade compliance of commodities such as cocoa, sugar, and coconut-based commodities on Fair Trade USA standards, and delves into environmental and social impacts of the Fair Trade model on farming communities worldwide. Nikita has previously worked with the Indo-German Environment Programme on e-waste management policies and instruments of climate change adaptation in coastal communities of India. She has also published papers that explore the role of municipalities in e-waste management, and that emphasize on the integration of climate change adaptation and disaster management. Nikita's master's thesis investigated the effect of e-waste management policy on the informal sector size in India.

Farhat Naz is Endeavour Scholar, Crawford School of Public Policy, College of Asia & the Pacific, The Australian National University, Canberra, Australia. She is a development sociologist by academic training and earned her PhD in development studies from the Centre for Development Research (ZEF), University of Bonn, Germany. She has master's and MPhil degrees in sociology from Jawaharlal Nehru University, New Delhi, India. Farhat has recently completed her two-year postdoctoral studies on gender with the World Agroforestry Centre (ICRAF), CGIAR, Vietnam. Her research interests include gender, agroforestry, water governance, food security, political ecology, and natural resource management. Apart from publishing her research in international journals and books, Farhat has led research projects on gender in Southeast Asia and South Asia.

Jayanthi A. Pushkaran is a New Delhi-based sociologist working on issues related to gender equality, environment, science policy, and child rights for a range of organizations including government think tanks, grassroots charities, international aid agencies, and social enterprises in India. She is also a doctoral candidate at the Centre for Studies in Science Policy, Jawaharlal Nehru University, New Delhi, India.

James Rajanayagam is Senior Project Advisor with the Centre for Social Innovation & Entrepreneurship at the Indian Institute of Technology Madras, Chennai, India. He is an innovation consultant in the development sector with 10 years of experience in promoting a number of small industries, enterprises, and other nonprofit organizations that provide social welfare in areas such as health, livelihood, and jobs. His core expertise is in formulating market-based solutions for the development of social welfare indicators. In this capacity, he has helped raise grants and equity for for-profit and nonprofit enterprises in sectors such as manufacturing, agriculture, construction materials, dairy, and healthcare. He has also mentored these enterprises for business development support. He brings extensive network from financing institutions, for-profit and nonprofit organizations, educational institutions, and government. Earlier, James was with Reliance Industries, India, in the petrochemical sector for 10 years. James is a textile technologist with an MBA in technology and innovation management obtained from Germany.

Naresh Chandra Sahu is currently working as Assistant Professor at the School of Humanities, Social Sciences & Management, Indian Institute of Technology Bhubaneswar, Odisha, India. After receiving his PhD in economics from the Indian Institute of Technology Kanpur, India, in 2008, he has taught at many reputed institutes of India. His research interests largely include forest resource management, mining, climate change, poverty, migration, financial inclusion, rural development, and energy. He has several research papers published in both national and international journals. He has authored couple of books published by reputed international publishers. In addition, he has presented several research papers in both international and national conferences in his research area. Furthermore, he has organized workshops on 'Environmental Impact Assessment' in 2012 and 'Productive

Efficiency: Theory and Practice' in 2016 at the Indian Institute of Technology Bhubaneswar.

Sakshi Saini is a development practitioner and urban researcher with around ten years of professional experience working in citizen-engagement-need-based program planning and applied participatory research methodologies. She is currently working at Participatory Research in Asia (PRIA) as Assistant Program Manager. Her prime area of work is citizen-centric program planning, and implementation and provision of policy level inputs to facilitate the process. Her prime work has been in the sector of water, sanitation, and hygiene (WASH), gender, climate change, development communication, and urban governance. She completed her doctoral research on enhancing the adaptive capacity of urban poor women to climate change from Delhi University, India, in 2014.

Manjeet Saluja is National Professional Officer (Environment and Public Health), WHO Country Office for India. Manjeet has been working in the area of environment and development for the last 15 years. He currently works on the issues of air pollution, climate change, water, and sanitation, and their effect on public health. He has previously worked as a technical advisor for the Indo-German Environment Partnership program of GIZ. Before joining GIZ, he has worked on issues of decentralization, livelihoods, and environment with the United Nations and international and national NGOs in various backward states of India. His research interests include the role of political economy on decentralized resource management. He has completed his master's in business administration from the Indian Institute of Forest Management (IIFM), Bhopal, India, and has a master's in environmental sciences.

Arushi Sen is Senior Communications Officer at the Center for Study of Science, Technology and Policy (CSTEP), Bangalore, India. Arushi has completed her master's degree in global studies from the University of Leipzig, Germany, and University of Wroclaw, Poland. She holds a bachelor's degree in journalism from Lady Sri Ram College, New Delhi, India. She has been working in the development and public policy space in India as a communications and media liaison officer for the last three years. Prior to joining CSTEP, she has worked with GIZ India, Yoda Press, and IORA Ecological Solutions. With over five years of work experience, she has acquainted herself with India's public policy space in the areas of climate policy, clean energy technologies, and sustainable development options. She is capable of understanding and developing communication strategies that gain traction within the policy-making space. The areas of work that interest her include strategic communication, editing, social media engagement, event management, and studying the intersection of communication, think tanks, and public policy.

Manasi Seshaiah is currently working as Associate Professor at the Centre for Research in Urban Affairs, Institute for Social and Economic Change, Bangalore, India. She has long-time research experience in natural resource management, livelihoods, and pollution abatement with specific focus on water, sanitation, and waste management. Her other research interests are urban ecology, environmental governance, and climate change. She has worked on collaborative research studies and authored several articles in journals and edited books.

Kartikeya Singh is Deputy Director and Fellow of the Wadhwani Chair in US–India policy studies at the Center for Strategic and International Studies (CSIS), Washington, DC, USA. He has many years of experience working with research institutions, start-ups, governments, businesses, and intergovernmental organizations. His research interests include climate change and energy policy, innovation, and the geopolitics of energy use. At CSIS, he is charged with leading the Wadhwani Chair's work on India's states, including a new project on energy. Prior to joining CSIS, Kartikeya worked at the US Department of Energy, supporting and managing US–India and US–Pakistan bilateral energy cooperation. He was also a consultant with the Environmental Defense Fund and is founder of the Indian Youth Climate Network (IYCN). Kartikeya received his BS in ecology and sustainable development from Furman University SC, USA; his MESc from the Yale School of Forestry and Environmental Studies, CT, USA; and his PhD from the Fletcher School of Law and Diplomacy, Tufts University, MA, USA.

Kanekanti Chandrashekar Smitha is Research Scientist, Centre for Research in Urban Affairs (CRUA), Institute for Social and Economic Change, Bengaluru, India. Her research interests are urban governance, service delivery, gender and urban poverty, urban informality, urban political economy, and urban political ecology. She has a PhD in political science and public administration from the Institute for Social and Economic Change, Bengaluru, through the University of Mysore. Her doctoral thesis examined urban governance and service delivery of water supply and sanitation in Bengaluru, exploring post-neoliberal urban sector reforms and their impact. Over the years, she has been actively engaged in varied urban research issues such as structure of Indian metropolises, rural–urban migration, urban deprivation, urban infrastructure, urban poor, and urban land and transport governance from an interdisciplinary perspective. She has several book chapters and articles published in peer-reviewed journals and very recently co-edited a volume, *Entrepreneurial Urbanism in India: Politics of Spatial Restructuring and Local Contestation* (2017).

Nidhi Yadav is Research Scholar at the School of Humanities, Social Sciences & Management, Indian Institute of Technology Bhubaneswar, India. She has done her postgraduation in economics from Banasthali University, Rajasthan, India, and was awarded UGC-NET and JRF in economics during her master's itself in 2011. She has papers published in international journals. In addition, she has presented research papers at international conferences. She has been a technical committee member of International Conference on Business and Management (ICBM) 2016, Singapore, as well as in forthcoming ICBM, 2017. Her current areas of interest are natural resources management, welfare economics, environmental economics, public finance, sustainable development, and ecosystem valuation.

Index